Current Topics in Microbiology and Immunology

198

Editors

Pathways for Cytolysis

Edited by G. M. Griffiths and J. Tschopp

With 45 Figures

Springer-Verlag
Berlin Heidelberg New York
London Paris Tokyo
Hong Kong Barcelona
Budapest

Gillian M. Griffiths
Basel Institute of Immunology
Grenzacherstr. 487
4005 Basel
Switzerland

Jürg Tschopp
University of Lausanne
Department of Biochemistry
Ch. Boveresses 155
1066 Epalinges/Lausanne
Switzerland

Cover Illustration: The cover shows a human cytotoxic T lymphocyte (CTL) clone which has recognised an EBV transformed B cell target. The lytic granules of the CTL are stained with an antibody to granzyme A (red), and visualised using confocal microscopy. Courtesy of S. Issaz and G.M. Griffiths.

Cover design: Künkel+Lopka Werbeagentur, Ilvesheim

ISSN 0070-217X
ISBN-13:978-3-642-79416-2 e-ISBN-13:978-3-642-79414-8
DOI: 10.1007/978-3-642-79414-8

Softcover reprint of the hardcover 1st edition 1995

Library of Congress Catalog Card Number 15–12910

Production: PRODUserv Springer Produktions-Gesellschaft, Berlin
Typesetting: Thomson Press (India) Ltd, New Delhi
SPIN: 10479798 27/3020-5 4 3 2 1 0 – Printed on acid-free paper.

Preface

At first glance the destruction of a target cell by a killer cell seems to be a simple endeavor. A closer look, however, reveals the complex mechanisms underlying this task. Killer cells are able to specifically recognize altered or infected cells. A transient contract with target cells has to be established to allow the delivery of lethal molecules or signals. The killer cell then disengages from the damaged cell and moves away to kill other target cells. After the eradication of the target cells, the number or activity of activated killer cells has to be reduced to avoid nonspecific killing of innocent cells.

In 1992, Herman Eisen concluded, in his introductory remarks in the most recent volume on lymphocyte cytotoxicity (Eisen 1993): "Given the immense amount of effort made in the past decade to understand these (lytic) mechanisms, it seems surprising that a consensus about the principal mechanisms has not been reached." Since that time, advances made in the field of cell-mediated cytolysis are so significant that our knowledge regarding the lytic mechanisms has been considerably augmented. Much of this is due to studies with transfectants and mutants–either naturally occurring or generated by gene targeting.

It is now clear that there are two predominant pathways which act hand in hand and lead to efficient target cell destruction. These pathways can be distinguished by the molecules which mediate them, i.e., Fas, a member of the tumor necerosis factor (TNF) receptor family, and perforin. The two pathways are complementary, one requiring cell-cell contact and using a regulated ligand able to lyse only Fas receptor-bearing cells, and a second, perforin-granzyme-based, mechanism whose lytic activity is dependent on the presence of phosphorylcholine, thus allowing killing of virtually any target cell. It is possible that, during evolution, the two cytolytic systems may have converged in the same cytolytic cell to render it optimally equipped to lyse all possible target cells.

The aim of this volume was to collect reviews covering the different pathways for cytolysis: from the importance of coreceptors in generating cytotoxic T lymphocytes (CTLs) to the molecular details of the effects of lytic proteins on the cell cycle which lead to apoptosis. The factors which induce the different pathways and the cell biology of thier storage and release are outlined. The results of the perforin knock-out mouse, the biology of the TNF family of surface proteins, and the interactions of the different lytic proteins, and their roles in contributing to an efficient killing mechanism are discussed. What is known about the mode of action of the different effector molecules on the target cells is presented and the potential of the different pathways in various disease and pathological states is explored. We have attempted to present the reader with a current, comprehensive overview of the different factors which contribute to killing. It was definitely a gratifying task to put this volume together in view of the new important findings described herein.

G.M. Griffiths and J. Tschopp

Reference

Eisen H (1993) Introduction. In: Sitkowsky MV, Heukart PA (eds) Cytotoxic cells: recognition, effector function, generation and methods. Birkhäuser, Boston.

List of Contents

List of Contributors

(Their addresses can be found at the beginning of their respective chapters.)

Perforin and Granzymes: Crucial Effector Molecules in Cytolytic T Lymphocyte and Natural Killer Cell-Mediated Cytotoxicity

B. Lowin, M.C. Peitsch*, and J. Tschopp

1 Lymphocytes and Natural Killer Cells

The killing mediated by cytotoxic T lymphocytes (CTLs) and natural killer (NK) cells represents an important mechanism in the immune defense against tumors, virus-infected cells, parasites and other foreign invaders. The deleterious effects observed in the absence of a properly working cellular immune response have been characterized extensively in immunodeficient patients (for a review see Matsumoto et al. 1992) and naturally occurring animal models such as SCID (severe combined immunodeficiency) or nude mice. In general, immunodeficient individuals are predisposed to develop severe opportunistic infections which ultimately can lead to their death. Recently, gene targeting technology has been used repeatedly to generate mice with defined mutations in genes implicated in T and NK cell function (for a review see Yeung et al. 1993), in an attempt to dissect the individual pathways involved in the lymphocyte-mediated immune response.

Institute of Biochemistry, University of Lausanne, Ch. des Boveresses, 155, 1066 Epalinges, Switzerland
* Present address: Glaxo Institute for Molecular Biology, 1021 Plan-Les-Ouates, Switzerland

Both β_2-microglobulin-deficient mice (KOLLER et al. 1990; ZIJLSTRA et al. 1990) and CD8-deficient mice (FUNG-LEUNG et al. 1991) clearly showed the importance of $CD8^+$ T cells in virus clearance via their cytotoxic activity. In the initiation and development of immunopathological disease processes, as in autoimmune diseases, $CD4^+$ T cells play a crucial role. In addition, $CD8^+$ lymphocytes are also found in autoimmune lesions. Their role and significance in the development of autoimmunity, however, is still controversial; however, using mice lacking CD8 molecules, PENNINGER et al. (1993) demonstrated that $CD8^+$ T cells may regulate the disease severity in experimental autoimmune myocarditis.

T lymphocytes recognize antigens by means of T cell receptor (TCR) glycoprotein complexes, in which the antigen-binding chains are encoded for by rearranging gene families. Approximately 95% of blood T cells express the α/β TCR, up to 5% have the γ/δ TCR. For T cells, the TCR/CD3 complex is necessary yet not sufficient for antigen recognition. Another set of receptors, namely CD4 and CD8, is required. The generation of mice, deficient either in CD8 (FUNG-LEUNG et al. 1991) or CD4 (RAHEMTULLA et al. 1991), has shown convincingly that CTLs, which respond to antigenic challenge by lysis of their target, are recruited predominantly from the $CD8^+$ pool. CTLs are acquired by the host within a week after infection and form a secondary immune defense. They play a central role in antiviral immunity, but may also be of major importance in many bacterial and protozoan infections. CD8-mediated protection has been shown for *Listeria monocytogenes, Mycobacterium tuberculosis, Plasmodium* spp., *Trypanosoma cruzi* and *Leishmania major*. It is, however, not yet fully understood if protection is always associated with cytolytic activity or if it is provided through Interferon-γ (IFN-γ) which is produced by most $CD8^+$ T cells. Allograft rejection is another situation in which T cell-mediated mechanisms are involved. A number of experiments suggest a central role for $CD4^+$ T lymphocytes in the rejection of vascularized organ and skin allografts. Since endocrine cells, such as islet allografts, do not express MHC class II molecules, their rejection is thought to be dependent on $CD8^+$-mediated mechanisms (DESAI et al. 1993).

NK cells are a subpopulation of lymphocytes distinct from T and B cells that have the ability to mediate cytotoxicity and cytokine production. They represent 5%–15% of lymphocytes in the blood and are associated with a particular morphology, that is, the large granular lymphocyte phenotype. NK cells are $CD3^-$ (except for the ζ chain), and they do not productively rearrange and express TCR genes (for a review see TRINCHIERI 1989). They express the CD16 Fc receptor for IgG on their cell surface, through which they can mediate ADCC (antibody-dependent cell cytotoxicity) of antibody-coated target cells. In vivo, NK cells mediate rejection of transplanted tumor cells which is well reflected in in vitro assays. NK cells kill a variety of tumor targets, while normal cells are usually resistant. NK cells can also reject nonautologous bone marrow cells and lymphocytes (YU et al. 1992). The first phase of certain virus infections (such as Cytomegalovirus) represents a third case in which NK cells can mediate natural resistance in vivo. NK cells have also been implicated in resistance to certain bacterial and parasite infections (TRINCHIERI 1989).

2 Target Cell Death

For nearly 20 years efforts have been made to elucidate the molecular mechanisms underlying the killing mediated by CTLs and NK cells. Data obtained in both cell systems suggested a common death pathway. In contrast to killing by complement and bacterial toxins, target cells of CTLs/NK cells do not succumb to osmotic lysis, but to a suicidal process known as apoptosis. Apoptosis is characterized morphologically by cytoplasmic compaction, membrane blebbing, chromatin condensation and DNA fragmentation with an initial preservation of the integrity of cytoplasmic organelles (KERR et al. 1972). The affected cells separate into membrane-limited fragments, so-called apoptotic bodies, and are rapidly phagocytosed by adjacent cells. On agarose gels, DNA fragmentation, a hallmark of apoptosis, can be seen as the formation of the so-called DNA ladder. This pattern is generated by the preferential cleavage of genomic DNA in the linker region of nucleosomes giving rise to nucleosomal fragments (multiples of 180 bp).

Apoptosis is not restricted to lymphocyte-mediated cytotoxicity, but is also characteristic of physiological tissue turnover, cell death during embryonic development, maturation of the immune system through clonal deletion and many more situations. In contrast to the "classical" form of cell death, lymphocyte-mediated DNA fragmentation occurs more rapidly (in minutes instead of hours or days), suggesting that lymphocytes provide the target cells with the relevant effector molecules or activators triggering an intrinsic death program. Recently, a number of putative apoptosis-inducing molecules have been characterized in different systems (GAGLIARDINI et al. 1994; MIURA et al. 1993b; YUAN and HORVITZ 1992; YUAN et al. 1993). The nematode *Caenorhabditis elegans* has been widely used as a model system for apoptosis since development of the adult worm is accompanied by the generation and subsequent death of 131 well defined cells. Under the many genes affecting *C.elegans* cell death, the activities of two genes, *ced*-3 and *ced*-4, are directly required for cell death induction. CED-3 is an Asp-specific cysteine protease and may control cell death by proteolytical activation or inactivation of other proteins (MIURA et al. 1993a). Interestingly, Asp-specific proteases have also been described in mammalian apoptosis systems. ICE, the IL-1 β converting enzyme, shows a 43% amino acid identity with the active domain of CED-3 and is possibly involved in neuronal cell death.

3 The Perforin/Granule Exocytosis Model

Cytoplasmic granules have been described as a common feature of both activated CTLs and constitutively cytolytic NK cells. Based on electron Microscopy (EM) studies showing that granules localize selectively to the contact sites of effector-target cell conjugates and eventually fuse with the effector cell mem-

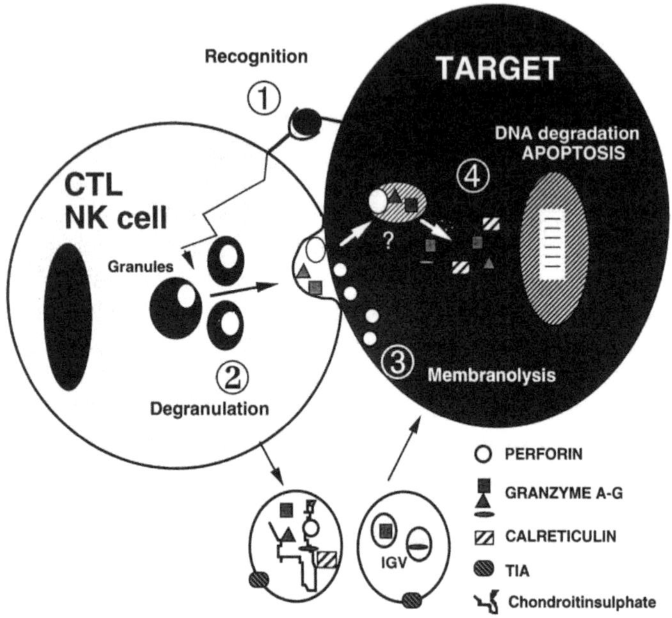

Fig. 1. The perforin/granule exocytosis model of lymphocyte-mediated cytotoxicity proposes that, upon effector-target cell conjugation (*1*), cytoplasmic granules, containing perforin, granzymes, calreticulin, TIA-1 and proteoglycans, are exocytosed into the intercellular space at the contact site (*2*). In the presence of Ca^{2+}, perforin binds to the target cell membrane and polymerizes to form tubular transmembrane lesions (*3*). Perforin also allows the access of granzymes and calreticulin to the cytoplasm of the target cell where these molecules induce apoptosis (*4*). The molecular mechanisms of the delivery process are unknown. Most likely, the exocytosed proteins or intragranular vesicles (*IGV*) are initially endocytosed and then translocate into the cytoplasm

brane, Henkart and colleagues (Henkart and Henkart 1982) put forth the perforin/granule exocytosis model of lymphocyte-mediated cytotoxicity (Fig. 1). According to this model, granules are redirected towards the target cell upon effector-target cell contact, whereby they fuse with the plasma membrane and release their content into the intercellular cleft, resulting in cell death by osmotic lysis and DNA degradation. At the EM level, granules appear as membrane-delimited organelles, containing an electron-dense core and small intragranular vesicles (IGVs) (Peters et al. 1989). Although it was initially assumed that the granule-associated proteins are released in their free form, recent studies show that at least a portion of the proteins are encapsulated into IGVs and transferred to the target cell using vesicles as vehicles (Schaerer et al. 1995).

Granules isolated from various types of killer lymphocytes by nitrogen cavitation and subsequent Percoll gradient centrifugation were shown to be cytolytic for red blood cells and a variety of nucleated cells. It was therefore speculated that they harbor the cytolytic effector molecules involved in CTL-and NK cell-mediated killing. In addition to classical lysosomal proteins, purification of granule-stored proteins led to the discovery of perforin, a membranolytic protein (Masson and Tschopp 1985; Podack et al. 1985); granzymes a set of serine

Table 1. Main molecules found in cytoplasmic granules of mouse CTL/NK cells

Protein	Distinctive features	Function
Perforin	Pore-forming protein	Osmotic lysis
Granzyme A	Protease, tryptase	Apoptosis
Granzyme B	Protease, Asp-ase	Apoptosis
Granzyme C,D,E,F,G	Proteases	Unknown
Calreticulin	Ca^{2+}-binding protein	Chaperone, modulation of gene expression
TIA-1	RNA-binding protein	Unknown
Proteoglycans	Chondroitin sulfate A	Carrier molecule

CTL, cytotoxic T lymphocyte; NK, natural killer.

esterases (Jenne et al. 1988); calreticulin, a calcium-binding protein (Dupuis et al. 1993b), *TIA-1,* a RNA-binding protein (Anderson et al. 1990) and proteoglycans (MacDermott et al. 1985) (Table 1). Not all granule-stored molecules are directly implicated in the death pathway, but may play an important role in the maintenance of granule structure, storage and repression of the lytic effector molecules and other "housekeeping" functions. Proteoglycans are thought to condensate proteins inside the granules to allow a higher density. Furthermore, packaging of granular proteases by proteoglycans may minimize autolysis of the enzymes or degradation of other cellular components (Stevens et al. 1987). Calreticulin, referring to its calcium-binding capacities, could function as a Ca^{2+} buffer, thereby preserving perforin, which is easily inactivated by low concentrations of calcium, in its lytic form (Dupuis et al. 1993b; Young and Cohn 1987). In addition, calreticulin may be implicated in the direct stabilization of perforin in a chaperone-like manner. Recently, two groups provided evidence that calreticulin can modulate the regulation of gene transcription by nuclear hormone receptors through direct interaction with the receptor's DNA binding site (Burns et al. 1994; Dedhar et al. 1994). Whether this activity also plays a role in the induction of target cell death remains to be seen.

4 Perforin

Perforin was the first granule-stored protein to which cytotoxic activity could be attributed. It was identified as the protein responsible for the tubular lesions found on target cells after CTL/NK cell attack (Dourmashkin et al. 1980). In a complement-like manner, monomeric perforin, in the presence of calcium, binds to membranes, inserts into the lipid bilayer and finally polymerizes into pore-like complexes with an average diameter of about 16 nm (Podack and Hengartner 1989; Tschopp and Nabholz 1990; Young 1989) (Fig. 2). Pore formation leads eventually to target cell lysis by osmotic effects, calcium influx and energy depletion. In addition, such pores allow the passage of dyes and even large proteins into the target cell (Peters et al. 1990).

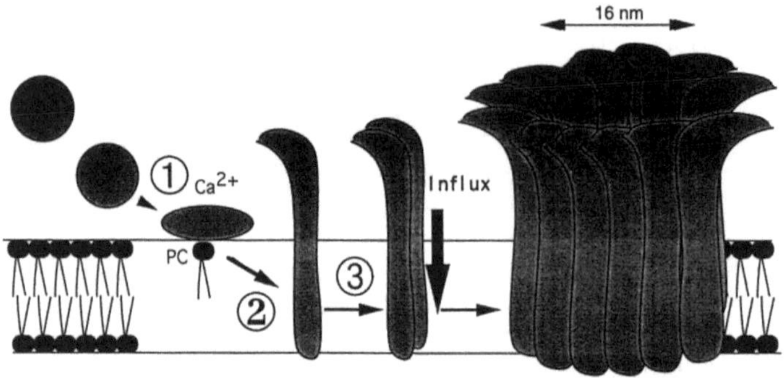

Fig. 2. Insertion of perforin. Monomeric perforin binds to phosphorylcholine headgroups of lipid membranes. This interaction is strictly Ca^{2+}-dependent and reversible (*1*). At physiological temperature, perforin undergoes a drastic conformational change and inserts into the lipid bilayer (*2*). Polymerization of the perforin protomers ensues (*3*). Cylindrical polymers of up to 18 monomers are formed. The largest tubules have an inner diameter of > 10 nm

4.1 Structure

cDNA clones for human, mouse and rat perforin have been isolated and revealed a 555 residue long transcript, of which the first 20–21 residues are the signal peptide. The amino acid sequence of perforin was shown to be related to the terminal complement components C6–C9. The region of similarity, however, does not span the complete sequence; only residues 170–408 of human perforin are related to residues 295–540 of human C9 (Fig. 3a,b). This region comprises the potential membrane spanning domain and the LDL class B domain, or epidermal growth factor (EGF)-like module. In this segment, 20% of the amino acids are shared by all family members, suggesting its involvement in the common biological function of pore formation. The NH_2-terminal and COOH-terminal exhibit unique sequences. No homologous proteins have been described as yet.

Proteins involved in pore formation contain a region(s) forming an amphipathic structure(s) to ensure the formation of a hydrophilic channel through the lipophilic environment. This amino acid sequence is not hydrophobic enough to avoid protein trapping in the endoplasmic reticulum during membrane translocation. Furthermore, the primary structures of perforin and the terminal

Fig. 3. a Structural organization of human perforin and the terminal complement component C9. Here the arrangement of the various structural motifs present in perforin and C9 are shown: *TS*, thrombospondin module; *L*, LDL receptor class A module; *EGF*, EGF receptor module; *MB*, candidate lipid binding domain. **b** Sequence alignment of perforin and human complement component C9 (*Co9*). The amino acid sequences of human, mouse and rat perforin were aligned with the human complement component C9; of the latter, only the region of similarity to perforin is shown. The regions corresponding to the signal peptide, the two potential amphipathic helices proposed to represent the transmembrane domain, and the epidermal growth factor (EGF)-like module are indicated *below the alignment*. The sequences are identified by their SwissProt (Release 28) identification codes (except for rat perforin which was deduced from Genbank: RATCYTA)

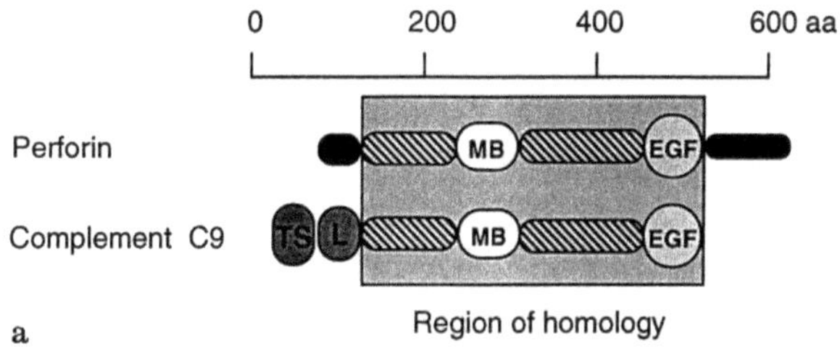

a

```
Perf_human:   MAARLLLLGILLLLLPLPVPAPCHTAARSECKRSHKFVPGAWLAGEGVDVTSLRRSGSFP   60
Perf_mouse:   -MATCLFLLGLFLLLPRPVPAPCYTATRSECKQKHKFVPGVWMAGEGMDVTTLRRSGSFP   59
Perf_rat  :   -MAAYLFLLGLFLLLPRPVPAPCYTATRSECKQNHKFVPGVWAAGEGVDVTTLRRSSSFP   59
              <- Signal Peptide ->

Perf_human:   VDTQRFLRPDGTCTLCENALQEGTLQRLPLALTNWRAQGSGCQRHVTRAKVSSTEAVARD  120
Perf_mouse:   VNTQRFLRPDRTCTLCKNSLMRDATQRLPVAITHWRPHSSHCQRNVAAAKVHSTEGVARE  119
Perf_rat  :   VNTGKFLRPDRTCTLCKNALMNDGIQRLPVAIAHWRPHGSHCQRNVATTKVSSTEGVARE  119

Perf_human:   AARSIRNDWKVGLDVTPKPTSNVHVSVAGSHSQAANFAAQKTHQDQYSFSTDTVECRFYS  180
Perf_mouse:   AAANINNDWRVGLDVNPRPEANMRASVAGSHSKVANFAAEKTYQDQYNFNSDTVECRMYS  179
Perf_rat  :   AAANINNDWRAGLDVNPKPEANVHVSVAGSHSKIANFAAEKAHQDQYNFNTDTVECRMYS  179
Co9_human :   ................................................FLHVKGEIHLGR  306

Perf_human:   FHVVHTPPLHPDFKRALGDLPHHFNASTQPAYLRLISNYGTHFIRAVELGGRISALTALR  240
Perf_mouse:   FRLVQKPPLHLDFKKALRALPRNFNSSTEHAYHRLISSYGTHFITAVDLGGRISVLTALR  239
Perf_rat  :   FRLAQKPPLHPDFRKALKNLPHNFNSSTEHAYRRLISSYGTHFITAVDLGGRVSVLTALR  239
Co9_human :   FVMRNRDVLTTTFVDDIKALPTTYEKG---EYFAFLETYGTHYSSSGSLGGLYELIYVLD  363
                    <---- Helix 1 --->       <----- Helix 2 ----->

Perf_human:   TCELALEGLTDNEVEDCLTVEAQVNIGIHG-SISAE--AKACEEKKKKHKMTASFHQTYR  297
Perf_mouse:   TCQLTLNGLTADEVGDCLNVEAQVSIGAQA-SVSSE--YKACEEKKKQHKMATSFHQTYR  296
Perf_rat  :   TCQLTLDGLTADEVGDCLSVEAQVSIGAQA-SVSSE--YKACEEKKKQHKIATSFHQTYR  296
Co9_human :   KASMKRKGVELKDIKRCLGYHLDVSLAFSEISVGAEFNKDDCVKRGEGRAVNITSENLID  423

Perf_human:   ERHSEVVGGHHT---SIND-LLFGIQAGPEQYSAWVNSLPGSPGLVDYTLEPLHVLL---  351
Perf_mouse:   ERHVEVLGGPLD---STHD-LLFGNQATPEQFSTWTASLPSNPGLVDYSLEPLHTLL---  350
Perf_rat  :   ERHVEVLGGPLD---SSND-LLFGNQATPEHFSTWIASLPTRPDVVDYSLEPLHILL---  350
Co9_human :   DVVSLIRGGTRKYAFELKEKLLRGTVIDVTDFVNWASSINDAPVLISQKLSPIYNLVPVK  483

Perf_human:   -DSQDPRREALRRALSQYLTDRARWRDCSRPCPPGRQKSPRDP-CQCVCHGSAVTTQDCC  408
Perf_mouse:   -EEQNPKREALRQAISHYIMSRARWQNCSRPCRSGQHKSSHDS.CQCECQDSKVTNQDCC  407
Perf_rat  :   -EDSDPKREALRQAISHYVMSRARWRDCNRPCRAGQHKSSRDS.CQCVCQDSNVTNQDCC  407
Co9_human :   MKNAHLKKQNLERAIEDYINE-FSVRKCHT-CQNGGTVILMDGKCLCACPFK-FEGIACE  540
                                         <--------- EGF-Like module --------->

Perf_human:   PRQRGLAQLEVTFIQAWGLWGDWFTATDAYVKLFFGGQELRTSTVWDNNNPIWSVRLDFG  468
Perf_mouse:   PRQRGLAHLVVSNFRAEHLWGDYTTATDAYLKVFFGGQEFRTGVVWNNNNPRWTDKMDFE  467
Perf_rat  :   PRQRGLAKLMVRNFQAKGLWGDYITSTDAYLKVFFGGQEIRTGVVWNNNHPSWSDKMDFG  467

Perf_human:   DVLLATGGPLRLQVWDQDSGRDDDLLGTCDQAPKSGSHEVRCNLNHGHLKFRYHARCLPH  528
Perf_mouse:   NVLLSTGGPLRVQVWDADYGWDDDLLGSCDRSPHSGFHEVTCELNHGRVKFSYHAKCLPH  527
Perf_rat  :   NVLLSTGGPLRVQVWDADNGWDDDLLGTCDKSPKSGFHEVNCPLNHGSIKFIYQANCLPD  527

Perf_human:   LGGGTCLDYVPQMLLGEPPGNRSGAVW   555
Perf_mouse:   LTGGTCLEYAPQGLLGDPPGNRSGAVW   554
Perf_rat  :   LTGETCLEYAPQGLLGDPRGNRSGAVW   554
```

b

complement components lack a continuous stretch of hydrophobic amino acids. Yet, experimental and theoretical consideration have allowed definition of the membrane interacting regions within C9 and perforin (PEITSCH et al. 1990). The membrane binding motif (residues 267–377 in human C9) was determined by photoaffinity labeling experiments, using polymerized C9 previously assembled in artificial liposomes (AMIGUET et al. 1985), and consists of two amphipathic α-helices separated by a tight turn, located at the site of highest similarity among all members of the perforin/terminal complement family (Fig. 3B). Subsequent mutagenesis experiments performed within this region revealed that any mutation only slightly affecting the amphipathic character had a strongly disturbing effect on C9 secretion and cytotoxicity (DUPUIS et al. 1993a). Perforin is believed to behave in a similar manner.

The COOH-terminal LDL class B domain of C9 is the second domain shared by perforin and the terminal complement components. This motif of

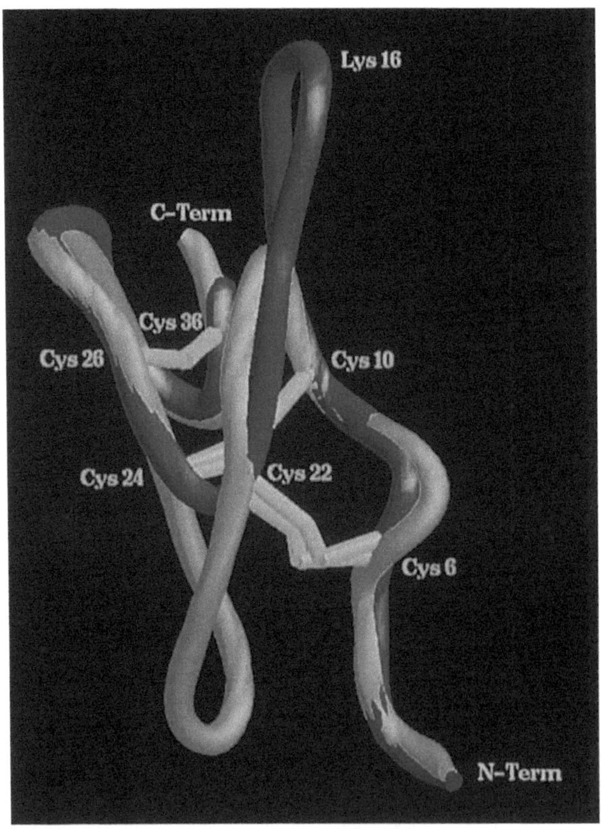

Fig. 4. Molecular model of the LDL class B module of human perforin. A model of the human perforin EGF-like module was designed based on the known 3-D structure of the EGF-like module of human coagulation factor IX using the automated protein modeling package *ProMod*. The modeled structure *(red)* is superimposed onto the NMR-based molecular structure of human coagulation factor IX *(blue)*. The disulphide bonds are shown in *yellow*, and were kept in topologically similar places throughout the modeling procedure

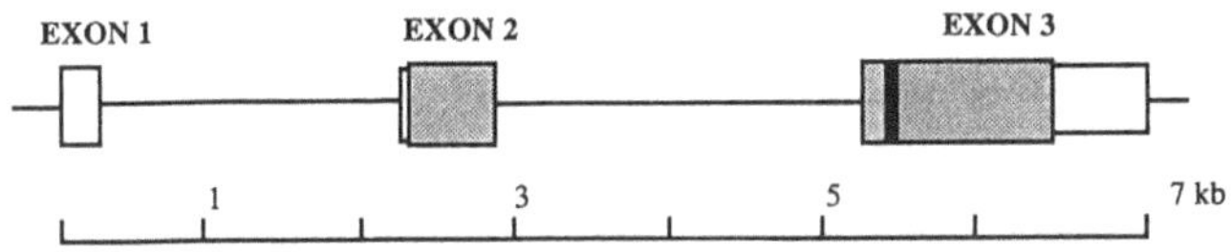

Fig. 5. Genomic organization of the mouse perforin gene. *Shaded boxes* represent coding exon sequences, *open boxes* untranslated exon sequences. The putative membrane spanning domain is indicated by a *black box*. Introns are represented by *thin lines*

approximately 40 amino acids shares the cysteine pattern with a region found in EGF and other proteins such as the low density lipoprotein (LDL) receptor (STANLEY et al. 1985). The conserved cysteines form three disulfide bonds which stabilize four β strands. The structure of the LDL class B module of perforin was modeled based on the NMR-based structure of the coagulation factor IX EGF-like module (Fig. 4). The striking difference between the factor IX EGF-like module and the model of the corresponding region of perforin is the length of the first and second loops. While the first loop consists of five and 11 residues in factor IX and perforin, respectively, the second loop is largely absent except for one residue in perforin (eight residues in factor IX). The great variability in residue composition and loop lengths encountered in this module, its conservation throughout the perforin/ terminal complement family and its known recognition/binding function reported in other systems suggest its involvement in subunit polymerization.

Both mouse and human perforin are encoded by a single copy gene, located on chromosome 10 (FINK et al. 1992; TRAPANI et al. 1990). In both species, the gene is small, containing only three exons (LICHTENHELD and PODACK 1989; TRAPANI et al. 1990) (Fig. 5). Exon 1 codes for the majority of the 5' untranslated sequence, exon 2 for the remainder of 5' untranslated sequence, the translational start signal, the signal peptide, and the NH_2-terminal part of the molecule up to the putative membrane binding site. Exon 3 encodes the remainder of the molecule including the 3' untranslated sequence.

4.2 Expression

Expression of perforin correlates with cytotoxicity and is restricted to the immune system. In circulating nonactivated lymphocytes perforin is only expressed in NK cells and γ/δ T cells. Both cell populations express perforin constitutively. Upon activation and acquisition of cytolytic activity, all $CD8^+$ T cells express perforin. In contrast to early reports (ALLBRITTON et al. 1988; BERKE and ROSEN 1988), perforin mRNA was also found in primary CTLs in peritoneal exudates (PELs) after intraperitoneal immunization with allogeneic spleen cells (NAGLER-ANDERSON et al. 1989) indicating that perforin-dependent killing is used by this cell population. In pregnant rodents, perforin is expressed in NK-like granulated metrial gland cells which are thought to be involved in immune surveillance during pregnancy (ZHENG et al. 1991). Under pathological conditions, perforin expression has been found in infiltrating lymphocytes during viral infections such as lymphocytic choriomeningitis virus infection (MULLER et al. 1989) and viral myocarditis (SEKO et al. 1993), during syngeneic and allogeneic tumor rejection (DOETSCHMAN et al. 1985; NAGLER-ANDERSON et al. 1989), transplant rejection (CLEMENT et al. 1991; GRIFFITHS et al. 1991), and autoimmune diseases such as nonobese diabetes (NOD) (YOUNG et al. 1989), experimental allergic encephalomyelitis (EAE) (HELD et al. 1993) and systemic lupus erythematosus (SLE) (NAKAMURA et al. 1992). In vivo expression of perforin appears to correlate with cytotoxic activity and tissue injury, although Bugeon et al. recently questioned the role of perforin in transplant rejection (BUGEON et al. 1993). They showed the presence of perforin expressing cells in both rejected and tolerated heart allografts.

4.3 Role in the Killing Process

Over the last few years many experiments have focused on the importance of perforin in lymphocyte-mediated killing. ACHA et al. showed that treatment of CTLs with perforin antisense oligonucleotides partially (up to 65%) inhibited their cytolytic activity (ACHA ORBEA et al. 1990). Whether the remaining activity was due to insufficient elimination of perforin mRNA or reflects the presence of alternative killing mechanisms remained unproved. Although purified perforin is capable of lysing red blood cells, DNA fragmentation, a prerequisite for the killing of nucleated cells, can only be induced by whole granule preparations (DUKE et al. 1989), implicating other granule-contained proteins in the induction of apoptosis. SHIVER and HENKART (1992) used a nonlytic rat basophilic leukemia (RBL) cell line to show lytic activity for the IgE-sensitized (IgE) receptor-driven exocytic granule pathway on red blood cells, following transfection with a perforin expression vector. In contrast to lymphocytes, however, these artificial killer cells were not able to lyse nucleated cells, most likely a consequence of insufficient release of perforin, as nucleated targets generally require higher concentrations of perforin for lysis than anucleated cells, or due to the need of additional cytotoxic factors.

Some investigators have questioned whether perforin is the primary mechanism of CTL-induced killing, since most of the evidence in favor of the granule exocytosis model has been obtained with CTL lines cultured in vitro in the presence of high levels of interleukin-2 (IL-2) (for a debate, see KRAHENBUHL and TSCHOPP 1991). Furthermore, this controversy has been heightened by the following unexplained observations: Some CTL lines were obtained which seem to lack detectable amounts of perforin, but are still cytolytic (BERKE et al. 1993), while other cell lines kill target cells in the absence of calcium (OSTERGAARD et al. 1987; TRENN et al. 1987); yet the lytic activity of perforin and successful exocytosis of granular proteins are strictly Ca^{2+}-dependent.

4.3.1 Perforin Knock-Out Mice

To elucidate the role of perforin in cytotoxicity, three separate groups (in Los Angeles, Zürich and Lausanne) have generated mice with a null mutation in the perforin gene, using the strategy of gene targeting in embryonic stem cells (KÄGI

Table 2. Effector activities of perforin knock-out mice

	Perforin (–/–)/Zürich	Perforin (–/–) Lausanne
Maintenance of pregnancy	Normal	Normal
Activation and expansion of T cells	Normal	Normal
Degranulation of granzyme A	n.d.	Normal
Alloreactivity	Absent	Impaired
P815 target cells	Impaired [a]	Impaired [b]
D2 fibroblasts	Absent [a]	n.d.
3T3 fibroblasts	n.d.	Impaired [a]
L929	n.d.	Impaired [a]
MEFs	Absent [a]	Impaired [a]
MEFs LPR	n.d.	Absent
Activated T cells	n.d.	Impaired
Activated T cells LPR	n.d.	Absent
NK Activity (Poly IC)	Absent [a]	Impaired [b]
Antiviral activity	Absent	n.d.
In vitro	Absent	
In vivo (LCMV)	Absent	
Tumor rejection (fibrosarcoma)	Highly impaired	n.d.

Results shown are a synopsis of the study of Kägi (KÄGI, et al. 1994) and of our own (LOWIN et al. 1994). MEF, mouse embryonic fibroblast; LPR, lymphoproliferation; NK, natural killer; LCMV, lymphocytic choriomeningitis virus.

[a] Target cell death was assessed by the ^{51}Cr release assay.

[b] Target cell death assessed by the DNA fragmentation ^{125}-IUdR test.

et al. 1994; LOWIN et al. 1994). In our hands, the perforin –/– mice are healthy and thriving (LOWIN et al. 1994). They reproduced normally, although it was shown that perforin is expressed during pregnancy in granulated metrial gland cells in the mouse uterus (ZHENG et al. 1991). These cells are described as NK-like and speculated to be required for the successful completion of pregnancy. This presumptive function therefore, is not mediated by a perforin-dependent mechanism. Expansion and activation of T cells in perforin-mice are normal. No difference between normal and perforin (–/–) mice was measured either in proliferation rate and T cell subtype pattern or in INF-γ release upon lymphocyte activation. The cytolytic activity of CD8$^+$ cells derived from perforin (–/–) mice was, however, greatly impaired in several cytolytic assay systems (Table 2).

Using P815 cells and determining membrane damage by the ^{51}Cr release assay, at least a 30-fold higher ratio of killer to target cells was required to achieve the same degree of lysis as with T cells derived from +/– or +/+ mice using alloreactive T cells. The difference was less pronounced when 3T3 fibroblasts were used as targets. Only a threefold reduction of the cytotoxic activity was observed (LOWIN et al. 1994). Different susceptibilities of fibroblasts towards perforin-deficient CTLs have been reported for the Zürich and Lausanne perforin knock-out mouse. A complete absence of lysis in allogeneic target cells derived from epithelial, neuroectodermal and mesodermal origin was observed in the study by KÄGI et al. (1994), whereas fibroblasts were still killed in the LOWIN et al. (1994) study. The reason for this discrepancy remains unknown. T cells from heterozygous (+/–) mice were consistently less active than the wild-type T cells, indicating that in primary cultures of lymphocytes perforin represents a limiting factor. This is in agreement with the antisense oligonucleotide study, in which a twofold reduction of perforin directly correlated with a decrease in cytolytic activity (ACHA ORBEA et al. 1990).

Almost identical results were obtained with NK cells, thus providing evidence that the two effector cells use similar, if not identical, killing mechanisms. Complete absence of NK target cell lysis was detected in the Zürich knock-out mouse, whereas cytotoxicity was only impaired in our hands The disagreement in the results may be simply explained by the difference in the lysis assays used.

The crucial role of perforin in the cytotoxic process may be interpreted in several ways. The most simple of these is that the absence of perforin results in the lack of transmembrane pores in the target cell membrane. Incorporation of isolated perforin into membranes induces massive ion conductance (YOUNG et al. 1986) and allows dyes and macromolecules to pass the plasma membrane (BLUMENTHAL et al. 1984; PETERS et al. 1990). Since in perforin-lacking CTLs, no toxic lethal massive Ca^{2+} influx ensues, target cells resist the attack of the killer cells.

Other explanations cannot be excluded a priori. Not only may the absence alter the biogenesis of the secretory granules in a way in which they become incompetent for the storage of other proteins, but perforin may also play the role of a fusogen akin to viral proteins, allowing the fusion of granule and cell membrane. This supposition seems unlikely, as T cell receptor-induced exocytosis of granzyme A is not impaired.

Since all target cells tested up to date are still partially susceptible to the attack by perforin-deficient lymphocytes, (an) alternative killing mechanism(s) must exist. The most likely second lytic effector system is the one based on Fas (described elsewhere in this volume) which is active in the absence of extracellular Ca^{2+} (ITOH et al. 1991; MACLENNON et al. 1980; ROUVIER et al. 1993; SUDA and NAGATA 1994; SUDA et al. 1993; TRAUTH et al. 1989; YOUNG et al. 1987). Susceptible target cells carrying a defective Fas receptor become indeed resistant to lysis by perforin-free lymphocytes (Table 2, LOWIN et al. 1994b).

Perforin messenger RNA has been shown to be highly increased in T cells at the site of lesions in various disease models including murine choriomenengitis, herpes virus infection, nonobese diabetic mice, myocarditis, graft rejection, rheumatoid arthritis. It will therefore be interesting to analyze these in vivo models in perforin knock-out mice to evaluate the contribution of the pore forming protein.

5 Granzymes

5.1 Structure

To date, seven mouse (granzymes A–G) and four human lymphocyte serine esterases (granzyme A, B, H and possibly 3) have been identified (TALENTO et al. 1992; JENNE and TSCHOPP 1988). Granzymes show a high degree of homology to one another and are thought to originate from an ancestral gene which evolved by duplication and exon shuffling (JENNE et al. 1989).

Granzyme A has a structure which is unique among serine proteases in that it forms a disulfide-linked homodimer of 60 kDa, while granzymes B–G all consist of a single polypeptide chain and M_r ranging between 27 and 55 kDa. The granzymes are antigenically related and, except for granzyme C, highly glycosylated, although the carbohydrate content of each granzyme is heterogeneous, ranging from 10% to 50% of the total molecular mass.

Granzymes are highly homologous to each other (at least 39% amino acid identity in the most unfavorable case between granzymes A and D) (Fig. 6). All granzymes contain, at equivalent positions, three characteristic residues, His, Asp and Ser, which form the catalytic center of serine proteases. Adjacent residues are also highly conserved as are six cysteine residues which are involved in disulfide bond formation in a 1–2, 3–6 and 4–5 pattern by analogy to rat mast cell protease (RMCP) II. The structure of granzyme A, however, shows two peculiarities in that it contains a fourth cysteine bond at the COOH-terminal analogous to that of chymotrypsin, trypsin and elastase (WOODBURY and NEURATH 1980) and that it occurs as a disulfide-linked homodimer via a free cysteine residue at position 76. Granzyme A closely resembles factor XI of the coagulation system in that both catalytic centers of the granzyme A dimer appear to be accessible and actually cleave substrates (MASSON and TSCHOOPP 1988).

All predicted granzyme sequences start with a typical hydrophobic signal peptide indicating that granzymes are translocated across the lipid membrane into the rough endoplasmatic reticulum (Fig. 6). Two sequence features characterize the granzyme subfamily of serine proteases. First, the mature enzymes found in granules share a strictly conserved NH_2-terminal sequence from positions +1 to +4 (I I G G) and positions +9 to +16 (P H S R P Y M A). Second, they are synthesized as inactive precursor molecules with a very short acidic propeptide at the NH_2 terminal consisting of either Gly-Glu or Glu-Glu (Fig. 6). The propeptide Glu-Arg and Glu-Lys of human and mouse granzyme A, respectively, appears to be an exception since it ends with a positively charged residue. Recent studies have suggested that the lysosomal dipeptidyl peptidase I (DPPI) is the major posttranslational processing enzyme responsible for removing the dipeptide, generating active granule serine proteases (BROWN et al. 1993). In contrast to low levels of DPPI activity in unstimulated $CD4^+$ spleen T cells, both unstimulated $CD8^+$ spleen T cells and in vitro-activated CTL populations are several-fold enriched in DPPI activity. Depletion of DPPI-enriched cells by treatment with Leu-Leu-OMe results in loss of the active enzyme and cytolytic effector function.

5.2 Substrate Specificities

A panel of synthetic peptide substrates has been used to identify the substrate's granzymes. Both granzymes A and D show trypsin-like activity in that they cleave optimally after Arg or Lys residues, although the activity of granzyme D is very low compared to granzyme A (MASSON et al. 1986; ODAKE et al. 1991; SIMON

```
                                                                        1
Graa_human:  MRNSYRFLASSLSVVVSLLLIPEDVC EK IIGGNEVTPHSRPYMVLLSLDRKT----ICAG   56
Graa_mouse:  MRNASGPRGPSLATLLFLLLIPEGGC ER IIGGDTVVPHSRPYMALLKLSSNT----ICAG   56
Grab_human:  --------MQPILLLLAFLLLPRADA GE IIGGHEAKPHSRPYMAYLMIWDQKSLK-RCGG   51
Grab_mouse:  --------MKILLLLLTLSLASRTKA GE IIGGHEVKPHSRPYMALLSIKDQQPEA-ICGG   51
Grac_mouse:  --------MPPVLILLTLLLPLRAGA EE IIGGNEISPHSRPYMAYYEFLKVGGKKMFCGG   52
Grad_mouse:  --------MPPILILLTLLLPLRAGA EE IIGGHVVKPHSRPYMAFVMSVDIKGNRIYCGG   52
Grae_mouse:  --------MPPVLILLTLLLPLGAGA EE IIGGHVVKPHSRPYMAFVKSVDIEGNRRYCGG   52
Graf_mouse:  --------MPPILILLTLLLPLRAGA EE IIGGHEVKPHSRPYMARVRFVKDNGKRHSCGG   52
Grag_mouse:  --------MPPILILLTLLLPLRAGA EE IIGGHEVKPHSRPYMAFIKSVDIEGKKKYCGG   52
Grah_human:  --------MQPFLLLLAFLLTPGAGT EE IIGGHEAKPHSRPYMAFVQFLQEKSRK-RCGG   51
Sec. Str  :                                                 bbbbbb      bbbb

                        *1
Graa_human:  ALIAKDWVLTAAHCNLNK---RSQVILGAHSITREEPTKQIMLVKKEFPYPCYDPATREG  113
Graa_mouse:  ALIEKNWVLTAAHCNVGK---RSKFILGAHSINKEPE-QQILTVKKAFPYPCYDEYTREG  112
Grab_human:  FLIQDDFVLTAAHCWGS----SINVTLGAHNIKEQEPTQQFIPVKRPIPHPAYNPKNFSN  107
Grab_mouse:  FLIREDFVLTAAHCEGS----IINVTLGAHNIKEQEKTQQVIPMVKCIPHPDYNPKTFSN  107
Grac_mouse:  FLVRDKFVLTAAHCKGS----SMTVTLGAHNIKAKEETQQIIPVAKAIPHPDYNPDDRSN  108
Grad_mouse:  FLIQDDFVLTAAHCKNSSVQSSMTVTLGAHNITAKEETQQIIPVAKDIPHPDYNATIFYS  112
Grae_mouse:  FLVQDDFVLTAAHCRNR----TMTVTLGAHNIKAKEETQQIIPVAKAIPHPDYNATAFFS  108
Graf_mouse:  FLVQDYFVLTAAHCTGS----SMRVILGAHNIRAKEETQQIIPVAKAIPHPAYDDKDNTS  108
Grag_mouse:  FLVQDDFVLTAAHCRNR----SMTVTLGAHNIKAKEETQQIIPVAKAIPHPAFNRKHGTN  108
Grah_human:  ILVRKDFVLTAAHCQGS----SINVTLGAHNIKEQERTQQFIPVKRPIPHPAYNPKNFSN  107
Sec. Str  :  bbb  bbbbbb          bbbbbbbbbb       bbbbbbbb          bbb

             *                                  2
Graa_human:  DLKLLQLTEKAKINKYVTILHLPKKGDDVKPGTMCQVAGWGRTH-NSASWSDTLREVNIT  172
Graa_mouse:  DLQLVRLKKKATVNRNVAILHLPKKGDDVKPGTRCRVAGWGRFG-NKSAPSETLREVNIT  171
Grab_human:  DIMLLQLERKAKRTRAVQPLRLPSNKAQVKPGQTCSVAGWGQTA-PLGKHSHTLQEVKMT  166
Grab_mouse:  DIMLLKLKSKAKRTRAVRPLNLPRRNVNVKPGDVCYVAGWGRMA-PMGKYSNTLQEVELT  166
Grac_mouse:  DIMLLKLVRNAKRTRAVRPLNLPRRNAHVKPGDECYVAGWGKVT-PDGEFPKTLHEVKLT  167
Grad_mouse:  DIMLLKLESKAKRTKAVRPLKLPRSNARVKPGDVCSVAGWGSRSINDTKASARLREVQLV  172
Grae_mouse:  DIMLLKLESKAKRTKAVRPLKLPRPNARVKPGDVCSVAGWGSRSINDTKASARLREAQLV  168
Graf_mouse:  DIMLLKLESKAKRTKAVRPLKLPRPNARVKPGHVCSVAGWGRTSINATQRSSCLREAQLI  168
Grag_mouse:  DIMLLKLESKAKRTKAVRPLKLPRPNARVKPGDVCSVAGWGKTSINATKASARLREAQLI  168
Grah_human:  DIMLLQLERKAKWTTAVRPLRLPSSKAQVKPGQLCSVAGWGYVSM--STLATTLQEVLLT  165
Sec. Str  :  bbbbbbbbb                          bbbbbbbbb             bbbbbb

                   3                   3           # 4   *    2
Graa_human:  IIDRKVCNDRNHYNFNPVIGMNMVCAGSLRGGRDSCNGDSGSPLLCEGVFRGVTSFGLEN  232
Graa_mouse:  VIDRKICNDEKHYNFHPVIGLNMICAGDLRGGKDSCNGDSGSPLLCDGILRGITSFGGEK  231
Grab_human:  VQEDRKCESDL---RHYYDSTIELCVGDPEIKKTSFKGDSGGPLVCNKVAQGIVSYGRNN  223
Grab_mouse:  VQKDRECESYF---KNRYNKTNQICAGDPKTKRASFRGDSGGPLVCKKVAAGIVSYGYKD  223
Grac_mouse:  VQKDQVCESQF---QSSYNRANEICVGDSKIKGASFEEDSGGPLVCKRAAAGIVSYGQTD  224
Grad_mouse:  IQEDEECKKR----FRYYTETTEICAGDLKKIKTPFKGDSGGPLVCHNQAYGLFAYAKNG  228
Grae_mouse:  IQEDEECKKR----FRHYTETTEICAGDLKKIKTPFKGDSGGPLVCDNKAYGLLAYAKNR  224
Graf_mouse:  IQKDKECKKY----FYKYFKTMQICAGDPKKIQSTYSGDSGGPLVCNNKAYGVLTYGLNR  224
Grag_mouse:  IQEDEECKKL----WYTYSKTTQICAGDPKKVQAPYEGESGGPLVCDNLAYGVVSYGINR  224
Grah_human:  VQKDCQCERLF---HGNYSRATEICVGDPKKTQTGFKGDSGGPLVCKDVAQGILSYGNKK  222
Sec. Str  :  b                        bbbbb              bbbbb    bbbbbbbb

              4
Graa_human:  KCGDPRGPGVYILLSKKHLNWIIMTIKGAV   262
Graa_mouse:  -CGDRRWPGVYTFLSDKHLNWIKKIMKGSV   260
Grab_human:  GMP----PRACTKVS-SFVHWIKKTMKRY-   247
Grab_mouse:  GSP----PRAFTKVS-SFLSWIKKTMKSS-   247
Grac_mouse:  GSA----PQVFTRVL-SFVSWIKKTMKHS-   248
GRaD_mouse:  TIS----SGIFTKVV-HFLPWISWNMKLL-   252
Grae_mouse:  TIS----SGVFTKIV-HFLPWISRNMKLL-   248
Graf_mouse:  TIG----PGVFTKVV-HYLPWISRNMKLL-   248
Grag_mouse:  TIT----PGVFTKVV-HFLPWISTNMKLL-   248
Grah_human:  GTP----PGVYIKVS-HFLPWIKRTMKRL-   246
Sec. Str  :            bbbbbbb
```

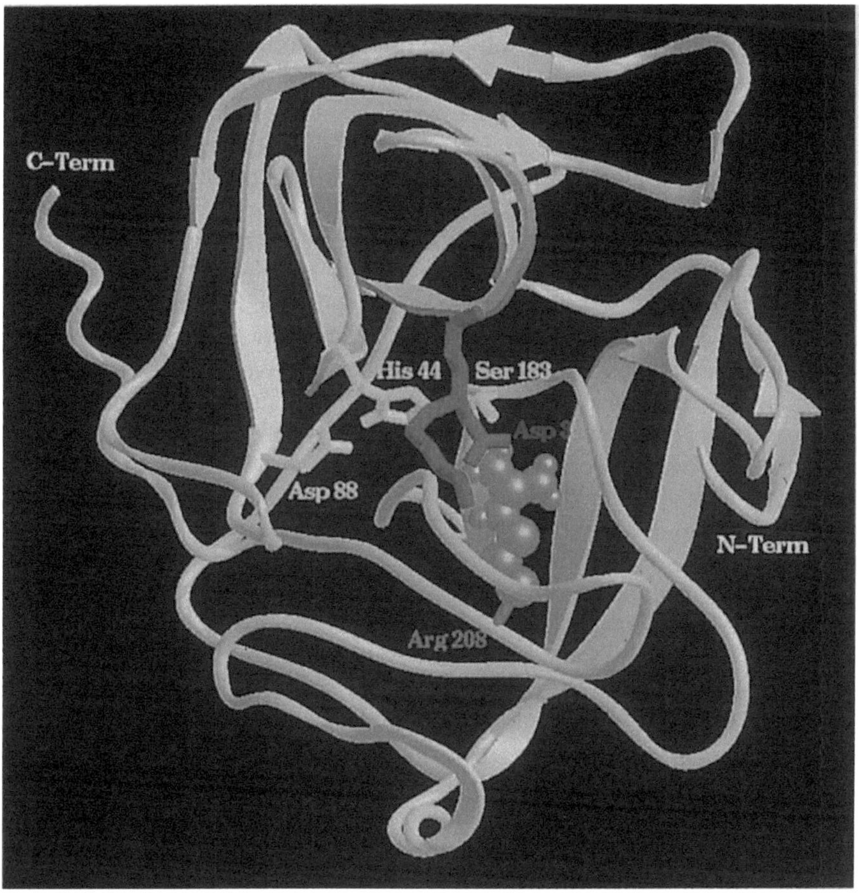

Fig. 7. Molecular model of human granzyme B. The structure of granzyme B is shown as a *brown ribbon*, while the catalytic triad (His-44, Asp-88 and Ser-183) is represented by *yellow rods*. The side chain of Arg-208 which confers Asp/Glu specificity to granzyme B is shown by *blue spheres*. The *red structure* represents the position of the modeled substrate (AADG), the side chain of Asp-3 pointing towards Arg-208

Fig. 6. Sequence alignment of human and mouse granzymes. The two residues corresponding to the propeptide are shown in *boldface*, the sequence at their *left* being the signal peptide. The three residues forming the catalytic triad (His, Asp, Ser) are shown in *boldface* and by an *asterix* above the alignment. The residue determining the nature of the S_1 substrate pocket is found at position −6 relative to the active site serine and is marked by a #; in addition, Arg-208, responsible for the Asp-ase activity of granzyme B, is shown in *boldface*. The positions of potential N-linked glycosylations is shown by *underlined asparagines*, and the cysteine bonding pattern is indicated by *numbers above the sequences*. The secondary structure assignment shown on the *last line* of the alignment was derived from rat mast cell protease II, the closest homologue with known 3-D structure (Brookhaven Protein Data Bank entry 3RP2). All sequences are identified by their SwissProt (Release 28) identification codes et al. 1986) and may be due to contaminating granzyme A. Granzyme A cleaves Pro-Phe-Arg-7-amino-4-methyl-coumarin, Pro-Phe-Arg-nitroanilide and N-α-benzyloxycarbonyl-Lys-thiobenzylester (BLT) most efficiently. In contrast, granzyme B has Asp-ase activity and efficiently hydrolyzes butoxycarbonyl (Boc)-Ala-Ala-Asp-thiobenzoyl (SBzl) (Odake et al. 1991; Poe et al. 1991). This preferential activity for Asp is unique among mammalian serine proteases and is only found in the cysteine protease IL-1 converting enzyme (ICE) (Molineaux et al. 1993). A model of human granzyme B has been designed based on six known 3 D structures of serine proteases (including RMCP II) using the automated protein modeling package *ProMod* (Fig. 7, M. Peitsch et al., unpublished). From this model it is apparent that the conserved side chain of Arg-208 confers Asp specificity to granzyme B. Granzyme B also has minor activity towards Boc-Ala-Ala-X-Sbzl substrates where X is Asn or Ser. The pH optimum of granzyme A and B for the cleavage of the various substrates is approximately pH 8.

Inhibitors of serine proteases like PMSF, aprotinin, DFP and benzamidine are all potent inhibitors for the esterolytic activity of granzyme A (Masson et al. 1986; Simon et al. 1986; Young et al. 1986). TLCK and TPCK, excellent inhibitors for trypsin and chymotrypsin, respectively, have no or little effect on granzyme A. The plasma inhibitor antithrombin III (Masson and Tschopp 1988) and the tissue inhibitor protease nexin-1 (Gurwitz et al. 1989) inhibit granzyme A by forming a heparin-dependent, covalently linked 2:1 complex with granzyme A, each subunit interacting with one molecule of antithrombin III or nexin-1. Two specific inhibitor of granzyme A, H-D-Pro-Phe-Arg-chloromethyl-ketone (PFR-CK) and a isocoumarin-based one, designed according to the preferred amino acid sequence cleaved by this protease, have been shown to inhibit granzyme A (Odake et al. 1994; Simon et al. 1989). The esterolytic activity of granzyme B is efficiently blocked by 3, 4-dichloroisocoumarin (Odake et al. 1991).

5.3 Expression

Like perforin, granzymes are expressed in NK cells, γ/δ T cells, activated $CD8^+$ and $CD4^+$ T cells of both killer and helper type, but not always at comparable expression levels (Garcia et al. 1990a; Nagler-Anderson et al. 1989). In vivo, granzyme expression is found in lymphocytes infiltrating transplants during rejection (Clement et al. 1991) and rheumatoid arthritis (Cowing and Gilmore 1992). In pregnant rodents, granzymes are coexpressed in the maternal metrial gland along with perforin.

5.4 Function

Recent publications suggest an important role for granzymes in lymphocyte-mediated killing (Table 3). Inactivation of granzymes by either specific protease inhibitors (Ewoldt et al. 1992) or transfection with antisense mRNA results in reduced cell lysis. It was therefore speculated that granzymes cleave and activate

Table 3. Substrates of granzymes

	Function	Substrate
Granzyme A		
Apoptosis	Proteolytic cascade	Pro-ICE? PARP
Neurodegenerative diseases	Neurite retration Reversal of astrocyte stellate formation	Thrombin receptor Thrombin receptor
Inflammation	Cytokine activation	Pro-IL-1, Pro-ICE?
Polymyositis	Degradative activity on skeletal muscle proteins	Dystrophin, myosin, and nebulin
Tissue infiltration	Destruction of extracellular matrix	Collagen, fibronectin, etc.
Granzyme B		
Apoptosis	Proteolytic cascade	Pro-ICE?

ICE, interleukin-1 converting enzyme; IL-1, interleukin-1; PARP, poly-ADP-ribose polymerase.

intragranular proteins such as perforin (Talento et al. 1992). Also, DNA breakdown can be blocked by serine esterase inhibitors. Following the DNA fragmentation activity during granule fractionation of mouse CTLs, Hayes et al., showed a role for granzyme A in target cell DNA breakdown (Hayes et al. 1989). Recently, Shi et al., reported the purification of three serine proteases, called fragmentins, from rat large granular lymphocytes (Shi et al. 1992). Fragmentin 2, identified as an Asp-ase and the rat orthologue of granzyme B, induced DNA fragmentation in a fast-acting manner, whereas the two other fragmentins, both tryptases and homologous to granzyme A and human granzyme 3, respectively, induced slow DNA breakdown. Inactivation of the granzyme B gene by homologous recombination demonstrated that this protease is indeed responsible for the rapid DNA degradation observed during CTL attack (Heusel et al. 1994). Nevertheless, apoptosis was not completely abolished. Longer incubation of the target cells with CTLs revealed a slow-acting DNA degrading activity, probably due to other granzymes. It has been argued that every kind of protease could promote DNA degradation by cleaving histones to facilitate the access of DNases to the nucleus (Ortaldo 1993).

In all the aforementioned experiments, DNA degradation was seen only when target cells were pretreated either with detergents or perforin, suggesting that granzymes act inside the target cell. Whether they are directly involved in DNA degradation or, more likely, function as activators of a cellular DNA degrading machinery has yet to be investigated. Pasternack et al. demonstrated that granzyme A binds selectively to nuclear proteins, particularly nucleolin, which it also cleaves (Pasternack et al. 1991). In addition, TIA-1, a poly (A)-binding protein found in granules, was suggested as a natural target of granzymes (Tian et al. 1991). We have evidence that granzyme A efficiently cleaves poly-ADP-ribose-polymerase (PARP). The processing of PARP renders the NADPH-consuming enzyme active in the absence of DNA single strand nicks, resulting in energy depletion (unpublished). Shiver et al. showed, in a reconstitution system, that cotransfection of perforin and granzyme A in RBL cells is necessary to generate killer cells which cause both target cell lysis and DNA degradation (SHIVER et al. 1992). As discussed previously, transfection with perforin alone (SHIVER and HENKART 1991) led only to target cell lysis, whereas single transfectants of granzyme A induced neither DNA breakdown nor cell lysis. Cytostasis, the arrest of cellular proliferation without death, has been proposed to be another function of granzymes,. Sayers et al. found that rat Asp-ase (granzyme B) inhibited the growth of three adherent tumor lines for several days, probably by the cleavage of Arg-Gly-Asp (RGD)-containing adhesion molecules (SAYERS et al. 1992). Lymphocyte proteases are also capable of degrading extracellular matrix (ZANOVELLO et al. 1990) and muscle proteins (NAKAMURA et al. 1993).

The interleukin-1 β-converting enzyme (ICE), the mammalian homologue of the *C. elegans ced*-3 protein, plays a crucial role in cell death. Overexpression of either ced-3 or ICE was shown to induce apoptosis (MIURA et al. 1993a). Unlike granzymes, ICE is a cysteine proteinase, yet it shares the rare substrate site Asp with granzyme B. This raises the possibility that granzymes and ICE share common substrates (Fig. 8). The bona fide substrate of ICE, pro-IL-1 β, is indeed processed by granzymes into the active cytokine (Irmler et al., unpublished). In spite of the common substrate specificity of ICE and granzyme B, the latter enzyme is unable to convert pro-IL-1 β. In turn, granzyme A is a highly active converting enzyme, cleaving the precursor-IL-1 β downstream of the authentic Asp cleavage site (Irmler et al., unpublished).

The pro-enzyme form of ICE must be proteolytically cleaved at four aspartic acids to generate two ICE peptides required for activity in vitro. Both granzyme A and B cleave the proenzyme form of ICE into large fragments similar in size to the one found in the active processed ICE (unpublished).

Granzyme A may also contribute to neurodegenerative diseases. If not only induces neurite retraction at nanomolar concentrations, but also reverses the stellation of astrocytes (SUIDAN et al. 1994). All these effects are critically dependent on granzyme A's esterolytic activity and are inhibited by the protein kinase inhibitor staurosporine. Neurite retraction is known to be induced by thrombin. A synthetic peptide spanning the NH_2-terminal thrombin receptor activation sequence is cleaved by granzyme A at the authentic thrombin cleavage site LDPR/S. Antibodies to the thrombin receptor inhibit both thrombin- and granzyme A-mediated neurite retraction. Therefore, T cell-released granzyme A is capable of activating the thrombin receptor. Moreover, basic myelin protein is an excellent in vitro substrate of granzyme A (VANGURI et al. 1993).

Uncontrolled release of granzyme A from lymphocytes in the brain may have profound consequences, since nexin-1, the serine protease inhibitor found in the brain, is an inefficient inhibitor of granzyme A and is probably not able to block its proteolytic activity sufficiently (GURWITZ et al. 1989). For example, EAE (ZAMVIL and STEINMAN 1990) is a neurological autoimmune disease reminiscent of multiple sclerosis in humans. T lymphocytes reactive against myelin basic protein were found to be the pathogenic mechanism of EAE. The etiology involves homing,

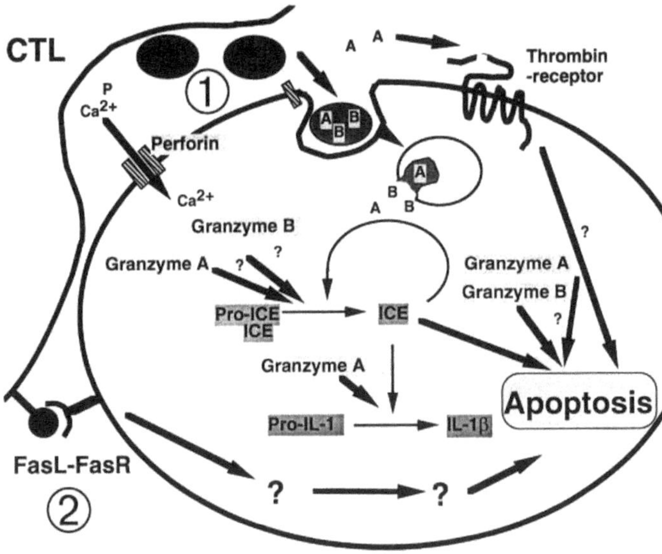

Fig. 8. Pathways of induction of apoptosis in target cells. *Pathway 1*: Granzymes are released from lymphocytes packed into vesicles. Upon endocytosis and fusion with the target cell membrane, granzymes gain access to the cytoplasm where they act on several poorly defined substrates whose cleavage induces the cell to collapse. It has been shown that granzyme A converts pro-interleukin-1 (IL-1) into the active form IL-1β, but granzymes may in addition cut IL-1 converting enzyme (ICE) and other substrates whose degradation lead to apoptosis. Signaling from the surface via granzyme A-induced activation of the thrombin receptor may also contribute to the cell death. Perforin forms transmembrane channels which allow uncontrolled Ca^{2+} entry. *Pathway 2*: The Fas ligand is exposed at the surface upon T cell receptor engagement. The ligand interacts with its respective receptor, which signals apoptosis via an unknown mechanism

extravasation and induction of tissue damage in the central nervous system. The encephalitogenic myelin basic protein (MBP)-specific T lymphocytes are, in most cases, $CD4^+$ and are MHC class II-restricted. Little is known about the mechanisms used by the encephalitogenic T lymphocytes. One of the earliest effects observed on target cells inflicted by EAE-causing T cells is loss of adhesion. Detachment of adherent astrocytes was suggested to be lymphocyte-specific. Granzyme A is found in activated T lymphocytes of both subtypes, $CD8^+$ or $CD4^+$ (GARCIA et al. 1990b). Brain infiltrating lymphocytes may therefore transport granzyme A into the brain, where its release from the autoreactive T cell population may exert its effect on neurons and astrocytes. It is noteworthy that astrocytes are believed to induce microcapillary and venule endothelial cells in the central nervous system to form the highly impermeable blood-brain barrier. Granzyme A could therefore impair brain function not only by disrupting the integrity of astroglial processes, but by damaging the blood-brain barrier.

How do granzymes enter the target cell to induce apoptosis? In the lymphocyte, perforin could stimulate the uptake of the granzymes into the target cell, although the exact mechanism is not yet defined. Similar to diphtheria toxin, SHI et al. (1992) proposed that the DNA damaging molecule could be pinocytosed by the target cells as they try to repair the membrane damage induced by perforin. Since at least a fraction of granzymes are released from CTLs in a vesicle enwrapped form, granzymes may enter cells as viruses do (Schärer et al., unpublished). Another, less probable, route of entry would be by direct passage through the pores created by perforin. Alternatively, granzymes may act from the outside by cleaving and activating surface receptors.

6 Perspectives

Cytotoxic T lymphocytes and NK cells kill their targets via a mechanism involving two independent lytic systems (Fig. 8). One is based on proteins which are stored in granules of activated CTLs and in NK cells. Upon specific contact with the target cells, the content of these storage organelle is secreted. Moreover, the release is directional, i.e., towards the interaction site with the target cells. Perforin and granzyme B are two proteins released from CTLs whose roles have been unequivocally described by the creation of mice deficient for the respective proteins. Perforin perforates the target cell membrane, while granzyme B is most likely involved in a proteolytic cascade leading to apoptosis. The two molecules work in concert. For granzyme B it has been shown that nuclear damage was initiated through premature activation of $p34^{cdc2}$, a serine-threonine kinase involved in cell cycle control (Heald et al. 1993; Shi et al. 1994).

The perforin knock-out mice have also elucidated a second lytic pathway which is functional in CTLs. Most, if not all, target cells tested display the Fas receptor at the surface and are lysed by perforin-free lymphocytes and NK cells. Fas-based killing accounts in most cases for approximately 25% of the observed lysis in a short-term assay. The challenge in the future will be to identify the mechanisms, which govern the activity of the Fas-based lytic machinery. What controls Fas ligand surface appearance? Why do not all cells die which bare the receptor? Why did two independent lytic systems evolve during evolution? Truly exciting years are yet to come.

References

Acha Orbea H, Scarpellino L, Hertig S, Dupuis M, Tschopp J (1990) Inhibition of lymphocyte mediated cytotoxicity by perforin antisense oligonucleotides. EMBO J 9: 3815–3819

Allbritton NL, Nagler AC, Elliott TJ, Verret CR (1988) Target cell lysis by cytotoxic T lymphocytes that lack detectable hemolytic perforin activity. J Immunol 141: 3243–3248

Amiguet P, Brunner J, Tschopp J (1985) The membrane attack complex of complement: lipid insertion of tubular and nontubular polymerized C9. Biochemistry 24: 7328–7334

Anderson P, Nagler AC, O BC, Levine H, Watkins S, Slayter HS, Blue ML, Schlossman SF (1990) A monoclonal antibody reactive with a 15-kDa cytoplasmic granule-associated protein defines a subpopulation of $CD8^+$ T lymphocytes. J Immunol 144: 574–582

Berke G, Rosen D (1988) Highly lytic in vivo primed cytolytic T lymphocytes devoid of lytic granules and BLT-esterase activity acquire these constituents in the presence of T cell growth factors upon blast transformation in vitro, J Immunol 141: 1429–1436

Berke G, Rosen D, Ronen D (1993) Mechanism of lymphocyte-mediated cytolysis: functional cytolytic T cells lacking perforin and granzymes. Immunology 78: 105–112

Blumenthal R, Millard PJ, Henkart MP, Reynolds CW, Henkart PA (1984) Liposomes as targets for granule cytolysin from cytotoxic large granular lymphocyte tumors. Proc Natl Acad Sci USA 81: 5551–5555

Bosma GC, Custer RP, Bosma MJ (1983) A severe combined immunodeficiency mutation in the mouse. Nature 301: 527

Brown GR, McGuire MJ, Thiele DL (1993) Dipeptidyl peptidase I is enriched in granules of in vitro- and in vivo-activated cytotoxic T lymphocytes. J Immunol 150: 4733–4742

Bugeon L, Cuturi MC, Paineau J, Anegon I, Soulillou JP (1993) Similar levels of granzyme A and perforin mRNA expression in rejected and tolerated heart allografts in donorspecific tolerance in rats. Transplantation 56: 405–408

Burns K, Duggan B, Atkinson EA, Famulski KS, Nemer M, Bleackley RC, Michalak M (1994) Modulation of gene expression by calreticulin binding to the glucocorticoid receptor Nature 367: 476–480

Clement MV, Soulie A, Legros MS, Guillet J, Gluckman E, Sigaux N, Sasportes M (1991) Perforin and granzyme B: predictive markers for acute GVHD or cardiac rejection after bone marrow or heart transplantation. Nouv Rev Fr Hematol 33: 465–470

Cowing C, Gilmore GL (1992) Allogeneic chimerism in SCID mice after neonatal transfer of bone marrow, J Immunol 148: 1072–1079

Dedhar S, Rennie PS, Shago M et al. (1994) Inhibition of nuclear hormone receptor activity by calreticulin. Nature 367: 480–483

Desai NM, Bassiri H, Kim J, Koller BH, Smithies O, Barket CF, Naji A, Markmann JF (1993) Islet allograft, islet xenograft, and skin survival in CD8$^+$ T lymphocyte-deficient mice. Transplantation 55: 718–722

Doetschman TC, Elstetter H, Kate M, Schmidt W, Kemler R (1985) The in vitro development of blastocyst-derived embryonic stem cell lines: formation of visceral yolk sac, blood islands, and myocardium, J Embryol Exp Morphol 87: 27–45

Dourmashkin RR, Deteix P, Simone CB, Henkart P (1980) Electron microscopic demonstration of lesions on target cell membranes associated with antibody-dependent cellular cytotoxicity. Clin Exp Immunol 43: 554–560

Duke RC, Persechini PM, Chang S, Liu CC, Cohen J, Young JD (1989) Purified perforin induces target cell lysis but not DNA fragmentation, J Exp Med 170: 1451–1456

Dupuis M, Peitsch MC, Hamann U, Stanley KK, Tschopp J (1993a) Mutations in the putative lipid-interaction domain of complement C9 result in defective secretion of the functional protein. Mol Immunol 30: 195–100

Dupuis M, Schaerer E, Krause KH, Tschopp J (1993b) The calcium-binding protein calreticulin is a major constituent of lytic granules of cytolytic T lymphocytes. J Exp Med 177: 1–7

Ewoldt GR, Winkler U, Powers JC, Hudig D (1992) Sulfonyl fluoride serine protease inhibitors inactivate RNK-16 lymphocyte granule proteases and reduce lysis by granule extracts and perforin. Mol Immunol 29: 713–721

Fink TM, Zimmer M, Weitz S, Tschopp J, Jenne DE, Lichter P (1992) Human perforin (PRF1) maps to 10q22, a region that is syntenic with mouse chromosome 10. Genomics 13: 1300–1302

Fung-Leung WP, Schilham MW, Rahemtulla A, Kündig TM, Vollenweider M, Potter, J, Van Ewijk W, Mak TW (1991) Immune response against lymphocytic choriomeningitis virus infection in mice without CD8 expression. J Exp Med 174: 443–449

Gagliardini V, Fernandez P-A, Lee RKK, Drexler HCA, Rotello RJ, Fishman MC, Yuan J (1994) Prevention of vertebrate neuronal death by the crmA gene. Science 263: 826–828

Garcia SJ, MacDonald HR, Jenne DE, Tschopp J, Nabholz M (1990a) Cell specificity of granzyme gene expression. J Immunol 145: 3111–3118

Garcia SJ, MacDonald HR, Jenne DE, Tschopp J, Nabholz M (1990b) Cell specificity of granzyme gene expression. J Immunol 145: 3111–3118

Gossler A, Doetschman TC, Korn R,Sterfling E, Kemler R (1986) Transgenesis by means of blastocyst-derived embryonic stem cell lines. Proc Natl Acad Sci USA 83: 9065–9069

Griffiths GM, Namikawa R, Müller C, Liu CC, Young JD, Billingham M, Weissman I (1991) Granzyme A and perforin as markers for rejection in cardiac transplantation. Eur J Immunol 21: 687–693

Gurwitz D, Simon MM, Fruth U, Cunningham DD (1989) Protease nexin-1 complexes and inhibits T cell serine proteinase-1. Biochem Biophys Res Commun 161: 300–304

Hayes MP, Berrebi GA, Henkart PA (1989) Induction of target cell DNA-release by the cytotoxic T lymphocyte granule protease granzyme A. J Exp Med 170: 933–946

Heald R, McLoughlin M, McKeon F (1993) Human Weel maintain mitotic timing by protecting the nucleus from cytoplasmically activated cdc2 kinase. Cell 74: 463–474

Held W, Meyermann R, Qin Y, Mueller C (1993) Perforin and tumor necrosis factor alpha in the pathogenesis of experimental allergic encephalomyelitis: comparison of autoantigen induced and transferred disease in Lewis rats. J Autoimmun 6: 311–322

Henkart MP, Henkart PA (1982) Lymphocyte mediated cytolysis as a secretory phenomenon. Adv Exp Med Biol 146: 227–242

Heusel JW, Wesselschmidt RL, Shresta S, Russell JH, Ley TJ (1994) Cytotoxic lymphocytes require granzyme B for the rapid induction of DNA fragmentation and apoptosis in allogeneic target cells. Cell 76: 977–987

Itoh N, Yonehara S, Ishii A, Yonehara M, Mizushima S, Sameshima M, Hase A, Seto Y, Nagata S (1991) The polypeptide encoded by the cDNA for human cell surface antigen Fas can mediate apoptosis. Cell 66: 233–243

Jenne DE, Tschopp J (1988) Granzymes: a family of serine proteases in granules of cytolytic T lymphocytes. In: Podack ER (ed) Cytotoxic effector mechanisms. Springer, Berlin Heidelberg New York, pp 33–47 (current topics in microbiology and immunology, vol 140)

Jenne D, Rey C, Haefliger JA, Qiao BY, Groscurth P, Tschopp J (1988) Identification and sequencing of cDNA clones encoding the granule-associated serine proteases granzymes, D, E, and F of cytolytic T lymphocytes. Proc Natl Acad Sci USA 85: 4814–4818

Jenne DE, Masson D, Zimmer M, Haefliger JA, Li WH, Tschopp J (1989) Isolation and complete structure of the lymphocyte serine protease granzyme G, a novel member of the granzyme multigene family in murine cytolytic T lymphocytes. Evolutionary origin of lymphocyte proteases. Biochemistry 28: 7953–7961

Kägi D, Ledermann B, Bürki K, Seller P, Odermatt B, Olson KJ, Podack ER, Zinkernagel RM et al. (1994) Cytotoxicity mediated by T cells and natural killer cells is greatly impaired in perforin-deficient mice. Nature 369: 31–37

Kerr J, Wyllie AH, Currie AR (1972) Apoptosis a basic biological phenomenon with wideranging implications in tissue kinetics. Br J Cancer 68: 239–257

Koller BH, Marrack P, Kappler JW, Smithies O (1990) Normal development of mice deficient in beta 2-microglobulin, MHC Class I proteins, and $CD8^+$ T cells. Science 248: 1227–1230

Krahenbuhl O, Tschopp J (1991) Debate: the mechanism of lymphocyte-mediated killing. Perforin-induced pore formation. Immunol Today 12: 399–402

Lichtenheld MG, Podack ER (1989) Structure of the human perforin gene. A simple gene organization with interesting potential regulatory sequences. J Immunol 143: 4267–4274

Lowin B, Beermann F, Schmidt A, Tschopp J (1994) A null mutation in the perforin gene abolishes cytolytic T lymphocyte- and NK-mediated cytotoxicity. Proc Natl Acad Sci USA 91: 11571–11575

Lowin B, Hahne M, Mattmann C, Tschopp J (1994b) Cytolytic T-cell cytotoxicity is mediated through perforin and Fas lytic pathways. Nature 370: 650–652

MacDermott RP, Schmidt RE, Caulfield JP, Hein A, Bartley GT, Ritz J, Schlossman SF, Austen KF, Stevens RL (1985) Proteoglycans in cell-mediated cytotoxicity. J Exp Med 162: 1771–1787

MacLennon ICM, Gotch FM, Golstein P (1980) Limited specific T-cell mediated cytolysis in the absence of extracellular calcium. Immunology 39: 109–116

Masson D, Tschopp J (1985) Isolation of a lytic, pore-forming protein (perforin) from cytolytic T-lymphocytes. J Biol Chem 260: 9069–9072

Masson D, Tschopp J (1988) Inhibition of lymphocyte protease granzyme A by antithrombin III. Mol Immunol 25: 1283–1289

Masson D, Nabholz M, Estrade C, Tschopp J (1986) Granules of cytolytic T-lymphocytes contain two serine esterases EMBO J 5: 1595–1600

Matsumoto S, Sakiyama Y, Ariga T, Gallagher R, Taguchi Y (1992) Progress in primary immuno-deficiency. Immunol Today 13: 4–5

Miura M, Zhu H, Rotello R, Hartwieg EA, Yuan J (1993a) Induction of apoptosis in fibroblasts by IL-1 beta-converting enzyme, a mammalian homolog of the C. elegans cell death gene ced-3. Cell 75: 653–660

Miura M, Zhu H, Rotello R, Hartwieg EA, Yuan J (1993b) Induction of apoptosis in fibroblasts by IL-1beta-converting enzyme, a mammalian homolog of the C. elegans cell death gene ced-3.Cell 75: 653–660

Molineaux SM, Casano FJ, Rolando AM, Peterson EP, Limjuco G, Chin J, Griffin PR, Calaycay JR, Ding GJ, Yamin TT et al. (1993) Interleukin 1 beta (IL-1 beta) processing in murine macrophages requires a structurally conserved homologue of human IL-1 beta conventing enzyme Proc Natl Acad Sci USA 90: 1809–1813

Muller C, Kagi D, Aebischer T, Odermatt B, Held W, Podack ER, Zinkernagel RM and Hengartner H (1989) Detection of perforin and granzyme A mRNA in infiltrating cells during infection of mice with lymphocytic choriomeningitis virus Eur J Immunol 19: 1253–1259

Nagler-Anderson C, Lichtenheld MG, Eisen HN, Podack ER (1989) Perforin mRNA in primary peritoneal exudate cytotoxic T lymphocytes, J Immunol 143: 3440–3445

Nakamura K, Arahata K, Ishiura S, Osame M, Sugita H (1993) Degradative activity of granzyme A on skeletal muscle proteins in vitro: a possible molecular mechanism for muscle fiber damage in polymyositis, Neuromuscul Disord 3: 303–310

Nakamura T, Ebihara I, Osada S, Okumura K, Tomino Y, Koide H (1992) Perforin gene expression in T lymphocytes correlates with disease activity in immunoglobulin A nephropathy. Clin Sci 82: 461–468

Odake S, Kam CM, Narasimhan L, Poe M, Blake JT, Krahenbuhl O, Tschopp J, Powers JC (1991) Human and murine cytotoxic T lymphocyte serine proteases: subsite mapping with peptide thioester substrates and inhibition of enzyme activity and cytolysis by isocoumarins. Biochemistry 30: 2217–2227

Odake S, Kam CM, Oleksyszyn J, Hernandez MA, Poe M, Tschopp J, Virca GD, Powers JC (1994) Substrate and inhibitor studies with human recombinant granzyme A, murine granzyme B and human Q31 chymase. Biochemistry (in press)

Ortaldo J (1993) Cell-mediated cytotoxicity. Summary of the 5th international workshop. Neve Ilan, Israel. Eur Cytokine Netw 4: 383–384

Ostergaard HL, Kane KP, Mescher MF, Clark WR (1987) Cytotoxic T lymphocyte mediated lysis without release of serine esterase. Nature 330: 71–72

Pasternack MS, Bleier KJ, McInerney TN (1991) Granzyme A binding to target cell proteins. Granzyme A binds to and cleaves nucleolin in vitro. J Biol Chem 266: 14703–14708

Peitsch MC, Amiguet P, Guy R, Brunner J, Maizel JJ, Tschopp J (1990) Localization and molecular modelling of the membrane-inserted domain of the ninth component of human complement and perforin. Mol Immunol 27: 589–602

Penninger JM, Neu N, Timms E, Wallace VA, Koh D-R, Kishihara K, Pummerer C, Mak TW (1993) Induction of experimental autoimmune myocarditis in mice lacking CD3 of CD8 molecules. J Exp Med 178: 1831–1842

Peters PJ, Geuze HJ, van der Donk HA, Slot JW, Griffith JM, Stam NJ, Clevers HC, Borst J (1989) Molecules relevant for T cell-target cell interaction are present in cytolytic granules of human T lymphocytes. Eur J Immunol 19: 1469–1475

Peters R, Sauer H, Tschopp J, Fritzsch G (1990) Transients of perforin pore formation observed by fluorescence microscopic single channel recording. EMBO J 9: 2447–2451

Podack ER, Hengartner H (1989) Structure of perforin and its role in cytolysis. Year Immunol 6: 245–261

Podack ER, Young JE, Cohn ZA (1985) Isolation and biochemical and functional characterization of perforin 1 from cytolytic T-cell granules. Proc Natl Acad Sci USA 82: 8629–8633

Poe M, Blake JT, Boulton DA, Gammon M, Sigal NH, Wu JK, Zweerink HJ (1991) Human cytotoxic lymphocyte granzyme B. Its purification from granules and the characterization of substrate and inhibitor specificity. J Biol Chem 266: 98–103

Rahemtulla A, Fung-Leung WP, Schilham MW, Kündig TM, Sambhara SR, Narendran A, Arabian A, Wakeham A, Paige CJ, Zinkernagel RM et al. (1991) Normal development and functional $CD8^+$ cells but markedly decreased helper cell activity in mice lacking CD4. Nature 353: 180–184

Rouvier E, Luciani MF, Golstein P (1993) Fas involvement in Ca^{2+}-independent T cell-mediated cytotoxicity. J Exp Med 177: 195–200

Sayers TJ, Wiltrout TA, Sowder R, Munger WL, Smyth MJ, Henderson LE (1992) Purification of a factor from the granules of a rat natural killer cell line (RNK) that reduces tumor cell growth and changes tumor morphology. Molecular identity with a granule serine protease (RNKP-1). J Immunol 148: 292–300

Schaerer E, Karapetian O, Adrian M, Tschopp J (1995) Cytotoxic T cell specific proteins aim at the target cell nucleus (in preparation)

Seko Y, Shinkai Y, Kawasaki A, Yagita H, Okumura K, Yazaki Y (1993) Evidence of perforin-mediated cardiac myocyte injury in acute murine myocarditis caused by Coxsackie virus B3. J Pathol 170: 53–58

Shi L, Kam CM, Powers JC, Aebersold R, Greenberg AH (1992) Purification of three cytotoxic lymphocyte granoule serine proteases that induce apoptosis through distinct substrate and target cell interactions. J Exp Med 176: 1521–1529

Shi L, Nishioka WK, Th'ng J, Bradbury EM, Litchfield DW, Greenberg AH (1994) Premature p34 cdc2 activation required for apoptosis. Science 263: 1143–1145

Shiver JW, Henkart PA (1991) A noncytotoxic mast cell tumor line exhibits potent IgE-dependent cytotoxicity after transfection with the cytolysin/perforin gene. Cell 64: 1175–1181

Shiver JW, Su L, Henkart PA (1992) Cytotoxicity with target DNA breakdown by rat basophilic leukemia cells expressing both cytolysin and granzyme A. Cell 71: 315–322

Simon MM, Fruth U, Simon HG, Kramer MD (1986) A specific serine proteinase is inducible in Lyt-2+, L3T4- and Lyt-2-, L3T4+ T cells in vitro but is mainly associated with Lyt-2+, L3T4- effector cells in vivo. Eur J Immunol 16: 1559–1568

Simon MM, Prester M, Kramer MD, Fruth U (1989) An inhibitor specific for the mouse T-cell associated serine proteinase 1 (TSP-1) inhibits the cytolytic potential of cytoplasmic granules but not of intact cytolytic T cells. J Cell Biochem 40: 1–13

Stanley KK, Kocher HP, Luzio JP, Jackson P, Tschopp J (1985) The sequence and topology of human complement component C9. EMBO J 4: 375–382

Stevens RL, Otsu K,Weis JH, Tantravahi RV, Austen KF, Henkart PA, Galli MC, Reynolds CW (1987) Co-sedimentation of chondroitin sulfate A glycosaminoglycans and proteoglycans with the cytolytic secretory granules of rat large granular lymphocyte (LGL) tumor cells, and identification of a mRNA in normal and transformed LGL that encodes proteoglycans. J Immunol 139: 863–868
Suda T, Nagata S (1994) Purification and characterization of the Fas-ligand that induces apoptosis. J Exp Med 179: 873–879
Suda T, Takahashi T, Golstein P, Nagata S (1993) Molecular cloning and expression of the Fas ligand, a novel member of the tumor necrosis factor family. Cell 75: 1169–1178
Suidan HS, Bouvier J, Schaerer E, Stone SR, Monard D, Tschopp J (1994) Granzyme A released upon stimulation of cytotoxic T lymphocytes activates the thrombin receptor on neuroanal cells and astrocytes. Proc Natl Acad Sci USA 91: 8112–8116
Talento A, Nguyen M, Law S, Wu JK, Poe M, Blake JT, Patel M, Wu TJ, Manyak CL, Silberklang M et al. (1992) Transfection of mouse cytotoxic T lymphocyte with an antisense granzyme A vector reduces lytic activity. J Immunol 149: 4009–4015
Tian Q, Streuli M, Saito H, Schlossman SF, Anderson P (1991) A polyadenylate binding protein localized to the granules of cytolytic lymphocytes induces DNA fragmentation in target cells. Cell 67: 629–639
Trapani JA, Kwon BS, Kozak CA, Chintamaneni C, Young JD, Dupont B (1990) Genomic organization of the mouse pore-forming protein (perforin) gene and localization to chromosome 10. Similarities to and differences from C9. J Exp Med 171: 545–557
Trauth BC, Klas C, Peters AM, Matzku S, Moller P, Falk W, Debatin KM, Krammer PH (1989) Monoclonal antibody-mediated tumor regression by induciton of apoptosis. Science 245: 301–305
Trenn G, Takayama H, Sitkovsky MV (1987) Exocytosis of cytolytic granules may not be required for target cell lysis by cytotoxic T-lymphocytes Nature 330: 72–74
Trinchieri G (1989) Biology of natural killer cells. Adv Immunol 47: 187–375
Tschopp J, Nabholz M (1990) Perforin-mediated targt cell lysis by cytolytic T lymphocytes. Annu Rev Immunol 8: 279–302
Vanguri P, Lee E, Henkart P, Shin ML (1993) Hydrolysis of myelin basic protein in myelin membranes by granzymes of large granular lymphocytes. J Immunol 150: 2431–2439
Woodbury RG, Neurath H (1980) Structure, specificity and localization of the serine preoteases of connective tissue. Febs Lett 114: 189–196
Yeung RSM, Penninger J, Mak TW (1993) Genetically modified animals and immunodeficiency, Curr Opin Immunol 5: 585–594
Young JE (1989) Killing of target cells by lymphocytes: a mechanistic view. Physiol Rev 69: 250–314
Young JE, Cohn ZA (1987) Cellular and humoral mechanisms of cytotoxicity: structural and functional analogies. Adv Immunol 41: 269–332
Young JE, Nathan CF, Podack ER, Palladino MA, Cohn ZA (1986) Functional channel formation associated with cytotoxic T-cell granules. Proc Natl Acad Sci USA 83: 150–154
Young JE, Clark WR, Liu CC, Cohn ZA (1987) A calcium- and perforin-independent pathway of killing mediated by murine cytolytic lymphocytes. J Exp Med 166: 1894–1899
Young LH, Peterson LB, Wicker LS, Persechini PM, Young JD (1989) In vivo expression of perforin by $CD8^+$ lymphocytes in autoimmune disease. Studies on spontaneous and adoptively transferred diabetes in nonobese diabetic mice. J Immunol 143: 3994–3999
Yu YYL, Kumar V, Bennett M (1992) Murine natural killer cells and marrow graft rejection. Annu Rev Immunol 10: 189
Yuan J, Horvitz HR (1992) The Caenorhabditis elegans cell death gene ced-4 encodes a novel protein and is expressed during the period of extensive programmed cell death. Development 116: 309–320
Yuan J, Shaham S, Ledoux S, Ellix HM, Horvitz HR (1993) The C. elegans death gene ced-3 encodes a protein similar to mammalian interleukin-1 beta-converting enzyme. Cell 75: 641–652
Zamvil SS, Steinman L (1990) The T lymphocyte in experimental allergic encephalomyelitis, Annu Rev Immunol 8: 579–621
Zanovello P, Bronte V, Rosato A, Pizzo P, Di Virgilio F (1990) Responses of mouse lymphocytes to extracellular ATP. II. Extracellular ATP causes cell type-dependent lysis and DNA fragmentation. J Immunol 145: 1545–1550
Zheng LM, Liu CC, Ojcius DM, Young JD (1991) Expression of lymphocyte perforin in the mouse uterus during pregnancy. J Cell Sci 99: 317–323
Zijlstra M, Bix M, Simister NE, Loring JM, Raulet DH, Jaenisch R (1990) beta 2-microglobulin deficient mice lack $CD4\text{-}8^+$ cytolytic T cells. Nature 344: 742–746

Fas-Based T Cell-Mediated Cytotoxicity

P. GOLSTEIN

1 Introduction

Phenomenological studies on T cell-mediated cytotoxicity in vitro have been extensively reviewed (PERLMANN and HOLM 1969; CEROTTINI and BRUNNER 1974; GOLSTEIN and SMITH 1977; HENNEY 1977; MARTZ 1977; BERKE 1980; SANDERSON 1981; HENKART 1985; BERKE 1989a; YOUNG 1989; TSCHOPP and NABHOLZ 1990; DUKE 1991; GOLSTEIN et al. 1991; SITKOVSKY and HENKART 1993). Based on these studies, two major molecular pathways for the lethal hit stage of T cell-mediated cytotoxicity have been proposed. The first pathway involves granule exocytosis and perforin release and was detected about 10 years ago (HENKART 1985; PODACK 1985). When a cytotoxic T lymphocyte (CTL) encounters a target cell, the channel-forming molecule perforin and serine-esterases from CTL granules are released at the interface between the CTL and the target cell. Perforin molecules then form channels within the target cell membrane, which, together with some contribution from serine-esterases (SHIVER et al. 1992; SHI et al. 1992), lead to target cell death. The second pathway involves direct effector cell-target cell membrane interactions during which transduction of a signal within the target cell occurs

Centre d'Immunologie INSERM-CNRS de Marseille-Luminy, Case 906, 13288 Marseille Cedex 9, France leading to its death. Striking recent progress has led to demonstration of the use of *both* pathways by cytotoxic T cells.

Here we shall consider only those pathways and molecules involved in T cell-mediated cytotoxicity after the recognition stage, as detected in short-term (4h) ^{51}Cr release tests in vitro.

2 Cytotoxicity Pathways

2.1 A Perforin-Based Mechanism

The formal demonstration of a role for perforin in a major pathway of T cell-mediated cytotoxicity came from experiments using perforin gene knock-out (P°) mice (KÄGI et al. 1994a). In these mice, most of the antiviral and anti-allogeneic cytotoxicity was lost. Also, there was a loss of antiviral protection, at least in the lymphocyte choriomeningitis virus (LCMV) system. It was therefore concluded that: (1) perforin is required for a major cytotoxicity pathway and (2) this very same pathway is probably necessary for protection against at least some viral infections. A few questions remained, however. For instance, it is still not completely clear how perforin contributes to target cell death. Furthermore, cells from the P° mice still expressed a fraction of the cytotoxicity of wild-type mouse cells. Thus, a perforin-based mechanism was not sufficient to explain T cell-mediated cytotoxicity.

Moreover, even in the perforin pathway, perforin alone was not enough, since when large amounts of purified perforin were added to target cells, nonapoptotic death ensued (GROMKOWSKI et al. 1988; DUKE et al. 1989) instead of the apoptotic death (KERR et al. 1972) caused by CTLs (COHEN 1991; GOLSTEIN et al. 1991). Perforin cell-mediated cytotoxicity (P-CMC) seems to also require other molecules, notably certain serine-esterases (SHIVER et al. 1992; SHI et al. 1992). The requirement for granzyme B in a mechanism of T cell-mediated cytotoxicity was indeed recently demonstrated (HEUSEL et al. 1994).

Several lines of evidence suggested that, in addition to the perforin-based mechanism, at least another mechanism was at play in T cell-mediated cytotoxicity (CLARK et al. 1988; BERKE 1989b; YOUNG et al. 1989; JU et al. 1990; CHANG et al. 1990; DUKE 1991; LANCKI et al. 1991; JU 1991; SMYTH et al. 1992; FITCH et al. 1993). For example, although cytolysis by most CTL populations or clones requires extracellular calcium, in line with the Ca^{2+} requirements of a perforin-based mechanism (PODACK et al. 1985; YOUNG et al. 1987b; ISHIURA et al. 1990), part of the cytotoxic activity they exert is often independent of extra-cellular Ca^{2+} (MACLENNAN et al. 1980; TIROSH and BERKE 1985; TRENN et al. 1987; OSTERGAARD et al. 1987; YOUNG et al. 1987a). These considerations led to a search for a nonperforin-based mechanism(s).

3 The Search for Nonperforin-Based Mechanisms

Some ligands can bind receptors at the surface of a cell and induce the transduction of a signal. This signal can be interpreted as a cell death signal and then lead to the death of this cell. For instance, antigens would bind B cell surface immunoglobulins or T cell receptors and lead in certain cases to death of the cells bearing these receptors, or tumor necrosis factor (TNF) would bind cell surface TNF receptors and often lead to the death of the corresponding cells. We thought a CTL pathway might operate in a similar way, that is via a target cell surface, signal transducing molecule. An immediate problem was that CTLs were suspected and are now known to kill by more than one pathway, which could have blurred the search for molecules involved in one of them. We therefore tried to isolate a cytotoxic cell that killed via only one molecular mechanism.

A mouse CTL line derived by H. von Boehmer and Coworkers from C57Bl/6 anti-male killer cells (VON BOEHMER et al. 1979) was fused by Markus Nabholz and Coworkers in Lausanne with a rat T lymphoma (Fig. 1). The resulting PC60 mouse x rat CTL hybridoma (CONZELMANN et al. 1982) was cytotoxic and was also unstable in culture. Taking advantage of the instability, we grew this PC60 hybridoma in vitro and subcloned it serially 10 times, selecting each time the most cytotoxic

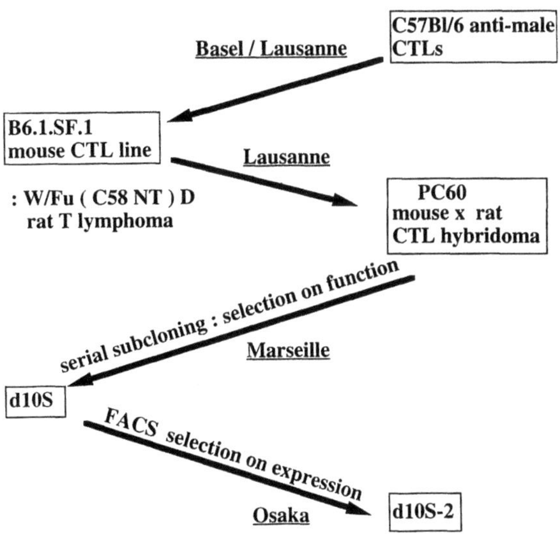

Fig. 1. Pedigree of PC60 and PC60-derived clones. The mouse T cell clone used as one of the parental cells for the construction of the PC60 hybridoma had initial anti-male D^b and cross-reactive anti-D^d specificity, as tested in Basel and Lausanne (VON BOEHMER et al. 1979). It was fused in Lausanne with a rat T lymphoma (CONZELMANN et al. 1982). The resulting PC60 hybridoma cells were serially subcloned in Marseille with constant selection of the clone most cytotoxic for YAC target cells. The product of the tenth serial subcloning, called d10S, was used for cytotoxicity experiments showing the Fas pathway (ROUVIER et al. 1993; GOLSTEIN et al. 1994) and selected by FACS using a soluble Fas molecule to obtain the Fas ligand-overexpressing d10S-2 cell (SUDA et al. 1993), which was used to clone the Fas ligand cDNA (SUDA et al. 1993)

clone. The resulting d10S clone had about 53 chromosomes compared to the initial 79 chromosomes in PC60 cells. However, by design, the cytotoxic function was conserved. In fact the d10S cells appeared more cytotoxic than its ancestor PC60 hybridoma cells. A detailed account of the derivation and properties of d10S cells is given elsewhere (GOLSTEIN et al. 1994).

3.1 The Fas-Based Mechanism: Requirement for Fas at the Target Cell Surface

Upon incubation with a mixture of PMA and ionomycin (PI), d10S cells were particularly cytotoxic for a range of tumor cells, such as YAC cells, in an MHC-unrestricted manner and at relatively low ratios (ROUVIER et al. 1993; GOLSTEIN et al. 1994). These d10S cells activated with PI were also able to significantly lyse thymocytes at ratios as low as 0.1 : 1 (ROUVIER et al. 1993). We compared these findings to recently published work on the Fas molecule. The Fas cell surface molecule (YONEHARA et al. 1989; TRAUTH et al. 1989; ITOH et al. 1991; WATANABE-FUKUNAGA et al. 1992b) can transduce a cell death signal when engaged by antibodies (YONEHARA et al. 1989; TRAUTH et al. 1989) and is either not expressed or abnormally expressed (WATANABE-FUKUNAGA et al. 1992a) in the *lpr* mutant mouse (reviewed in DAVIDSON et al. 1986), which exhibits a lymphoproliferative disorder. The latter observation immediately suggested that Fas was involved in the control of lymphocyte development, a point that we shall consider in more detail below. With regard to the mechanism of d10S-mediated lysis, the observation (WATANABE-FUKUNAGA et al. 1992a) that Fas was abundantly expressed in the thymus of normal but not of *lpr* mice provided us with an hypothesis and a way to test it.

PI-activated d10S cells were found to lyse wild-type but not *lpr* thymocytes (ROUVIER et al. 1993). In these experiments, a useful control was provided by the use of target thymocytes from *gld* mice; these mutant mice have the same cellular phenotype as *lpr* mice, but have an altered Fas ligand (TAKAHASHI et al. 1994); *gld* thymocytes were lysed by PI-activated d10S cells, strongly indicating that the inability to lyse *lpr* thymocytes was due to lack of Fas and not a secondary consequence of the mutant phenotype (ROUVIER et al. 1993). We also tested the cytotoxicity of PI-activated d10S cells for a number of conventional tumor target cells. YAC, EL-4 and P815 cells were sensitive, while L1210 cells were either not sensitive or were only barely lysed. L1210 cells were found by northern blot to express no or little of the main Fas transcript. We then transfected these L1210 cells with Fas cDNA and obtained L1210-Fas cells, which expressed the main Fas transcript and were sensitive to PI-activated d10S cells. Thus, by two independent approaches (mutants and transfectants) target cell expression of Fas was found to be required for d10S cell-mediated lysis. We also showed that this mechanism was not particular to d10S, but was used by practically all of the CTL clones and populations we tested (data not shown), and also accounted for part of the antigen-specific cytotoxicity by alloimmune MLC and PEL killer cells (ROUVIER et al. 1993).

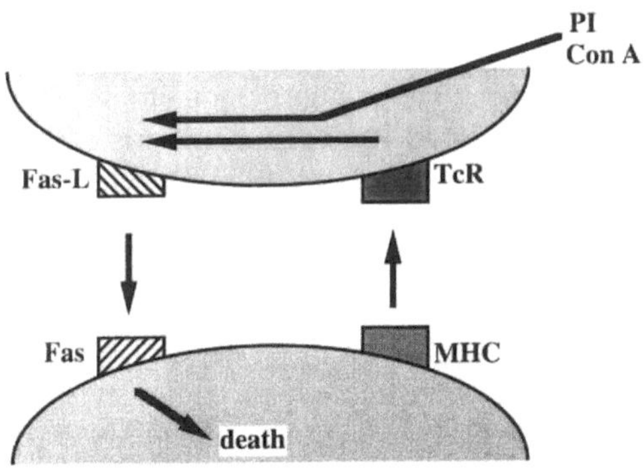

Fig. 2. Fas-based T cell-mediated cytotoxicity. Fas was shown to be required for the transduction of a death signal into the target cell (ROUVIER et al. 1993). This implied the existence of a Fas ligand, which was secondarily cloned (SUDA et al. 1993). This Fas ligand is expressed, either nonspecifically upon activation with PMA and ionomycin (PI) (ROUVIER et al. 1993; VIGNAUX and GOLSTEIN 1994) or concanavalin A (ConA) (KÄGI et al. 1994b), or antigen-specifically upon encounter with alloantigen (ROUVIER et al. 1993), which implies interactions in the effector cell between the T cell receptor (*TcR*) and the Fas ligand. In this scheme, at least two molecules are involved on the target cell (Fas and the MHC) and on the effector cell (Fas-L and the TCR). Also, this pathway of cytotoxicity is molecularly defined through both Fas on the target cell and Fas-L on the effector cell

Thus, Fas was required at the target cell level, presumably at the target cell surface, which implied the functional availability of a Fas ligand at the effector cell surface. This functional availability could be obtained in two different ways: (1) either by nonantigen-specific activation with PI, or (2) by antigen-specific stimulation of the T cell receptor (TCR). The latter would occur upon recognition by the TCR of MHC antigens at the target cell surface. Thus, at the target cell surface this mechanism involved both Fas for the transduction of the cell death signal and MHC to ensure specific recognition (Fig. 2). In molecular terms, Fas defined this cytotoxicity pathway at the target cell level, while a definition at the effector cell level required cloning of the Fas ligand.

3.2 Requirement for a Fas Ligand at the Effector Cell Surface

Since d10S cells exert cytolysis via the Fas pathway, they could be suspected to express a cell surface ligand for Fas. Nagata and his colleagues in Osaka selected from d10S cells variants expressing high amounts of this Fas ligand. The selection was done by FACS after binding of a Fas-Fc chimeric molecule (Fig. 1). Variants were obtained which overexpressed the Fas ligand. From these variant clones the Fas ligand cDNA was cloned by expression (SUDA et al. 1993). It is remarkable how flexible the PC60-derived cells proved to be, allowing the several rounds of selection depicted in Fig. 1. This flexibility might be related to the fact that these cells are an heterospecific, naturally unstable hybridoma.

It was in fact the rat Fas ligand which was cloned from the d10S cells. This raised the interesting problem of a control in *trans* of the expression of this Fas ligand, which was probably initially expressed by the mouse component of the original PC60 hybridoma.

The sequence of the rat Fas ligand was found to be homologous to the TNFs and to the CD40, CD27 and CD30 ligands (SUDA et al. 1993). Thus, there is a family of homologous ligands including the Fas ligand, the TNFs and others, corresponding to a family of homologous receptors including the NGF receptor, Fas, the TNF receptors, CD40, CD27 and CD30.

The Fas ligand cDNA was transfected into Cos cells. The transfectants were highly cytotoxic against Fas-bearing target cells (SUDA et al. 1993). Thus, and spectacularly, transfection of the Fas ligand was enough to confer cytotoxic activity to Cos cells. Since these are fibroblast-like cells, the results showed that Fas-mediated cytotoxicity could occur in the absence of other molecules preferentially expressed in activated lymphocytes. Moreover, purified Fas ligand protein exhibited strong cytotoxic activity against Fas-bearing target cells (SUDA and NAGATA 1994), showing that effector cell molecules other than the Fas ligand were not required for *execution* of Fas cell-mediated cytotoxicity (F-CMC) (while, for instance, the TCR is required for *induction* of antigen-specific F-CMC, and other, unknown molecules may modulate the Fas pathway).

Together with other results indicating that some granzymes are not expressed in d10S cells (GOLSTEIN et al. 1994), the experiments above showed that at least some components of the perforin pathway of cytotoxicity are not required for the Fas pathway. In the same vein, P° effectors could lyse via the Fas pathway (KÄGI et al. 1994a,b), and *gld* effectors, known not to lyse via the Fas pathway (VIGNAUX and GOLSTEIN 1994; RAMSDELL et al. 1994) because of a mutation of the Fas ligand (TAKAHASHI et al. 1994), could lyse via the perforin pathway (VIGNAUX and GOLSTEIN 1994). Thus, no molecular cross-talk i.e., no common molecular component, has been detected so far between these mechanisms, making it all the more remarkable that these mechanisms can coexist in the same effector cells (unpublished experiments).

Thus, F-CMC is now defined in molecular terms not only through the requirement for Fas at the target cell surface, but through the requirement for a Fas ligand at the effector cell surface. Both Fas and Fas ligand cell surface molecules interact when an effector cell encounters a target cell.

4 Properties and Extent of Fas-Based Cell-Mediated Cytotoxicity

Fas-based Cell-mediated Cytotoxicity is rapid, leading to detectable ^{51}Cr release (a late indicator of cell death) after 2–3 h of incubation of YAC target cells with d10S cells, even those not activated with PI (GOLSTEIN et al. 1994). Also, the characteristic apoptotic ladder pattern of DNA fragmentation, that occurred even in the absence of extracellular Ca^{2+}, could be detected in YAC target cells after 2h of incubation with d10S PI (unpublished results). Moreover, F-CMC required no extracellular Ca^{2+}, since, in contrast to P-CMC, F-CMC could still be detected in the presence of EGTA-Mg^{2+} (Rouvier et al. 1993; Golstein et al. 1994), accounting for about 10%–20% (MacLennan et al. 1980; Rouvier et al. 1993) of specific T cell-mediated cytotoxicity. More recent results suggested that this may well be an underestimate: direct evaluations in P° mice (Kägi et al. 1994b) and observations that induction, as opposed to execution, of F-CMC was Ca^{2+}-dependent (F. Vignaux et al., unpublished results) indicate that F-CMC and P-CMC are of the same order of magnitude in the presence of the appropriate target cells (Kägi et al. 1994b).

Effector d10S cells were activated with PI, which as mentioned above is known to induce in T killer cells significant levels of cytotoxic activity (Russell 1986; Lancki et al. 1987). For d10S cells this activation led to transcription of Fas ligand message (Suda et al. 1993) and could be inhibited by macromolecular synthesis inhibitors (cycloheximide, actinomycin D, DRB) for the first 3 h of the induction period (Luciani and Golstein 1994). Macromolecular synthesis may be required not only for the synthesis of sufficient Fas ligand (Suda et al. 1993), but also for the synthesis or functional availability of any molecule other than Fas that is also required for lysis by d10S cells.

In apparent contrast, cytotoxicity by already activated d10S cells was not altered in the presence of macromolecular synthesis inhibitors. Indeed, while T cell-induced target cell death shares with classical models of apoptosis characteristic morphological features (Sanderson 1976; Don et al. 1977; Liepins et al. 1977; Matter 1979) and DNA fragmentation (Russell et al. 1982; Russell 1983; Duke et al. 1983; Cohen et al. 1985; Schmid et al. 1986) (but see Zychlinsky et al. 1991), it is not affected by macromolecular synthesis inhibitors, which is one of the features making this type of apoptosis apparently different from several others. The d10S cell-mediated death of thymocytes did not require synthesis of any of the molecules that are required for the dexamethasone-induced apoptotic death of the same thymocytes (Luciani and Golstein 1994). Thus, in F-CMC macromolecular synthesis is required for d10S effector cell activation, but not for lysis by already activated effector cells nor for target cell death.

Interestingly, Fas-transduced cell death occurring upon engagement of Fas with antibodies has been reported to increase upon inhibition of macromolecular synthesis (Yonehara et al. 1989; Itoh et al. 1991). F-CMC may not be equivalent to anti-Fas antibody-mediated cell death: cytotoxic T cells may contribute quantitative or qualitative factors other than the mere presence of the Fas ligand.

5 Evidence for Only Two Mechanisms for T Cell-Mediated Cytotoxicity In Vitro

Are there T cell-mediated cytotoxicity mechanisms other than P-CMC and F-CMC? Using specifically sensitized antiallogeneic mixed leukocyte culture cells and peritoneal exudate cells, or anti-viral spleen cells, we recently showed that

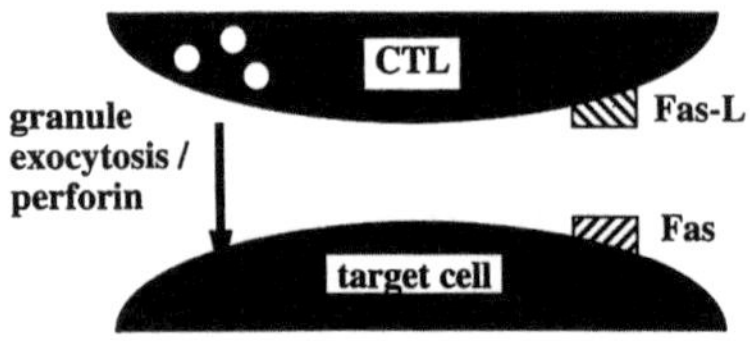

Fig. 3. Two major molecular pathways of cytotoxicity. The Fas-based and perforin-based mechanisms, clearly quite different, can coexist in the same effector cell (unpublished results)

F-CMC could be readily detected in P° mice and seemed to account for all of the cytotoxicity, i.e., within the experimental systems used no major mechanism of T cell-mediated cytotoxicity other than P-CMC and F-CMC could be detected (KÄGI et al. 1994b).

The detailed mechanism of P-CMC is not known and may itself be heterogenous, for instance when involving perforin alone or perforin plus granzyme. In addition, other mechanisms may be involved in vitro, in tests that are of longer duration or that make use of other target cells, and in vivo situations may also resort to other mechanisms. Nonetheless, it may thus well be that most, if not all, of T cell-mediated cytotoxicity in vitro is accounted for by these two mechanisms and by these two mechanisms only.

The existence of two distinct cytotoxicity pathways (Fig. 3) raises the interesting problem of *how* each of them appeared during evolution, since the two are quite different. For instance, the Fas-based mechanism uses a regulated ligand and can lyse only Fas-bearing target cells, while the perforin-based one does not seem to require any particular ligand-receptor system for lysis, and, accordingly, may lyse any target cell. It is thus quite extraordinary that two such different mechanisms should have converged in the same cells to exert cytotoxicity. It could be that the granule exocytosis mechanism shares a common origin with, say, mast cell degranulation. The Fas-based mechanism may be more related to developmental cell death and may have been borrowed from developmental pathways by the immune system.

Also, *why* two mechanisms? It may be that two mechanisms optimizes cytotoxic efficiency, at least against some target cells. It may also be that these two mechanisms have distinct roles, as discussed below.

6 From Cytotoxicity In Vitro to a Regulatory Role In Vivo

In addition to the involvement of Fas in cytotoxicity, what could its function be in vivo in the immune system (a possible role for Fas outside of the immune system will not be considered here)? Fas is required for the control of lymphocyte development, as indicated by the lymphoaccumulation and autoimmune diseases observed in *lpr* mice (COHEN and EISENBERG 1991), which are abnormal for Fas expression (WATANABE-FUKUNAGA et al. 1992a). These two roles of Fas, in cytotoxicity and in lymphocyte control, may well be causally related: F-CMC might ensure Fas-based control. In other words, some cytotoxic T cells may destroy syngeneic activated lymphocytes via a Fas-based mechanism. Thus, Fas may contribute to peripheral deletion of cells engaged in an immune response, thereby down-regulating this response.

We indeed showed that in vitro activated lymphocytes, restimulated with PI, could ensure the lysis of syngeneic activated lymphocytes via F-CMC (VIGNAUX and GOLSTEIN 1994). These results, while not proving a causal relationship between Fas-based lymphoproliferation control and F-CMC, demonstrated existence of the machinery required for such a relationship. F-CMC may thus be involved in part in the clonal elimination of peripheral T cells (D'ADAMIO et al. 1993) and perhaps also B cells (HARTLEY et al. 1993), as LPS-activated B cell blasts were found to be sensitive to F-CMC (results not shown). This is in line with alteration of the efficiency of peripheral deletion in in vitro models using cells from *lpr* or *gld* mice (RUSSELL et al. 1993; RUSSELL and WANG 1993).

Fas-based cytotoxicity, thus directed against activated self, can account at least in part for Fas-based lymphoproliferation control. It may represent one mechanism, most probably among several, for the control of immune responses. CTLs may thus exert two roles, the classical defense role against non-self and a Fas-based regulatory role acting on activated self and down-regulating immune responses.

F-CMC implies cell death, inflicted through direct cell contacts, presumably in close proximity to a strong ongoing immune reaction. The present results thus point at down-regulation through F-CMC. However, since Fas has also been shown to stimulate, not kill, certain cells (MIYAWAKI et al. 1992; MAPARA et al. 1993; ALDERSON et al. 1993), one cannot exclude the possiblity of other modes of Fas-based regulation. In this sense, at least some CTLs should also be considered regulatory cells. It is interesting to speculate whether some of the "suppressor cell" phenomenology might not be due in part to these killer/regulator cells.

In regulatory F-CMC, activated lymphocytes would be restimulated upon TCR triggering following antigen recognition. This would lead to expression of the Fas ligand on some cells and of Fas on the same or other cells. Fas ligand-Fas interactions would lead to cell death. This phenomenon would thus provide an example of social control of life or death (RAFF 1992), as also shown in other systems (SULSTON et al. 1980; LANG and BISHOP 1993; GRAHAM et al. 1993), involving in this case very similar cells killing each other through direct cell membrane interactions.

One may wonder whether similar Fas-based mechanisms might be at play in certain acquired or congenital pathological processes involving the immune system. If deficient, these mechanisms may lead to, e.g., poor limitation of immune reactions and autoimmunity, as in the *lpr* or *gld* mice. If exacerbated, they may lead in particular to lymphocyte depletion and immunodeficiencies and might be anticipated to play a role in strong and chronic T cell activation situations.

7 Conclusions

A major pathway of T cell-mediated cytotoxicity involves perforin along with other molecules, most probably serine-esterases. It is not known exactly how these molecules cause target cell death. Another mechanism involves an effector cell Fas ligand interacting with target cell Fas. These two mechanisms seem to entirely account for the whole of T cell-mediated cytotoxicity in vitro at least in short-term assays. Both mechanisms were formally demonstrated within the same period of a few months, providing a double answer to the long-standing question of how T cells kill.

To investigate whether other mechanisms may be at play in vivo, we are currently preparing P° x *gld* mice. These hybrid mice should express neither the perforin-based mechanism nor the Fas-based mechanism of cytotoxicity. It will be worthwhile to investigate whether they are able, for instance, to reject grafts or to defend themselves against some types of infection.

In vivo, Fas-based cytotoxicity pathways must be limited by both expression of a functional Fas ligand among effector cells and expression of Fas among putative target cells. In mice, Fas expression has been reported (WATANABE-FUKUNAGA et al. 1992b) in thymus, liver, ovary and heart, but could not be detected in brain, spleen, bone marrow, kidney, testis and uterus. In humans (TRAUTH et al. 1989; YONEHARA et al. 1989; DEBATIN et al. 1990; KOBAYASHI et al. 1990; ITOH et al. 1991; OWEN-SCHAUB et al. 1992), Fas (identical to APO-1, OEHM et al. 1992) was found expressed on various cell types such as myeloid cells and fibroblasts, on some but not all B and T lymphoid cell lines, on activated more than on resting lymphocytes and on some virus-infected lymphocytes. The Fas ligand was found expressed mostly in activated T cells and in the testis (SUDA et al. 1993). Thus, marked expression of both Fas and its ligand has been found so far only within activated lymphocyte populations, emphasizing the possible immunoregulatory role of Fas-based cytotoxicity.

Acknowledgements. This work was supported by institutional grants from INSERM and CNRS and by additional grants from ARC and LNFC. We thank Arnold H. Greenberg for comments on the manuscript.

References

Alderson MR, Armitage RJ, Maraskovsky E, Tough TW, Roux E, Schooley K, Ramsdell F, Lynch DH (1993) Fas transduces activation signals in normal human T lymphocytes. J Exp Med 178: 2231–2235

Berke G (1980) Interaction of cytotoxic T lymphocytes and target cells. Prog Allergy 27: 69–133

Berke G (1989a) Functions and mechanisms of lysis induced by cytotoxic T lymphocytes and natural killer cells. In: Paul WE (ed) Fundamental immunology, 2nd edn Raven, New York, pp 735–764

Berke G (1989b) The cytolytic T lymphocyte and its mode of action Immunol Lett 20: 169–178

Cerottini J-C, Brunner KT (1974) Cell-mediated cytotoxicity, allograft rejection, and tumor immunity Adv Immunol 18: 67–132

Chang JCC, Zhang L, Edgerton TL, Kaplan AM (1990) Heterogeneity in direct cytotoxic function of L3T4 T cells: TH1 clones express higher cytotoxic activity to antigen-presenting cells than TH2 clones. J Immunol 145: 409–416

Clark WR, Ostergaard HL, Gorman K, Torbett BE (1988) Molecular mechanisms of CTL-mediated lysis: a cellular perspective. Immunol Rev 103: 37–51

Cohen JJ (1991) Programmed cell death in the immune system. Adv Immunol 50: 55–83
Cohen JJ, Duke RC, Chervenak R, Sellins KS, Olson LK (1985) DNA fragmentation in targets of CTL: an example of programmed cell death in the immune system. Adv Exp Med Biol 184: 493–508
Cohen PL, Eisenberg RA (1991) *Lpr* and *gld*: single gene models of systemic autoimmunity and lymphoproliferative disease. Annu Rev Immunol 9: 243–269
Conzelmann A, Corthésy P, Cianfriglia M, Silva A, Nabholz M (1982) Hybrids between rat lymphoma and mouse T cells with inducible cytolytic activity. Nature 298: 170–172
D'Adamio L, Awad K M, Reinherz EL (1993) Thymic and peripheral apoptosis of antigen-specific T cells might cooperate in establishing self tolerance. Eur J Immunol 23: 747–753
Davidson WF, Dumont FJ, Bedigian HG, Fowlkes BJ, Morse HC III (1986) Phenotypic, functional, and molecular genetic comparisons of the abnormal lymphoid cells of C3H-*lpr/lpr* and C3H-*gld/gld* mice. J Immunol 136: 4075–4084
Debatin K-M, Goldmann CK, Bamford R, Waldmann TA, Krammer PH (1990) Monoclonal-antibody-mediated apoptosis in adult T-cell leukaemia. Lancet 335: 497–500
Don MM, Ablett G, Bishop CJ, Bundesen PG, Donald KJ, Searle J, Kerr JFR (1977) Death of cells by apoptosis following attachment of specifically allergized lymphocytes in vitro. Aust J Exp Biol 55: 407–417
Duke RC (1991) Apoptosis in cell-mediated immunity. In: Tomei LD, Cope FO (eds) Apoptosis: the molecular basis of cell death. Cold Spring Harbor Laboratory Press, New York, pp 209–226
Duke RC, Chervenak R, Cohen JJ (1983) Endogenous endonuclease-induced DNA fragmentation: an early event in cell-mediated cytolysis. Proc Natl Acad Sci USA 80: 6361–6365
Duke RC, Persechini PM, Chang S, Liu C-C, Cohen JJ, Young JD-E (1989) Purified perforin induces target cell lysis but not DNA fragmentation. J Exp Med 170: 1451–1456
Fitch FW, McKisic MD, Lancki DW, Gajewski TF (1993) Differential regulation of murine T lymphocyte subsets. Annu Rev Immunol 11: 29–48
Golstein P, Smith ET (1977) Mechanism of T cell-mediated cytolysis: the lethal hit stage. Contemp Top Immunobiol 7: 273–300
Golstein P, Ojcius DM, Young JD-E (1991) Cell death mechanisms and the immune system. Immunol Rev 121: 29–65
Golstein P, Mattéi M-G, Foa C, Luciani M-F (1994) Molecular mechanisms of T lymphocyte cytotoxicity, with emphasis on the Fas pathway. In: Gregory CD (ed) Apoptosis and the immune response. Wiley, New York (in press)
Graham A, Heyman I, Lumsden A (1993) Even-numbered rhombomeres control the apoptotic elimination of neural crest cells from odd-numbered rhombomeres in the chick hindbrain. Development 119: 233–245
Gromkowski SH, Brown TC, Masson D, Tschopp J (1988) Lack of DNA degradation in target cells lysed by granules derived from cytolytic T lymphocytes. J Immunol 141: 774–778
Hartley SB, Cooke MP, Fulcher DA, Harris AW, Cory S, Basten A, Goodnow CC (1993) Elimination of self-reactive B lymphocytes proceeds in two stages: arrested development and cell death. Cell 72: 325–335
Henkart PA (1985) Mechanism of lymphocyte-mediated cytotoxicity. Annu Rev Immunol 3: 31–58
Henney CS (1977) T-cell-mediated cytolysis: an overview of some current issues. Contemp Top Immunobiol 7: 245–272
Heusel JW, Wesselschmidt RL, Shresta S, Russell JH, Ley TJ (1994) Cytotoxic lymphocytes require granzyme B for the rapid induction of DNA fragmentation and apoptosis in allogeneic target cells. Cell 76: 977–987
Ishiura S, Matsuda K, Koizumi H, Tsukahara T, Arahata K, Sugita H (1990) Calcium is essential for both the membrane binding and lytic activity of pore-forming protein (perforin) from cytotoxic T-lymphocyte. Mol Immunol 27: 803–807
Itoh N, Yonehara S, Ishii A, Yonehara M, Mizushima S-I, Sameshima M, Hase A, Seto Y, Nagata S (1991) The polypeptide encoded by the cDNA for human cell surface antigen Fas can mediate apoptosis. Cell 66: 233–243
Ju S-T (1991) Distinct pathways of CD4 and CD8 cells induce rapid target DNA fragmentation. J Immunol 146: 812–818
Ju S-T, Ruddle NH, Strack P, Dorf ME, Dekruyff RH (1990) Expression of two distinct cytolytic mechanisms among murine CD4 subsets. J Immunol 144: 23–31
Kägi D, Ledermann B, Bürki K, Seiler P, Odermatt B, Olsen KJ, Podack ER, Zinkernagel RM, Hengartner H (1994a) Cytotoxicity mediated by T cells and natural killer cells is greatly impaired in perforin-deficient mice. Nature 369: 31–37
Kägi D, Vignaux F, Ledermann B, Bürki K, Depraetere V, Nagata S, Hengartner H, Golstein P (1994b) Fas and perforin pathways as major mechanisms of T cell-mediated cytotoxicity. Science 265: 528–530

Kerr JFR, Wyllie AH, Currie AR (1972) Apoptosis: a basic biological phenomenon with wide-ranging implications in tissue kinetics. Br J Cancer 26: 239–257
Kobayashi N, Hamamoto Y, Yamamoto N, Ishii A, Yonehara M, Yonehara S (1990) Anti-Fas monoclonal antibody is cytocidal to human immunodeficiency virus-infected cells without augmenting viral replication. Proc Natl Acad Sci USA 87: 9620–9624
Lancki DW, Weiss A, Fitch FW (1987) Requirements for triggering of lysis by cytolytic T lymphocyte clones. J Immunol 138: 3646–3653
Lancki DW, Hsieh C-S, Fitch FW (1991) Mechanisms of lysis by cytotoxic T lymphocyte clones: lytic activity and gene expression in cloned antigen-specific $CD4^+$ and $CD8^+$ T lymphocytes. J Immunol 146: 3242–3249
Lang RA, Bishop JM (1993) Macrophages are required for cell death and tissue remodeling in the developing mouse eye. Cell 74: 453–462
Liepins A, Faanes RB, Lifter J, Choi YS, De Harven E (1977) Ultrastructural changes during T-lymphocyte-mediated cytolysis. Cell Immunol 28: 109–124
Luciani M-F, Golstein P (1994) Fas-based d10S-mediated cytotoxicity requires macromolecular synthesis for effector cell activation but not for target cell death. R Soc Philos Trans B 345: 303–309
MacLennan ICM, Gotch FM, Golstein P (1980) Limited specific T-cell mediated cytolysis in the absence of extracellular Ca^{++}. Immunology 39: 109–117
Mapara MY, Bargou R, Zugck C, Döhner H, Ustaoglu F, Jonker RR, Krammer PH, Dörken B (1993) APO-1 mediated apoptosis or proliferation in human chronic B lymphocytic leukemia: correlation with bcl-2 oncogene expression. Eur J Immunol 23: 702–708
Martz E (1977) Mechanism of specific tumor-cell lysis by alloimmune T lymphocytes: resolution and characterization of discrete steps in the cellular interaciton. Contemp Top Immunobiol 7: 301–361
Matter A (1979) Microcinematographic and electron microscopic analysis of target cell lysis induced by cytotoxic T lymphocytes. Immunology 36: 179–190
Miyawaki T, Uehara T, Nibu R, Tsuji T, Yachie A, Yonehara S, Taniguchi N (1992) Differential expression of apoptosis-related Fas antigen on lymphocyte subpopulations in human peripheral blood. J Immunol 149: 3753–3758
Oehm A, Behrmann I, Falk W, Pawlita M, Maier G, Klas C, Li-Weber M, Richards S, Dhein J, Trauth BC, Ponstingl H, Krammer PH (1992) Purification and molecular cloning of the APO-1 cell surface antigen, a member of the tumor necrosis factor/nerve growth factor receptor superfamily. Sequence identity with the Fas antigen. J Biol Chem 267: 10709–10715
Ostergaard HL, Kane KP, Mescher MF, Clark WR (1987) Cytotoxic T lymphocyte mediated lysis without release of serine esterase. Nature 330: 71–72
Owen-Schaub LB, Yonehara S, Crump WL, III, Grimm EA (1992) DNA fragmentation and cell death is selectively triggered in activated human lymphocytes by Fas antigen engagement. Cell Immunol 140: 197–205
Perlmann P, Holm G (1969) Cytotoxic effects of lymphoid cells in vitro. Adv Immunol 11: 117–193
Podack ER (1985) Molecular mechanism of lymphocyte-mediated tumor cell lysis. Immunol Today 6: 21–27
Podack ER, Young JD-E, Cohn ZA (1985) Isolation and biochemical and functional characterization of perforin 1 from cytolytic T-cell granules. Proc Natl Acad Sci USA 82: 8629–8633
Raff M C (1992) Social controls on cell survival and cell death. Nature 356: 397–400
Ramsdell F, Seaman MS, Miller RE, Tough TW, Alderson MR, Lynch DH (1994) *gld/gld* mice are unable to express a functional ligand for Fas. Eur J Immunol 24: 928–933
Rouvier E, Luciani M-F, Golstein P (1993) Fas involvement in Ca^{++}-independent T cell-mediated cytotoxicity. J Exp Med 177: 195–200
Russell JH (1983) Internal disintegration model of cytotoxic lymphocyte-induced target damage. Immunol Rev 72: 97–118
Russell JH (1986) Phorbol-ester stimulated lysis of weak and non-specific target cells by cytotoxic T lymphocytes. J Immunol 136: 23–27
Russell JH, Masakovski V, Rucinsky T, Phillips G (1982) Mechanisms of immune lysis. III. Characterization of the nature and kinetics of the cytotoxic T lymphocyte-induced nuclear lesion in the target. J Immunol 128: 2087–2094
Russell JH, Wang R (1993) Autoimmune *gld* mutation uncouples suicide and cytokine/proliferation pathways in activated, mature T cells. Eur J Immunol 23: 2379–2382
Russell JH, Rush B, Weaver C, Wang R (1993) Mature T cells of autoimmune *lpr/lpr* mice have a defect in antigen-stimulated suicide. Proc Nat Acad Sci USA 90: 4409–4413
Sanderson CJ (1976) The mechanism of T cell mediated cytotoxicity. II. Morphological studies of cell death by time-lapse microcinematography. Proc R Soc Lond 192: 241–255

Sanderson CJ (1981) The mechanism of lymphocyte-mediated cytotoxicity. Biol Rev 56: 153–197
Schmid DS, Tite JP, Ruddle NH (1986) DNA fragmentation: manifestation of target cell destruction mediated by cytotoxic T-cell lines, lymphotoxin-secreting helper T-cell clones, and cell-free lymphotoxin-containing supernatant. Proc Natl Acad Sci USA 83: 1881–1885
Shi L, Kraut RP, Aebersold R, Greenberg AH (1992) A natural killer cell granule protein that induces DNA fragmentation and apoptosis. J Exp Med 175: 553–566
Shiver JW, Su L, Henkart PA (1992) Cytotoxicity with target DNA breakdown by rat basophilic leukemia cells expressing both cytolysin and granzyme A. Cell 71: 315–322
Sitkovsky M, Henkart P (1993) Cytotoxic cells. Birkhauser, Boston
Smyth MJ, Norihisa Y, Ortaldo J R (1992) Multiple cytolytic mechanisms displayed by activated human peripheral blood T cell subsets. J Immunol 148: 55–62
Suda T, Nagata S (1994) Purification and characterization of the Fas-ligand that induces apoptosis. J Exp Med 179: 873–879
Suda T, Takahashi T, Golstein P, Nagata S (1993) Molecular cloning and expression of the Fas ligand: a novel member of the tumor necrosis factor family. Cell 75: 1169–1178
Sulston JE, Albertson DG, Thomson JN (1980) The *Caenorhabditis elegans* male: postembryonic development of nongonadal strucutres. Dev Biol 78: 542–576
Takahashi T, Tanaka M, Brannan CI, Jenkins NA, Copeland NG, Suda T, Nagata S (1994) Generalized lymphoproliferative disease in mice, caused by a point mutation in the Fas ligand. Cell 76: 969–976
Tirosh R, Berke G (1985) T lymphocyte-mediated cytolysis as an excitatory process of the target. I. Evidence that the target may be the site of Ca^{++} action. Cell Immunol 95: 113–123
Trauth BC, Klas C, Peters AMJ, Matzku S, Moller P, Falk W, Debatin K-M, Krammer PH (1989) Monoclonal antibody-mediated tumor regression by induction of apoptosis. Science 245; 301–305
Trenn G, Takayama H, Sitkovsky MV (1987) Exocytosis of cytolytic granules may not be required for target cell lysis by cytotoxic T-lymphocytes. Nature 330: 72–74
Tschopp J, Nabholz M (1990) Perforin-mediated target cell lysis by cytolytic T lymphocytes. Annu Rev Immunol 8: 279–302
Vignaux F, Golstein P (1994) Fas-based lymphocyte-mediated cytotoxicity against syngeneic activated lymphocytes: a regulatory pathway? Eur J Immunol 24: 923–927
Von Boehmer H, Hengartner H, Nabholz M, Lernhardt W, Schreier MH, Haas W (1979) Fine specificity of a continuously growing killer cell clone specific for H-Y antigen. Eur J Immunol 9: 592–597
Watanabe-Fukunaga R, Brannan CI, Copeland NG, Jenkins NA, Nagata S (1992a) Lymphoproliferation disorder in mice explained by defects in Fas antigen that mediates apoptosis. Nature 356: 314–317
Watanabe-Fukunaga R, Brannan CI, Itoh N, Yonehara S, Copeland NG, Jenkins NA, Nagata S (1992b) The cDNA structure, expression, and chromosomal assignment of the mouse Fas antigen. J Immunol 148: 1274–1279
Yonehara S, Ishii A, Yonehara M (1989) A cell-killing monoclonal antibody (Anti-Fas) to a cell surface antigen co-downregulated with the receptor of tumor necrosis factor. J Exp Med 169: 1747–1756
Young JD-E (1989) Killing of target cells by lymphocytes: a mechanistic view. Physiol Rev 69: 250–314
Young JD-E, Clark WR, Liu C-C, Cohn ZA (1987a) A calcium- and perforin-independent pathway of killing mediated by murine cytolytic lymphocytes. J Exp Med 166: 1894–1899
Young JD-E, Damiano A, Dinome MA, Leong LG, Cohn ZA (1987b) Dissociation of membrane binding and lytic activities of the lymphocyte pore-forming protein (perforin). J Exp Med 165: 1371–1382
Young LHY, Klavinskis LS, Oldstone MBA, Young JD-E (1989) In vivo expression of perforin by $CD8^+$ lymphocytes during an acute viral infection. J Exp Med 169: 2159–2171
Zychlinsky A, Zheng L-M, Liu C-C, Young JD-E (1991) Cytotoxic lymphocytes induce both apoptosis and necrosis in target cells. J Immunol 146: 393–400

Structure and Biogenesis of Lytic Granules

G.M. GRIFFITHS[1] and Y. ARGON[2]

1 Introduction: The Lytic Cycle

The major mechanism by which natural killer (NK) cells and cytotoxic lymphocytes (CTLs) kill their targets is regulated exocytosis of specialized granules, termed lytic granules (HENKART 1985). Each activated CTL or NK cell contains on the order of 30–50 of these lytic granules, and they are the hallmark of specialized killer cells. Many studies have demonstrated that the lytic granules

[1] Basel Institute for Immunology, Grenzacherstrasse 487, 4005 Basel, Switzerland
[2] Department of Immunology, Duke Medical Center, Durham, NC 27710, USA

contain proteins which are expressed only by CTL and are involved in target cell destruction. The most conclusive evidence that these proteins are essential for the process of cell-mediated cytotoxicity comes from transfection (SHIVER and HENKART 1991; SHIVER et al. 1992) and gene disruption experiments (HEUSEL et al. 1994; KÄGI et al. 1994) which affirm the central role of granule exocytosis in killing by CTL and NK cells.

When a cytolytic lymphocyte recognizes a suitable target, such as a virus-infected cell or a tumor cell, cross-linking of surface receptors triggers a rapid reorganization of the killer cell cytoskeleton and secretory apparatus. Within 5 min, the microtubule organizing center (MTOC) and the Golgi complex reorient to face the bound target, and the lytic granules concentrate at the area of membrane contact (BYKOVSKAJA et al. 1978; FREY et al. 1982; PETERS et al. 1989a; ZAGURY et al. 1975). In this way the lymphocyte assumes polarity with respect to its target. As soon as this polarity has been established, and in the presence of Ca^{2+}, some of the granules fuse with the plasma membrane, releasing their contents in the space between the killer cell and the target (BYKOVSKAJA et al. 1978; PETERS et al. 1989a; ZAGURY et al. 1975). Granule release has only been seen to occur in an area of tight membrane contact between the killer and target cells. The proteins released from the granules then induce perforations in the target cell membrane and degradation of the target cell's DNA (HAYES et al. 1989a; MILLARD et al. 1984; PODACK and KONIGSBERG 1984). Such regulated secretion within a restricted area of membrane contact ensures a high degree of target specificity during killing.

Interestingly, the lytic granules are not found in resting T cells, but appear only after activation via the T cell receptor (TCR). The biogenesis of these organelles requires the correct and selective sorting of proteins involved in target cell lysis to the same intracellular compartment. Once formed, these organelles must then be able to move to the site of membrane contact during killing and then fuse with the plasma membrane. How is the biogenesis of the granules linked to T cell activation? How are the lytic proteins sorted to the granules? What controls granule movement? How is their polarized exocytosis regulated? In the last few years answers to some of these questions have begun to emerge. In this chapter we review what is currently known about the biogenesis, the composition, movement and functions of lytic granules.

2 Ultrastructure of Lytic Granules

By light microscopy the granules of CTL, NK or lymphokine activated killer (LAK) cells seem homogeneous. However, at the electron microscopic (EM) level there is clearly ultrastructural heterogeneity among these granules. In each type of killer cell there are three kinds of granules, termed type I, type II and intermediate granules (NEIGHBOUR et al. 1982). The three types of granules are built of two structural domains, and the classification into types reflects the relative

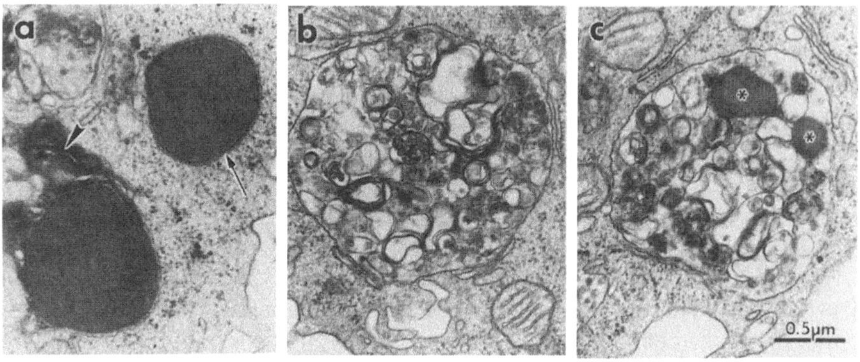

Fig. 1a–c. Electron micrographs showing the three different types of granules found in CTL and NK cells: **a** type I, **b** type II, **c** intermediate (*arrow*, thick membrane around the dense core of a type I granule; *arrowhead*, a whorl within the cortical rim; *asterisks*, small dense core regions of intermediate vesicles). Reproduced from *The Journal of Cell Biology* 1990, III, 2327–2340, by copyright permission of the Rockefeller University Press)

Table 1. Structural characterstics of lytic granules

Granule type	Dense core domain	Multivesicular domain
Type I	Large	Small rim
Type II	No	Yes
Intermediate type	Smaller	Yes

proportion of these two domains in each granule (Table 1, Fig. 1). Type I granules are mostly comprised of an electron dense core of homogeneous appearance, surrounded by a thin cortex of vesicles or lamellae. Type II granules are larger (up to 1.1 μm), more irregularly shaped, and are extensively multivesicular. Their volume is filled with membrane whorls and small (30–150 nm) vesicles (Burkhardt et al. 1990; Frey et al. 1982; Neighbour et al. 1982; Zarcone et al. 1987). The intermediate type granules contain both structural domains but most of their volume is the multivesicular domain and their dense cores are smaller on average, than the cores in type I granules.

The ultrasturcture of the multivesicular domain is very similar to that of late endosomes or prelysosomes of fibroblasts (Griffiths et al. 1988). Not only do the multivesicular domains look like prelysosomes, they are also as acidic as late endosomes or prelysosomes. The acidic intragranule pH was determined from histochemical staining with weak basic dyes (Shau and Dawson 1985) and more quantitative analyses using the microscopic DAMP method (Burkhardt et al. 1990; Masson et al. 1990) showed that the pH is 5.1–5.4. This pH value is higher than that of mature lysosomes (4.6–5.0) and more acidic than the typical pH of early endosomes (5–6) (Mellman et al. 1986).

An important structural point is the lack of membrane surrounding the dense cores and separating them from the multivesicular domains around them. The lack of membrane delimiting the dense core places certain constraints on how the granules can function during cytolysis (see Sect. 8).

3 Lytic Granule Contents

The proteins known to reside in lytic granules are summarized in Table 2. They comprise both proteins which are expressed specifically by killer lymphocytes and proteins which are also found in other cell types.

3.1 Lysis-Associated Proteins

The proteins which are expressed specifically in killer cells fall into three groups: perforin, granzymes and T1-A. Perforin (or cytolysin) is a membrane insertion protein which in the presence of Ca^{2+} undergoes a conformational change and forms homopolymers which can insert into the target cell membrane and destroy its integrity. The second group are serine proteases, the granzymes. In addition to their expression in CTL and residence in granules, the granzymes are distinct among all serine proteases in having a signature NH_2-terminal sequence (JENNE and TSCHOPP 1989). Common to all of them is the IIGG tetrapeptide, followed four residues later by a common octapeptide, amino acids 9–16. Granzymes A and B have been shown to be involved in triggering apoptosis in target cells (HAYES et al. 1989b; HEUSEL et al. 1994; SHIVER and HENKART 1991; SHIVER et al. 1992), although the exact mechanism is not clear. Another role suggests that granzymes act as auxilliary proteins to perforin (HUDIG et al. 1993) and in DNA fragmentation, one of the first events in the induced apoptosis of the target cells. Presumably, they act in both aspects by proteolytically activating other proteins. Together, the proteolytic spectrum of the granzymes is very broad: the group comprises tryptases (granzyme A, tryptase 3 and possibly granzyme D), putative chymases (granzyme E,F,G and H; ODAKE et al. 1991), an Aspase (granzyme B), a Metase (HUDIG et al. 1991; SMYTH et al. 1992), and one which cleaves after Ser or Asn (HUDIG et al. 1991; ODAKE et al. 1991). The third type of lysis-associated granule protein is TIA-1 and its relative TIAR. TIA-1 is a membrane associated protein with RNA binding domains, which when cleaved proteolytically yields a 15 kDa protein capable of causing DNA fragmentation (ANDERSON et al. 1990; KAWAKAMI et al. 1992; TIAN et al. 1991). Its role in cytolysis is discussed in more detail in the chapter by Anderson.

Proteoglycans are granule components which are not expressed exclusively in killer lymphocytes but are typical components of other regulated secretory granules. The predominant proteoglycan found in lytic granules is chondroitin sulfate. The proteoglycans of lytic granules have been shown to complex with both perforin and the granzymes in a pH-dependent manner (MASSON et al. 1990). Therefore, the proteoglycans seem to play important structural roles. First, they serve as a means of packaging lysis-specific proteins at high concentrations, and second, they may maintain these proteins in inactive forms until the pH and other environmental conditions change upon exocytosis.

Table 2. Components of lytic granules

Protein	Characteristics	Resides in	References
Perforin (cytolysin)	Pore forming protein; soluble	Core	(BURKHARDT et al. 1990; GROSCURTH et al. 1990; PETERS et al. 1991)
Granzyme A	Protease; tryptase; DNA fragmentin activity; soluble	Core	(BURKHARDT et al. 1989; GRIFFITHS and ISAAZ 1993)
Granzyme B	Protease; Asp-ase; DNA fragmentin activity; soluble	Core	(PETERS et al. 1991)
Granzyme C	Protease; mouse only; predicted Ser/Asn-ase; nonglycosylated; soluble	Unknown	(JENNE et al. 1988)
Granzyme D	Protease; mouse only; unknown specificity soluble	Core	(PETERS et al. 1991)
Granzyme E	Protease; mouse only; predicted chymase; soluble	Core	(PETERS et al. 1991)
Granzyme F	Protease; mouse only; predicted chymase; soluble	Core	(PETERS et al. 1991)
Granzyme G	Protease; mouse only; predicted chymase; soluble	Unknown	(JENNE et al. 1991)
Granzyme H	Protease; mouse only; unknown specificity; soluble	Unknown	(HADDAD et al. 1991)
Met-ase	Protease; human and rat; soluble	Unknown	(SMYTH et al. 1992)
Tryptase 3	Protease; human and rat; soluble	Unknown	(HAMEED et al. 1988)
Chondroitin sulfate proteoglycan	Complexes with other granule proteins; soluble	Core	(BURKHARDT et al. 1990; MACDERMOTT et al. 1985)
Calreticulin	Ca^{2+} binding protein; soluble	Unknown	(BURNS et al. 1992; DUPUIS et al. 1993)
Arylsulphatase	Lysosomal enzyme; soluble	Cortex	(HARGROVE et al. 1993)
β-glucuronidase	Lysosomal enzyme; soluble	Cortex	(ORYE et al. 1984)
Acid phosphatase	Lysosomal enzyme; soluble	Cortex	(BURKHARDT et al. 1990; FREY et al. 1982)
Cathepsin B	Lysosomal enzyme; soluble	Cortex	(GRIFFITHS and ISAAZ et al. 1993)
Cathepsin D	Lysosomal enzyme; soluble	Cortex	(BURKHARDT et al. 1990; PETERS et al. 1991)
α-glucosidase	Lysosomal enzyme; soluble	Cortex	(BURKHARDT et al. 1990; FREY et al. 1982)
CD 63	Lysosomal membrane protein	Cortex	(PETERS et al. 1989a)
LAMP1 and LAMP2	Lysosomal membrane protein; membrane-bound	Cortex	(BURKHARDT et al. 1990; PETERS et al.1991)
TIA-1	RNA binding; DNA fragmentation; membrane-bound	Cortex	(TIAN et al. 1991)
Leukophysin	Membrane-bound granule protein	Cortex	(ABDELHALEEM et al. 1991)
CD8	Receptor; membrane bound	Cortex	(PETERS et al. 1989a)
T cell receptor	Receptor; membrane bound	Cortex	(PETERS et al. 1989a)
MHC class I	Receptor; membrane bound	Cortex	(PETERS et al. 1989a)
MHC class II	Receptor; membrane bound		(G.M. Griffiths, unpublished)
Mannose-6-phosphate receptor	Receptor; membrane bound	Cortex	(BURKHARDT et al. 1990; PETERS et al. 1991)

3.2 Lysosomal Proteins

The other components of the lytic granules are lysosomal/endosomal enzymes, receptors and membrane proteins. Both soluble lysosomal hydrolases and lysosomal membrane proteins have been shown to reside in the granules. Most notable among these are the lysosomal membrane proteins from the LAMP family: LAMP-1, LAMP-2 and lgp-120 (LEWIS et al. 1985). Almost all of the CTL LAMP proteins reside in the granules. Another important constituent of the granule membrane is the 300 kDa cation-independent mannose-6-phosphate receptor (MPR). This receptor, which is present in all cell types, cycles among the Golgi complex and endosomes.

The multivesicular domains are acidified, as discussed above and the local pH is more characteristic of prelysosomal compartments than of mature lysosomes (MELLMAN et al. 1986). This granule acidification implies the existence of a proton ATPase in the granule membrane. Although this protein has not been directly demonstrated, treatment of cells with inhibitors of proton ATPases has specific effects on granule ultrastructure (KATAOKA 1993). These effects are correlated with an impaired ability of the treated cells to kill targets, showing that the acidic granule pH is important for their function. The lysosomal granule enzymes are active mostly at the acidic intra-granular pH while the lytic components are active only at neutral pH. This dichotomy clearly indicates the dual function of the lytic granules, which serve as both the lysosomes and secretory granules of the killer lymphocytes.

Finally, several membrane proteins are found associated with lytic granules. Among them are proteins like TCR and MHC proteins, which mediate the recognition of appropriate targets. PETERS et al. (1989b) proposed a novel mechanism for secretion of lytic proteins, based on the observations that TCR and MHC molecules are associated with internal vesicles of the granules and with membranes around the core, and that such small vesicles are released during the killing cycle. They suggested that these recognition molecules serve to direct the small vesicles to the target cell membrane. The lack of a delimiting membrane around the cores leaves them without a mechanism to confer cytolytic specificity, as proposed by Peters et al., and thus they must rely on the specificity of interaction at the level of the cell membrane. Furthermore, if the cores are not surrounded by membrane, some other compartmental barrier must exist between their contents and the lysosomal domain which is engaged in endocytosis. Perhaps the physical properties of the core itself, being densely packed with highly charged macromolecules, serve as such a spatial compartmentation device.

3.3 Other Granule Membrane Proteins

The composition of membranes from lytic granules has not yet been studied in detail. In addition to the lysosomal membrane proteins mentioned above, only a few granule proteins are characterized. One such protein is leukophysin (ABDELHALEEM et al. 1991), a 28 kDa protein with homology to granulophysin and synaptophysin, proteins which are markers of neutrophil granules and of synaptic vesicles, respectively. The existence of other membrane proteins is predicted from functional studies. For example, the ATPase is predicted from acidification of the granules, and granule movement along microtubules predicts receptors for the cytoplasmic motor proteins kinesin and dynein (see Sect. 6). In addition, a macrophage activating factor in the granules has been detected but not yet biochemically characterized (ROUSSEL and GREENBERG 1991).

3.4 Segregation of Granule Components

A remarkable feature of the granule architecture is the spatial segregation of its component proteins. The soluble, secreted proteins which are killer cell-specific are all concentrated in the dense core domain of the granules, as demonstrated by immunogold labeling of thin frozen EM sections (BURKHARDT et al. 1990; PETERS et al. 1989b). Conversely, most of the lysosomal proteins are confined to the multivesicular domains. In addition to the type II and intermediate granules, these lysosomal proteins are found in the cortex of type I granules. The distribution of the lysosomal proteins mirrors that of the hydrolases. The segregation of granule components is presumably related to the functions that the granules fulfill, as discussed below.

4 Targeting of Granule Proteins

Each of the lytic granule proteins has to be selectively sorted to this compartment from the Golgi complex during biosynthesis. As far as can be judged, there is no targeting mechanism unique to the lytic granules. Rather, the granule proteins are targeted by several distinct mechanisms, all of which are variations on targeting in other systems. This is evident from the elegant experiments of Henkart and colleagues, who showed that both granzyme A and perforin can be correctly targeted to the granules of mast cells as well as of CTLs (SHIVER and HENKART 1991; SHIVER et al. 1992).

4.1 Mannose-6-Phosphate Receptor-Dependent Targeting

The first indication that lytic proteins might be sorted to the granules by lysosomal targeting signals came from the observation that granzyme A bears the mannose-6-phosphate (M-6-P) signal (BURKHARDT et al. 1989). Support for the M-6-P sorting mechanism was provided by the finding, from immunoelectron microscopy, that the 300 kDa MPR was present in the granules (BURKHARDT et al. 1989; PETERS et al. 1991). MPR is an intracellular receptor which recognizes proteins that bear M-6-P incorporated into their N-linked glycans (KORNFELD 1987). The modification

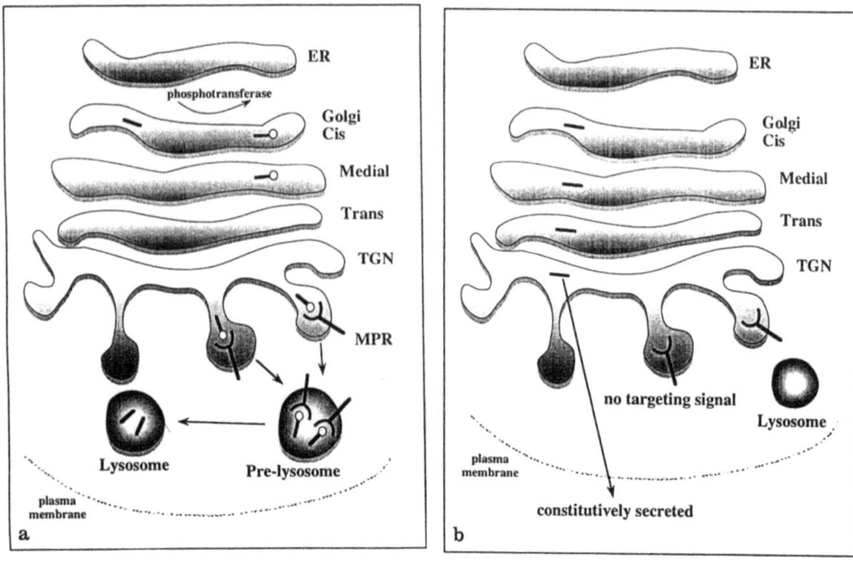

Fig. 2a, b. The sorting of soluble lysosomal enzymes in normal and I-cell disease cells. **a** In normal cells a phosphotransferase causes addition of a mannose-6-phosphate (*o*) to the enzyme as it passes through the Golgi during biosynthesis. This modification serves as the lysosomal targeting signal, being recognized by the mannose-6-phosphate receptor (*MPR*) in the *trans*-Golgi network (*TGN*). This receptor shuttles between the TGN and prelysosome, delivering the lysosomal enzymes to the lysosome; ER, endoplasmic reticulum. **b** In I-cell disease cells, the phosphotransferase is missing or inactive and the mannose-6-phosphate signal cannot be added to the newly synthesized proteins. Lacking any signal the proteins are secreted via the constitutive pathway. This same scheme applies to the sorting of granzymes A and B to the lytic granules (GRIFFITHS and ISAAZ 1993)

is added to passing lysosomal hydrolases by a phosphotransferase residing in the *cis* or mid-Golgi compartments and the receptor binds to its ligands in the *trans*-Golgi or *trans*-Golgi network (TGN) compartments (GRIFFITHS et al. 1988). The MPR-ligand complexes are sorted into Golgi clathrin-coated pits and vesicles and then move to late endosomes/prelysosomes. There, at a pH less than 6.0, the receptor and ligands dissociate, with the receptor recycling back to the loading compartment in the Golgi complex (Fig. 2a).

Functional evidence that the granzymes are sorted via this pathway was provided by studies on CTL clones from a patient with I-cell disease. In such patients, the M-6-P phosphotransferase is either absent or nonfunctional, and lysosomal enzymes are not sorted to lysosomes but are secreted as they are synthesized via the constitutive secretory pathway (Fig. 2b). I-cell disease derived CTL clones show such constitutive secretion of granzymes A and B. In addition, granzyme B in the I-cell CTL acquires complex type oligosaccharides, which are incompatible with the M-6-P modification (GRIFFITHS and ISAAZ 1993). The total levels of granzymes A and B are significantly lower in I-cell derived CTL clones than in their normal counterparts and MPR is not detected in their granules. Thus, granzymes are missorted in the absence of the M-6-P signal, showing that it is a major determinant in their normal routing to the granules.

4.2 Mannose-6-Phosphate Receptor-Independent Targeting of Soluble Proteins

Not all soluble granule proteins use the MPR-dependent pathway and there are several reasons to suggest the existence of a different sorting mechanism even for soluble proteins. At least one murine protease, granzyme C, is not glycosylated (JENNE and TSCHOPP 1988) and therefore cannot bear the M-6-P sorting signal, yet it is packaged into the granules. Second, even in CTLs from I-cell patients, 30% of the normal complement of granzymes A and B are found in the granules in the absence of the M-6-P signal (GRIFFITHS and ISAAZ 1993). Furthermore, the I-cell disease has no effect on the sorting of perforin, and the killing ability of these CTLs is unimpaired. Third, perforin does not bear M-6-P and its carbohydrates are processed to complex-type glycans (BURKHARDT et al. 1989), yet it is sorted to the lytic granule cores. Thus, perforin passes through the *trans*-Golgi compartment and then is sorted to the granules by a different mechanism. All these observations indicate that alternative sorting pathways must exist, but nothing is currently known about their characterstics. An MPR-independent pathway has recently been demonstrated for cathepsin D in I cell-disease B lymphoid cells, accounting for its residual lysosomal targeting in these cells (GLICKMAN and KORNFELD 1993). Significantly, the MPR-independent signal was found to overlap with the phosphotransferase recognition motif. The sequence homologies between granzymes and cathepsins are not great enough to allow an accurate prediction of this motif in the granzymes, but it is possible that a similar overlapping motif accounts for the MPR-independent sorting.

4.3 Targeting of Membrane-Bound Lytic Proteins

The COOH-terminal regions of the two TIA proteins are highly homologous to those found in the LAMP proteins and lgp-120 (TIAN et al. 1991). These regions contain a hexapeptide motif with a critical tyrosine on which the sorting of the lysosomal membrane proteins has been shown to depend. The shared motif may thus target this class of proteins to the membranes of lytic granules and to lysosomes and endosomes. Since other membrane proteins found in the granules lack this motif there must be other targeting signals. For proteins such as MHC and TCR it is not yet established whether those that reach the lytic granules are directed there from the biosynthetic pathway or via the endocytic route.

4.4 Proteoglycan Sorting

The mechanism by which proteoglycans are sorted to the granules is also unclear. Proteoglycans are not normally found in lysosomes but rather are a typical constituent of secretory granules. Inhibition of proteoglycan glycosylation with xylosides does not inhibit their sorting to the granules in hormone-producing cells (BURGESS and KELLY 1987) or in killer lymphocytes (CHRISTMAS et al. 1988; MASSON et al. 1990). In hormone-producing cells proteoglycans are sorted by a selective, pH-dependent condensation of proteins in the TGN (HUTTNER and TOOZE 1989). In CTL lines, chondroitin sulfate proteoglycan has been shown to condense with both perforin and granzymes in a pH-dependent fashion (MASSON et al. 1990). Similar associations are also found in other types of granules: mast cells proteases, cathepsin G in neutrophils (a homologue of granzyme B) and pancreatic chymotrypsinogen are all found complexed with proteoglycans (REGGIO and DAGORN 1978). These kind of selective associations may represent a common mechanism of sorting granule proteins by condensation with proteoglycans.

4.5 Calreticulin

A curious targeting situation is brought about by the finding of calreticulin in lytic granules (BURNS et al. 1992; DUPUIS et al. 1993). Calreticulin is normally a resident endoplasmic reticulum (ER) protein, with a COOH-terminal KDEL tetrapeptide, which serves as an ER retention/retreival signal (MUNRO and PELHAM 1987). Nonetheless, there are examples of KDEL bearing proteins reaching compartments in the secretory apparatus distal to the ER (PETER et al. 1992), and calreticulin in lytic granules would appear to be one such case. Considering the essential role of calcium in the cytolytic process one hypothesis is that calreticulin serves as a molecular chaperone by chelating calcium in the lytic granules and thus preventing premature conformational changes in lytic proteins.

4.6 Proteolytic Processing of Granule Proteins

In general, secretory granule proteins and many lysosomal proteins are synthesized as longer precursors (proenzymes) which are cleaved to yield the active form. Despite the evidence that proteases are necessary for perforin-mediated cytotoxicity (HUDIG et al. 1993), there is no evidence for proteolytic processing of perforin before it is packaged into the granules. Both the granzymes and the cathepsins of the lytic granules undergo proteolytic processing during their intracellular transport. In the case of granzymes, the propeptides are unusual: they are short, only a dipeptide, and the second residue is Glu (JENNE and TSCHOPP 1989). Of the known cathepsins, only cathepsin G has a similar propeptide (SALVESEN et al. 1987). The propeptides are apparently cleaved before the granzymes are packaged into the dense core granules: both sequencing of granzymes isolated from granules and immunolabeling with reagents that react with their active sites show that most of the granzymes in the granules are in the mature, processed form. The cleavage of the dipeptide is necessary for the enzymatic activity of the proteases (CAPUTO et al. 1993). The cleavage of the granzyme family after Glu suggests a processing endoprotease with restricted substrates and potentially a tissue distribution restricted to lymphoid and myeloid cells. One candidate which fits these criteria is the lysosomal thiol protease DPPI (cathepsin C) (MCGUIRE et al. 1993). The cellular compartment where this cleavage takes place is not yet established. By analogy to processing of proinsulin in pancreatic acinar cells, which occurs in the TGN or even in the immature condensing vacuoles (ORCI et al. 1987), the relevent compartment in lymphocytes is also a late one. Whether it is the TGN, or whether processing of granzymes happens upon arrival at the granules is currently unknown.

4.7 Sorting of Proteins Within the Granules

In the absence of a continuous membrane surrounding the dense cores, what accounts for the segregation of lysosomal and secretory proteins and for the apparent difference in pH? One explanation is that soluble proteins are packaged into the granules by a two-step targeting mechanism. First, they are diverted from the constitutive secretory pathway to the granules, likely as they pass through the *trans*-Golgi or TGN compartments. At this stage enzymes like cathepsin D and granzyme B both use the same receptor system and are likely to coexist in the same transport vesicles. Once in the granules, all dissociate from the MPR because of the pH shift and a second sorting stage separates proteins destined to the core from those targeted to the multivesicular domain. This second sorting stage need not involve a specific machinery. It is possible that in the ionic environment of the granule, physical or biochemical properties of perforin and granzymes cause them to aggregate, leaving cathepsin D and other hydrolases soluble. Newly arrived molecules then continue to condense onto this aggregate, and it eventually becomes the microscopic dense core.

5 Biogenesis of Lytic Granules

$CD8^+$ peripheral blood cells or splenocytes are by and large agranular. When cultured with allogeneic stimulating cells they develop granules over the course of several days (GRIFFITHS and ISAAZ 1993). Similarly, peritoneal exudate lymphocytes are agranular, but when cultured in the presence of lymphokines they both synthesize lytic proteins and develop granules (BERKE and ROSEN 1988).

Which type of granule develops first? We have used EM, the only available method to distinguish the various granules types, to study the differentiation of resting $CD8^+$ cells. Whether the precursor cells are derived from mouse spleen and cultured in the presence of interleukin-2 (IL-2) and ConA supernatant of splenocytes, or whether they are human peripheral blood cloned CTLs stimulated by exposure to allogeneic cells, the main features of granule biogenesis are the same. Both type I (dense core) granules and type II (multivesicular) granules start to appear simultaneousely during the process of expansion and maturation of killer cells. Their number per cell was highest when the cells reached their maximal killing capacity. A striking feature of this development is that the dense core granules grow in size. When initially detected, their average diameter is

0.1–0.3 μm, whereas when the cells reach their peak lytic capacity the average size of the dense core granules is 0.4–0.5 μm (our unpublished data). Therefore, the dense core granules increase in volume by about tenfold. Another interesting result from this study was the transition of the granules from MPR^+ to MPR^- organelles when CTLs were cultured on allogeneic stimulators. This suggests that, like lysosomes, the granules mature through an MPR^+ stage and that the MPR^- granules represent the mature organelles. This transition also explains the discrepancy between earlier observations in which MPR^+ granules were reported to represent 10% (PETERS et al. 1991) or 70% (BURKHARDT et al. 1990) of the lytic granule population.

6 Interactions of Granules with Microtubules

The exocytosis of lytic granules by CTLs represents one of the clearest cases of microtubule involvement in regulated secretion. As has been known for some time, agents that depolymerize microtubules, including colchicine, nocodazole, vinblastine and vincristine, inhibit cell-mediated cytotoxicity (KATZ et al. 1982; KUPFER et al. 1983). If the microtubules are allowed to repolymerize, cytolytic activity resumes (KUPFER et al. 1983). In addition, immunofluorescence microscopy indicates that elaboration of microtubules from the MTOC towards the contact area with the target during the programming for lysis phase provides "tracks" along which granules can move (KUPFER et al. 1983).

In lymphocytes, as in most cell types, most microtubules are organized with their fast growing (plus) ends towards the plasma membrane and their slow-growing (minus) ends towards the cell center (BERGEN et al. 1980). This provides a polarized cytoskeletal scaffold along which organelles can bind and move. Several proteins which mediate organelle-microtubule interactions have been identified, most notably two major motor proteins, kinesin and cytoplasmic dynein (LYE et al. 1987; PASCHAL and VALLEE 1987; VALE et al. 1985b). Kinesin has been shown to support the motility of organelles toward the microtubule plus-end (PORTER et al. 1987; VALE et al. 1985a), while cytoplasmic dynein is a minus-end directed motor (PASCHAL and VALLEE 1987; SCHROER et al. 1989).

The role of microtubules in directing movement of lytic granules was established by a cell-free system in which the binding and movement of purified dense core granules (MILLARD et al. 1984) on microtubules was reconstituted (BURKHARDT et al. 1993). Binding was energy-dependent and required intact granule membrane proteins and cytosolic factors. Foremost among these factors was the plus end-directed motor, kinesin. When bound to the microtubules, the granules translocated along them unidirectionally, at speeds of up to 1.2 μm/s. The direction and speed of granule movement are sufficient to account for the release of granules within a few seconds, well within the observed range for delivery of the "lethal hit" in cytolysis. Remarkably, although the CTL cytosol contains both the kinesin and the dynein motors, the lytic granules themselves dictate the direction of their movement in the assay by selectively utilizing kinesin. The interactions of granules with microtubules in vitro is regulated, and recently several kinesin-associated proteins were identified whose state of phosphorylation affects the extent of motor activity and granule motility (McIlvain et al. 1994).

In the intact T cell, lytic granules probably interact with microtubules in multiple ways, depending on the physiological state of the cell. Prior to stimulation of secretion, the granules cluster near the MTOC (Carpen et al. 1982; Gieger et al. 1982; Kupfer et al. 1985; McKinnon et al. I988). In fibroblasts, similar clustering of late endosomes has been attributed to the action of minus-end directed motors (Bomsel et al. 1990; Matteoni and Kreis 1987) and to other microtubule binding proteins (Mithieux and Rousset 1989). Lytic granules may share with other endosomes a common protein machinery for maintaining their resting distribution. After killer cell are stimulated by binding to their targets, the MTOC reorients to face the target, carrying the associated granules with it (Kupfer et al. 1983; Yannelli et al. 1986). The nature of interactions between granules and microtubules must then change so that granules can move toward the plasma membrane. Granule exocytosis is directed via microtubules radiating from the MTOC to the plasma membrane and a kinesin-like motor.

While the need for regulating granule-microtubule interactions is obvious, the molecular details of such regulation are not clear. After binding to the target cell, granule-microtubules association may be affected by the wave of Ca^{2+} that passes through the cell as a result of recognition (Poenie et al. 1987). At other times, granule pH may affect their binding to microtubules. Treatment of intact CTLs with the protonophore CCCP disrupts the reorientation of lytic granules to face a bound target (McKinnon et al. 1988), very possibly by disrupting their binding to microtubules. The regulation of granule motility by phosphorylation of motor complexes suggests a link with the lymphocyte signal transduction machinery: kinesin-associated proteins may be among the targets of Ser/Thr kinases in the cascade of phosphorylation events that is triggered by occupancy of antigen receptors or lymphokine receptors on the T cell surface.

7 Secretion

Upon cross-linking of the TCR by the target, the granules fuse with the plasma membrane and exocytose their contents. Like exocytosis of secretory granules by other cells, degranulation of killer lymphocytes is Ca^{2+}-dependent, and the second messengers which seem to be involved are the same as in other secretory cell types-diacylglycerol and inositol-triphosphoate. The involvement of these second messengers is inferred from the susceptibility of exocytosis to staurosporine, a protein Kinase C (PKC) antagonist (Wu et al. 1993); to neomycin, an inhibitor of inositol phospholipid hydrolysis; and from the enhancement of exocytosis by phorbol esters (Atkinson et al. 1990). Electron micrographs capturing granules fused with the plasma membrane show, that in addition to the dense cores being exocytosed, multivesicular (type II or intermediate) granules also are secreted. Indeed, the secretion of lysosomal hydrolases can be measured during cytolysis (ZUCKER et al. 1983) and small vesicles derived from the multivesicular domain have been shown in the space between killer and target cells (FREY et al. 1982; PETERS et al. 1989a).

7.1 The Directional Nature of Granule Exocytosis

Lytic granule exocytosis is not merely regulated with respect to timing the release, but is also spatially regulated. Unlike in mast cell exocytosis, killer lymphocyte granules are secreted in a polarized fashion towards the target cell. As discussed in Sect. 6, one important mechansism for the directional delivery of the granules is their ability to interact and move along microtubules. Microtubules position the granules so as to direct their release, maximizing the efficiency of the process and minimizing damage to bystander cells.

Once they underlie the plasma membrane contacting the target cell, what is the machinery behind this directional, regulated granule exocytosis? There must be proteins associated with the cytoplasmic aspect of the granules which mediate such signal sensing and membrane fusion events. Whether these are resident membrane proteins specific to lytic granules or cytosolic proteins associated with the granules transiently remains one of the more interesting unanswered questions. Rab proteins, small GTP binding proteins involved in transient fission and fusion reactions of many vesicles in the cell, are likely to be involved.

7.2 Secretion of Lytic Proteins During Killing

Another remarkable feature about T cell killing is the fact that a single CTL can kill multiple targets (ROTHSTEIN et al. 1978; ZAGURY et al. 1975) How does this occur? One possibility is that CTLs do not release all of their granules for a single target but rather use quantal release that can last for several killing cycles. Alternatively, CTLs may use other mechanisms (such as Fas-based killing, see Chapter by Golstein), even after all the granules have been secreted. Recent results suggest other mechanisms whereby granules can be refilled during killing and lytic proteins can be secreted via a nongranule-mediated pathway (G.M.G, submitted). Studies on the secretion of granzyme B from human CTLs show that cross-linking of TCR not only induces the exocytosis of stored lytic granules, but also results in a rapid new synthesis of granzymes A and B. The majority of the granzyme B synthesized acquires the M-6-P granule targeting signal and will refill the granules during killing. However, one third of the newly synthesized granzyme B fails to acquire the M-6-P targeting signal and is secreted by the constitutive pathway as the protein is synthesized (Fig. 3). The constitutive secretion of newly synthesized lytic proteins offers an alternative mechanism of killing which does not require the presence of preformed granules. This could

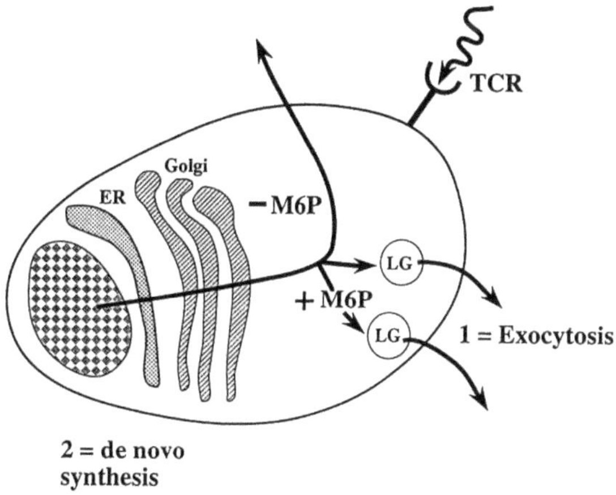

Fig. 3. Two events occur after T cell receptor (*TCR*) triggering during killing. **1** Exocytosis of the lytic granules (*LG*) and **2** de novo synthesis of lytic proteins. These may acquire the mannose-6-phosphate (*M6P*) targeting signal and refill the granules, or may fail to acquire this signal and are secreted directly via the constitutive pathway

explain findings in which killing is observed before the formation of granules (Berke and Rosen 1988).

8 Other Functions of the Granules

Clearly, lytic granules function as secretory organelles. They provide the means of packaging the lytic proteins, presumably in an inactive form which is not detrimental to the CTL itself. As discussed in Sects. 6 and 7, the granules also provide the means of coordinating and regulating secretion of lytic proteins onto the target cell.

The lysosomal nature of the granules (Sects. 2–4) suggests another, perhaps distinct, role for the lytic granules: that of the major endocytic/lysosomal compartment of NK cells and CTLs. There is considerable evidence to support this role. The vast majority of lysosomal hydrolases of killer lymphocytes reside in the granules and "conventional looking" lysosomes are very scarce. Exogenous ligands are taken up and routed to the granules in a process that has all the kinetic and pharmacological hallmarks of endocytosis as defined in fibroblasts (Burkhardt et al. 1990). The granules are the target for most of the endocytic traffic in these cells. Finally, although direct measurements are not yet available, the lysosomal hydrolases are active at the acidic pH of the granules. Thus, the lytic granules seem to perform the endocytic and catabolic functions common to all cells.

Several features place the lytic granules more precisely in the endocytic pathway, identifying them as the CTL equivalent of the prelysosomal compartment. In addition to the intermediate acidic pH, the codistribution of LAMP proteins with MPR, and the presence of rab5 and rab7, small GTP-binding proteins which normally associate with early and late endosomes (Burkhardt and Argon, unpublished), distinguish the granules from mature lysosomes.

Furthermore, the kinetics of uptake into the granules is also consistent with the view that they are late endosomes/prelysosomes.

It is interesting to speculate about possible specific roles which such a dual function organelle might serve. One possibility is that lytic granules are involved in antigen presentation by CTLs. In most species (mice being the exception) activated T cells express MHC class II molecules, and antigen presentation by T cells has been demonstrated (Lanzavecchia et al. 1988). The main peptide loading sites for class II molecules are endosomes and their presence within the lytic granules would be an obvious mechanism for antigen processing and presentation by CTLs. Indeed, isolated granules can process tetanus toxin for presentation (Lanzavecchia and Griffiths, unpublished).

Another possible role is in retreival of membrane after exocytosis. Since a considerable amount of membrane is added to the cell surface after each event of granule fusion, there is a need to recover this added surface area and the endocytic function of the granules is an obvious way to achieve this. Yet another role for this complex organelle may be protection of CTLs from viral infections. Each CTL comes in contact with multiple virus-infected cells (Martz 1976; Zagury et al. 1975) and is thus a prime target for infection by virus. Having a catabolic mechanism coupled to the exocytic mechanism may provide protection by efficiently destroying virions that may be taken up during the cellular engagement.

The dual nature of lytic granules is analogous to the nature of azurophilic granules of neutrophils. These granules also possess lysosomal enzymes, including elastase and cathpesin G, which are homologous to granzymes (Salvesen et al. 1987). Furthermore, these granules are "secreted" when the azurophilic granules fuse with phagosomes carrying pathogens. In both cell types the action of lysosomal hydrolases may augment membrane damage to the target and facilitate its degradation, increasing the efficiency with which it is cleared.

9 Summary

Lytic granules are specialized secretory organelles which appear after activation of CTLs and NK cells. The lytic granules contain a series of proteins that mediate target cell destruction after secretion from the cell. In addition, these organelles serve as the lysosomes of these lymphocytes.

At the EM level three types of granules with distinct regions are distinguished. Intriguingly, lytic and lysosomal proteins are localized in distinct regions. This is particularly interesting because lysosomal and lytic proteins can use the same sorting mechanisms to be targeted to this compartment. We favor the idea that a combination of sorting mechanisms result in this final segregation: the MPR receptor sorts both lysosomal proteins and granzymes from the Golgi complex, but a second event, such as selective aggregation with proteoglycans, then results in the segregation of lytic and lysosomal proteins in the granule.

Lytic granules provide a way to store and simultaneously secrete the lytic proteins in a highly specific fashion. The granules are able to move along microtubules using a kinesin-like motor, and thus can cluster at the site of membrane contact with a target cell. Once polarized, the granules exocytose their contents, using a molecular machinery that is as yet poorly defined. Understanding the machinery involved in both functions of the lytic granules will provide ways to control the action of cytotoxic lymphocytes, ultimately in clinical situations.

Acknowledgements. GMG and YA would like to thank Kristin Baetz, Witte Koopman and Jim McIlvain for critical reading of the manuscript. YA is the recipient of research grants from the American Cancer Society and the Arthritis Foundation. The Basel Institute for Immunology was founded and is supported by F. Hoffman-La Roche & Co. Ltd., Basel Switzerland.

References

Abdelhaleem MM, Hatskelzon L, Dalal BI, Gerrard JM, Greenberg AH (1991) Leukophysin: a 28 kDa granule membrane protein of leukocytes. J Immunol 147: 3053–3059

Anderson P, Nagler-Anderson C, O'Brien C, Levine H, Watkins S, Slayter HS, Blue ML, Schlossman SF (1990) A monoclonal antibody reactive with a 15-kDa cytoplasmic granule-associated protein defines a subpopulation of $CD8^+$ T lymphocytes. J Immunol 144: 574–582

Atkinson EA, Gerrard JM, Hildes GE, Greenberg AH (1990) Studies of natural killer degranulation and cytotoxicity. J Leukoc Biol 47: 39–48

Bergen LG, Kuriyama R, Borisy GG (1980) Polarity of microtubules nucleated by centrosomes and chromosomes of CHO cells in vitro. J Cell Biol 84: 151–159

Berke G, Rosen D (1988) Highly lytic in vivo primed cytolytic T lymphocytes devoid of lytic granules and BLT-esterase activity acquire these constituents in the presence of T cell growth factors upon blast transformation in vitro. J Immunol 141: 1429–1436

Bomsel M, Parton R, Kuznetsov SA, Schroer TA, Gruenberg J (1990) Microtubule- and motor-dependent fusion in vitro between apical and basolateral endocytic vesicles from MDCK cells. Cell 62: 719–731

Burgess TL, Kelly RB (1987) Constitutive and regulated secretion of proteins. Annu Rev Cell Biol 3: 243–293

Burkhardt JK, Hester S, Argon Y (1989) Two proteins targeted to the same lytic granule compartment undergo very different posttranslational processing. Proc Natl Acad Sci USA 86: 7128–7132

Burkhardt JK, Hester S, Lapham C K, Argon Y (1990) The lytic granules of natural killer cells are dual-function organelles combining secretory and prelysosomal compartments. J Cell Biol 111: 2327–2340

Burkhardt JK, McIlvain Jr, Sheetz MP, Argon Y (1993) Lytic granules from cytotoxic T cells exhibit kinesin-dependent motility on microtubules in vitro. J Cell Sci 104: 151–162

Burns K, Helgason CD, Bleackley RC, Michalak M (1992) Calreticulin in T- lymphocytes. Identification of calreticulin in T-lymphocytes and demonstration that activation of T cells correlates with increased levels of calreticulin mRNA and protein. J Biol Chem 267: 19039–19042

Bykovskaja SN, Rytenko AN, Rauschenbach MO, Bykovsky AF (1978) Ultrastructural alteration of cytolytic T lymphocytes following their interaction with target cells. I. Hypertrophy and change of orientation of the Golgi apparatus. Cell Immunol 40: 164–174

Caputo A, Garner RS, Winkler U, Hudig D, Bleackley RC (1993) Activation of recombinant murine cytotoxic cell proteinase-1 requires deletion of an amino-terminal dipeptide. J Biol Chem 268: 17672–17675

Carpen O, Virtanen I, Saksela E (1982) Ultrastructure of human natural killer cells: nature of the cytolytic contacts in relation to cellular secretion. J Immunol 128: 2691–2697.

Christmas SE, Steward WP, Lyon M, Gallagher JT, Moore M (1988) Chondroitin sulphate proteoglycan production by NK cells and T cells: effects of xylosidases on proliferation and cytotoxic function. Immunology 63: 225–231

Dupuis M, Schaerer E, Krause KH, Tschopp J (1993) The calcium-binding protein calreticulin is a major constituent of lytic granules in cytolytic T lymphocytes. J Exp Med 177: 1–7
Frey T, Petty TR, McConnell HM (1982) Electron microscopic study of natural killer cell-tumor cell conjugates. Proc Natl Acad Sci USA 79: 5317–5321
Gieger B, Rosen D, Berke G (1982) Spatial relationships of microtubule-organizing centers and the contact of cytotoxic T lymphocytes and target cells. J Cell Biol 95: 137–143
Glickman JN, Kornfeld S (1993) Mannose-6-phosphate-independent targeting of lysosomal enzymes in I-cell disease B lymphoblasts. J Cell Biol 123: 99–108
Griffiths GM, Isaaz S (1993) Granzymes A and B are targeted to the lytic granules of lymphocytes by the mannose-6 phosphate receptor. J Cell Biol 120: 885–896
Griffiths G, Hoflack B, Simons K, Mellman I, Kornfeld S (1988) The mannose-6-phosphate receptor and the biogenesis of lysosomes. Cell 52: 329–341
Groscurth P, Diener S, Stahel R, Jost L, Kagi D, Hengartner H (1990) Morphologic analysis of human lymphokine-activated killer LAK cells. Int J Cancer 45: 694–704
Haddad P, Jenne D, Tschopp J, Clement MV, Mathieu-Mahul D, Sasportes M (1991) Structure and evolutionary origin of the human granzyme H gene. Int Immunol 3: 57–66
Hameed A, Lowrey DM, Lichtenheld M, Podack ER (1988) Characterization of three serine esterases isolated from human IL-2 activated killer cells. J Immunol 141: 3142–3147
Hargrove ME, Wang J, Ting CC (1993) Regulation by glutathione of the activation and differentiation of IL-4-dependent activated killer cells. Cell Immunol 149: 433–434
Hayes MP, Berrebi GA, Henkart PA (1989a) Induction of target cell DNA release by the cytotoxic T lymphocyte granule protease granzyme A. J Exp Med 170: 933–946
Henkart PA (1985) Mechanism of Lymphocyte mediated cytotoxicity. Annu Rev Immunol 3: 31–58
Heusel JW, Wessselschmidt RL, Shrests S, Russell JH, Ley TJ (1994) Cytotoxic lymphocytes require granzyme B for the rapid induction of DNA fragmentation and apoptosis in allogeneic target cells. Cell 76: 977–987
Hudig D, Allison NJ, Pickett TM, Winkler U, Kam CM, Powers JC (1991) The function of lymphocyte proteases. Inhibition and restoration of granule-mediated lysis with isocoumarin serine protease inhibitors. J Immunol 147: 1360–1368
Hudig D, Ewoldt GR, Woodard SL (1993) Proteases and lymphocyte cytotoxic killing mechanisms. Curr Opin Immunol 5: 90–96
Huttner WB, Tooze SA (1989) Biosynthetic protein transport in the secretory pathway. Curr Opin Cell Biol 1: 648–654
Jenne DE, Tschopp J (1988) Granzymes, a family of serine proteases released from granules of cytolytic T lymphocytes upon T cell receptor stimulation. Immunol Rev 103: 53–71
Jenne DE, Tschopp J (1989) Granzymes: a family of serine proteases in granules of cytolytic T lymphocytes. In: Podack FR (ed) Cytotoxic effector mechanisms. Springer, Berlin Heidelberg New York, pp 33–48 (Current topics in microbiology and immunology, vol 140)
Jenne D, Rey C, Masson D, Stanley KK, Herz J, Plaetinck G, Tschopp J (1988) cDNA cloning of granzyme C, a granule-associated serine protease of cytolytic T lymphocytes. J Immunol 140: 318–323
Jenne DE, Masson D, Zimmer M, Haefliger JA, Li WH, Tschopp J (1989). Isolation and complete structure of the lymphocyte serine protease granzyme G, a novel member of the granzyme multigene family in murine cytolytic T lymphocytes. Evolutionary origin of lymphocyte proteases. Biochemistry 28: 7953–7961
Kagi D, Lederman B, Burki K, Seiler P, Odermatt B, Olsen KJ, Podack ER, Zinkernagel RM, Hengartner H (1994) Cytotoxicity mediated by T cell and natural killer cells is greatly impaired in perforin-deficient mice. Nature 369: 31–37
Kataoka T (1993) EMBO workshop on cell-mediated cytotoxicity. 29 August-1 September 1993, Weizmann, Institute of Science, Israel
Katz P, Zaytoun AM, Lee JHJ (1982) Mechanisms of human cell mediated cytotoxicity: III. Dependence of natural killing on microtubule and microfilament integrity. J Immunol 129: 2816–2825
Kawakami A, Tian Q, Duan X, Streuli M, Schlossman SF, Anderson P (1992) Identification and functional characterization of a TIA-1-related nucleolysin. Proc Natl Acad Sci USA 89: 8681–8685
Kornfeld S (1987) Trafficking of lysosomal enzymes. FASEB J 1: 462–468
Kupfer A, Dennert G, Singer SJ (1983) Polarization of the Golgi apparatus and the microtubule-organizing center within cloned natural killer cells bound to their targets. Proc Natl Acad Sci USA 80: 7224–7228
Kupfer A, Dennert G, Singer SJ (1985) The reorientation of the Golgi apparatus and the microtubule-organizing center in the cytotoxic effector cell is a prerequisite in the lysis of bound target cells. J Mol Cell Immunol 2:37–49

Lanzavecchia A, Roosnek E, Gregory T, Berman P, Abrignani S (1988) T cells can present antigens such as HIV gp120 targeted to their own surface molecules. Nature 334: 530–532
Lewis V, Green SA, Marsh M, Vikho P, Helenius A, Mellman I (1985) Glycoproteins of the lysosomal membrane. J Cell Biol 100: 1839–1847
Lye RJ, Porter ME, Scholey JM, McIntosh JR (1987) Identification of a microtubule-based cytoplasmic motor in the nematode Caenorhabditis elegans. Cell 51: 309–318
MacDermott RP, Schmidt RE, Caulfield JP, Hein A, Bartley GT, Ritz J, Schlossman SF, Austen KF, Stevens RL (1985) Proteoglycans in cell-mediated cytotoxicity. Identification, localization, and exocytosis of a chondroitin sulfate proteoglycan from human cloned natural killer cells during target cell lysis. J Exp Med 162: 1771–1787
Martz E (1976) Multiple target cell killing by the cytolytic T lymphocyte and the mechanism of cytotoxicity. Transplantation 21: 5–11
Masson D, Peters PJ, Geuze HJ, Borst J, Tschopp J (1990) Interaction of chondroitin sulfate with perforin and granzymes of cytolytic T-cells is dependent on pH. Biochemistry 29: 11229–11235
Matteoni R, Kreis TE (1987) Translocation and clustering of endosomes and lysosomes depends on microtubules. J Cell Biol 105: 1253–1265
McGuire MJ, Lipsky PJ, Thiele DL (1993) Generation of active myeloid and lymphoid granule serine proteases requires processing by the granule thiol protease dipeptidyl peptidase I. J Biol Chem 268: 2458–2467
McIlvain JM, Burkhardt J, Hamm-Alvarez S, Argon Y, Sheetz M (1994) Regulation of kinesin activity by phosphorylation of kinesin-associated proteins. J Biol Chem 269: 19176–19182
McKinnon K, Dunlevy JR, Dawson JR, Argon Y (1988) Cell-mediated cytotoxicity and the reorientation of lytic granules are inhibited by the protonophore CCCP. Hum Immunol 22: 81–95
Mellman I, Fuchs R, Helenius A (1986) Acidification of the exocytic and endocytic pathways. Annu Rev Biochem 55: 663–700
Millard PJ, Henkart MP, Reynolds C W, Henkart PA (1984) Purification and properties of cytoplasmic granules from cytotoxic rat LGL tumors. J Immunol 132: 3197–3204
Mithieux G, Rousset B (1989) Identification of a lysosome membrane protein which could mediate ATP-dependent stable association of lysosomes to microtubules. J Biol Chem 264: 4664–4668
Munro S, Pelham H (1987) A C-terminal signal prevents secretion of luminal ER proteins. Cell 48: 899–907
Neighbour DA, Huberman HS, Kress Y (1982) Human large granular lymphocytes and natural killing: ultrastructural studies of strontium induced degranulation. Eur J Immunol 12: 588–595
Odake S, Kam CM, Narasimhan L, Poe M, Blake JT, Krahenbuhl O, Tschopp J, Powers JC (1991) Human and murine cytotoxic T lymphocyte serine proteases: subsite mapping with peptide thioester substrates and inhibition of enzyme activity and cytolysis by isocoumarins. Biochemistry 30: 2217–2227
Orci L, Ravazzola M, Storch MJ, Anderson RG, Vassalli JD, Perrelet A (1987) Proteolytic maturation of insulin is a post-Golgi event which occurs in acidifying clathrin-coated secretory vesicles. Cell 49: 865–868
Orye E, Plum J, De Smedt M (1984) Beta-glucuronidase activity in human T and B lymphocytes and the Tmu and T gamma subpopulations. Histochemistry 81: 287–290
Paschal BM, Vallee RB (1987) Retrograde transport by the microtubule-associated protein MAP 1C. Nature 330: 181–183
Peter F, Nguyen VP, Soling HD (1992) Different sorting of Lys-Asp-Glu-Leu proteins in rat liver. J Biol Chem 267: 10631–10637
Peters PJ, Geuze HJ, Van d DH, Slot JW, Griffith JM, Stam NJ, Clevers HC, Borst J (1989a) Molecules relevant for T cell-target cell interaction are present in cytolytic granules of human T lymphocytes. Eur J Immunol 19: 1469–1475
Peters PJ, Geuze HJ, Van der Donk HA, Slot JW, Griffith JM, Stam NJ, Clevers HC, Borst J (1989b) Molecules relevant for T cell-target cell interaction are present in cytolytic granules of human T lymphocytes. Eur J Immunol 19: 1469–1475
Peters PJ, Borst J, Oorschot V, Fukuda M, Krahenbuhl O, Tschopp J, Slot JW, Geuze HJ (1991) Cytotoxic T lymphocyte granules are secretory lysosomes, containing both perforin and granzymes. J Exp Med 173: 1099–109
Podack ER, Konigsberg PJ (1984) Cytolytic T cell granules. Isolation, structural, biochemical, and functional characterization. J Exp Med 160: 695–710
Poenie M, Tsien RY, Schmitt-Verhulst AM (1987) Sequential activation and lethal hit measured by $[Ca^{2+}]i$ in individual cytolytic T cells and targets. EMBO J 6: 2223–2232
Porter ME, Scholey JM, Stemple DL, Vigers GPA, Vale RA, Scheetz MP, McIntosh JR (1987)

Characterization of the microtubule movement produced by sea urchin egg kinesin. J Biol Chem 74: 2794–2802
Reggio H, Dagorn JC (1978) Ionic interactions between bovine chymotrypsinogen A and chondroitin sulfate A.B.C. A possible model for molecular aggregation in zymogen granules. J Cell Biol 78: 951–957
Rothstein TL, Mage M, Jones G, McHugh LL (1978) Cytotoxic T lymphocyte sequential killing of immobilized allogeneic tumor target cells measured by time-lapse microcinematography. J Immunol 121: 1652–1656
Roussel E, Greenberg AH (1991) Granules of human CD3^{+} large granular lymphocytes contain a macrophage regulating factor(s) that induces macrophage H_2O_2 production and tumoricidal activity but decreases cell surface Ia antigen expression. Cell Immunol 134: 31–41
Salvesen G, Farley D, Shuman J, Przybyla A, Reilly C, Travis J (1987) Molecular cloning of human cathepsin G: structural similarity to mast cell and cytotoxic T lymphocyte proteinases. Biochemistry 26: 2289–2293
Schroer T, Steuer ER, Sheetz MP (1989) Cytoplasmic dynein is a minus end-directed motor for membranous organelles. Cell 56: 937–946
Shau H, Dawson JR (1985) Identification and purification of NK cells with lysosomotropic vital stains: correlation of lysosome content with NK activity. J Immunol 135: 137–140
Shiver JW, Henkart PA (1991) A noncytotoxic mast cell tumor line exhibits potent IgE-dependent cytotoxicity after transfection with the cytolysin/perforin gene. Cell 64: 1175–1181
Shiver JW, Su L, Henkart PA (1992) Cytotoxicity with target DNA breakdown by rat basophilic leukemia cells expressing both cytolysin and granzyme A. Cell 71: 315–322
Smyth MJ, Wiltrout JA, Trapani JA, Ottaway KS, Sowder R, Henderson LE, Kam CM, Powers JC, Young HA, Sayers TJ (1992) Purification and cloning of a novel serine protease, RNK-Met-1, of a rat natural killer cell leukemia. J Biol Chem 267: 24418–24425
Tian Q, Streuli M, Saito H, Schlossman SF, Anderson P (1991) A polyadenylate binding protein localized to the granules of cytolytic lymphocytes induces DNA fragmentation in target cells. Cell 67: 629–639
Vale RD, Schnapp BJ, Mitchison TS, Steuer E, Reese TS, Sheetz MP (1985a) Different axoplasmic proteins generate movement in opposite directions along microtubules in vitro. Cell 43: 623–632
Vale RD, Schnapp BJ, Reese TS, Sheetz MP (1985b) Organelle, bead, and microtubule translocations promoted by soluble factors from the squid giant axon. Cell 40: 559–569
Wu J, Shiver J, Hargrove ME, Ting CC (1993) Role of protein kinase C and cytokines on the function and production of cytolytic granules in alpha CD3-activated killer-cell-mediated killing of tumor cells. Int J Cancer 53: 973–977
Yannelli JR, Sullivan JA, Mandell GL, Engelhard VH (1986) Reorientation and fusion of cytotoxic T lymphocyte granules after interaction with target cells as determined by high resolution cinemicrography. J Immunol 136: 377–382
Zagury D, Bernard J, Thierness N, Feldman M, Berke G (1975) Isolation and characterization of individual functionally reactive cytotoxic T lymphocytes: conjugation, killing and recycling at the single cell level. Eur J Immunol 5: 818–822
Zarcone D, Prasthofer EF, Malavasi F, Pistoia V, LoBuglio AF, Grossi CE (1987) Ultrastructural analysis of human natural killer cell activation. Blood 69: 1725–1736
Zucker FD, Grusky G, Yang JS (1983) Arylsulfatase in natural killer cells: its possible role in cytotoxicity. Proc Natl Acad Sci USA 80: 6977– 6981

The Role of CD28 Costimulation in the Generation of Cytotoxic T Lymphocytes

M. AZUMA[1] and L.L. LANIER[2]

1 Introduction

Cell-mediated cytotoxicity is one of the most important effector functions produced during cellular immune responses. There are several types of effector cells that can mediate cellular cytotoxicity and they demonstrate distinct target cell specificity. Antigen-specific, major histocompatibility complex (MHC)-restricted cytotoxic T lymphocytes (CTLs) recognize peptide antigens bound to class I or class II MHC molecules through the CD3/T cell receptor (TCR) complex. Natural killer (NK) cells and certain T lymphocytes can lyse target cells using a mechanism that does not require recognition of MHC molecules. The membrane receptors on the effector cells and target cells that are responsible for MHC-unrestricted recognition are still unknown. In the addition to the first signal through the antigen-specific receptor, a second signal derived from adhesion or costimulatory molecules is required to activate efficient cytolytic function.

Several cell adhesion receptor-ligand combinations are involved in cell-mediated cytotoxicity. One of the most well characterized is the interaction between the CD11a / CD18 (lymphocyte function-associated antigen, LFA-1)

[1] Department of Immunology, Juntendo University School of Medicine, 2-1-1 Hongo, Bunkyo-ku, Tokyo 113, Japan
[2] Department of Human Immunology, DNAX Research Institute for Molecular and Cellular Biology, 901 California Avenue, Palo Alto, CA 94304-1104, USA

integrin on the effector cells with the intercellular adhesion molecules (ICAMs) on the target cells. Three ICAMs (ICAM-1/CD54, ICAM-2/CD50, and ICAM-3/CD102) have been identified which. differ in their tissue distribution and regulation. NK and CTL cytolytic function also can be augmented by the interaction between CD2 on the effector cell with CD58 (LFA-3) on the target. These two receptor-ligands combinations have been shown to participate directly in the effector function of CTLs and NK cells. Additionally, other receptor-ligand interactions are involved in the generation of CTLs from their nonlytic precursors. Interactions between CD28 on small, resting T cells with CD80 (also known as B7 or BB1) or B70 (also known as B7-2 and recently demonstrated to be identical to CD86; Caux et al. 1994; Engel et al. 1994) on an antigen-presenting cell can efficiently costimulate the generation of CTLs, provided that a primary stimulation is initiated through the CD3/TCR complex. Herein, we review recent developments implicating CD28 costimulation in cell-mediated cytotoxicity and discuss the potential application of CD28 costimulation in the treatment of cancer and human disease.

2 CD28 Costimulation in T Cell Activation

CD28 is a cell surface glycoprotein composed of two identical, disulfide-linked 44 kDa subunits and is a member of the immunoglobulin (Ig) superfamily. In the peripheral blood of human adults, CD28 is expressed on most $CD4^+$ T cells and two thirds of $CD8^+$ T cells (Yamada et al. 1985; June et al. 1990; Azuma et al. 1993c). A molecule related to CD28, designated CTLA-4, initially was identified in a cDNA library prepared from a murine CTL and a human homolog was cloned subsequently (Brunet et al. 1987; Dariavach et al. 1988). Unlike CD28 that is constitutively expressed on resting and activated T cells, CTLA-4 is a T cell surface protein expressed only after activation (Linsley et al. 1991a; Harper et al. 1991; Lindsten et al. 1993). Both CD28 and CTLA-4 specifically bind to the B cell activation antigen, B7 (Freeman et al. 1991), also known as BB1 (Yokochi et al. 1982) (hereafter referred to as CD80). CTLA-4 binds CD80 with an approximately 20-fold greater affinity than CD28 (Linsley et al. 1991b, 1992). Soluble fusion proteins containing the extracellular domain of human or mouse CTLA-4 and the Fc portion of human or mouse IgG efficiently block the interaction between CD28 and its ligands (Lenschow et al. 1992; Turka et al. 1992; Linsley et al. 1991b). The role of membrane CTLA-4 in T cell immune responses presently is not clear, although there are suggestions that interactions between CTLA-4 and its ligands may provide either positive or negative signals.

The ability of CD28 to "costimulate" T lymphocytes initially was demonstrated by the finding that anti-CD28 monoclonal antibody (mAb) together with mitogens or phorbol esters enhance proliferation and cytokine production (Hara et al. 1985). Additionally, soluble anti-CD28 mAb augments proliferative responses induced either by anti-CD3 or by anti-CD2 mAb (Damle et al. 1988; Martin et al. 1986; Pierres et al. 1988; van Lier et al. 1988; Yang et al. 1988; de Jong et al. 1991). While mAbs against several other cell surface antigens have been shown to increase T cell proliferation when used in conjunction with suboptimal amounts of anti-CD3, anti-CD28 mAb clearly demonstrates the most potent effect and apparently works by a distinct signaling mechanism (VAN LIER et al. 1991; LEDBETTER et al. 1990; CERDAN et al. 1992; SOMMER et al. 1993). CD28-induced T cell proliferation is interleukin-2 (IL-2)-dependent (MARTIN et al. 1986; HARA et al. 1985) and results both from induction of IL-2 transcription (FRASER et al. 1991) and stabilization of cytokine mRNA (LINDSTEN et al. 1989). Cyclosporin A does not affect IL-2 production induced via CD28, whereas it efficiently inhibits IL-2 transcription initiated by stimulation through the CD3/TCR complex (JUNE et al. 1987; HESS and BRIGHT 1991).

CD80 is a 50–60 kDa glycoprotein expressed on antigen-presenting cells such as activated B cells, macrophages and dendritic cells. The structure of CD80 is composed of an extracellular region containing two Ig-like domains, a transmembrane domain and a short cytoplasmic tail (FREEMAN et al. 1989). The binding of soluble CD80-Ig fusion proteins to CD28 on T cells augments mitogen-induced T cell proliferation, cytokine production and T cell-dependent Ig production (LINSLEY et al. 1991a).

Recently, we and other investigators have discovered a second counter-receptor for both CD28 and CTLA-4, i.e., B70 or B7-2 (AZUMA et al. 1993b; FREEMAN et al. 1993a,b; HATHCOCK et al. 1993) (hereafter referred to as CD86). CD86, a member of the Ig gene superfamily with limited homology to CD80, is constitutively expressed on monocytes and is induced on T, NK, and B lymphocytes after activation (AZUMA et al. 1993b). By contrast, CD80 is present on monocytes and lymphocytes only after activation. Transfectants expressing CD86 are able to costimulate T cell proliferation and cytokine production (our unpublished data; FREEMAN et al. 1993b), similar to the ability of CD80 transfectants to enhance T cell functions. In the context of a normal, physiological immune response, it is possible that interactions between CD28 and CD86 may be of critical importance given the constitutive expression of CD86 on monocytes and possibly dendritic cells. This would permit initiation of a primary immune response, before the induction and expression of CD80. The potential importance of CD86 is supported by the observations that T cell responses are mounted efficiently in CD80 knock-out mice (FREEMAN et al. 1993c) and by the observation that anti-CD86 mAb alone efficiently blocks a primary allogeneic mixed lymphocyte reaction (MLR) when allogeneic monocytes are used as the antigen-presenting cell (AZUMA et al. 1993b).

The reason for the existence of two ligands for the CD28 and CTLA-4 receptors presently is unclear. Both CD86 and CD80 have similar affinities for CTLA-4 and the binding sites on CD86 and CD80 for CTLA-4 seem to be very close (unpublished data). Additionally, both CD86 and CD80 bind with similar affinities to CD28, but approximately 20-fold lower than the affinity to CTLA-4 (LANIER et al., in press). Consistent with their similar binding properties, both CD86 and CD80 provide important costimulatory signals to T cells via the CD28 counterreceptor which results in the activation of T cell proliferation and cytokine production.

3 CD28-Dependent Cytotoxicity

3.1 CD28-Dependent Cytotoxicity Mediated by Small, Resting T Cells

The first indication that CD28 may be important in the generation of T cell-mediated cytotoxicity came from studies attempting to elicit CTLs against melanoma targets (JUNG et al. 1991). In these experiments, human peripheral blood lymphocytes (PBLs) and melanoma cells were cocultured for 3 days with heterobispecific mAb composed of anti-CD3 X anti-melanoma antigen and anti-CD28 X anti-melanoma antigen antibodies. CTLs were generated only in the presence of both bispecific mAb and melanoma cells, suggesting that costimulation via CD28 was involved. However, while these experiments demonstrated a role for CD28 in the induction of CTLs, they did not address whether CD28 participated in triggering cytotoxicity.

Prior studies have shown that CTLs can lyse Fc receptor bearing tumor cell targets in the presence of anti-CD3 mAb (LEEUWENBERG et al. 1985). This type of mAb redirected cytotoxicity assay permits identification of CTLs, without the necessity to determine the antigenic specificity of the effector cell population. CTL clones, in vitro activated T cells, and a small subset of possibly in vivo activated T cells identified by the phenotype $CD3^{+}56^{+}$ or $CD3^{+}57^{+}$ (PHILLIPS and LANIER 1986; SEGAL et al. 1990) mediate potent anti-CD3 redirected cytotoxicity. Previously, it was assumed that high-buoyant density T cells isolated from Percoll gradients (small, resting T cells) do not possess cytolytic activity, possibly as a consequence of lacking the necessary cellular machinery. However, normal PBL T cells can mediate anti-CD3 induced redirected cytotoxicity if costimulated via CD28 (AZUMA et al. 1992a). This was demonstrated by the ability of small, resting T cells to kill Fc receptor bearing murine P815 mastocytoma cells transfected with human CD80 cDNA in the presence of anti-CD3 mAb. Killing was measured using a 4 h ^{51}Cr radioisotope release assay. Cytotoxicity was inhibited by pretreating the effectors or targets with F(ab)'2 fragments of anti-CD28 or anti-CD80 mAb, respectively, demonstrating that triggering of lysis was dependent upon the CD28/CD80 interaction (AZUMA et al. 1992a, 1993a). Recently, we also have demonstrated that transfection of P815 target cells with CD86 allows small, resting T cells to mediate CTL function in the presence of anti-CD3 mAb (LANIER et al., in press).

Specificity in this anti-CD3 redirected cytotoxicity assay system was demonstrated using a panel of P815 transfectants expressing several different human adhesion receptors including human CD72, CD106 (vascular cell adhesion molecule-1; VCAM-1), CD40 and CD58 (LFA-3) (Fig. 1). Counterreceptors for CD72, CD106, and CD58 (respectively, CD5, CD49d, and CD2) are present on essentially all small, resting T cells and the CD40 ligand is rapidly induced on T cells after stimulation with anti-CD3 (HERMANN et al. 1993; LANE et al. 1992). In these studies, we observed no anti-CD3 redirected cytotoxicity against CD72, CD106 or CD40

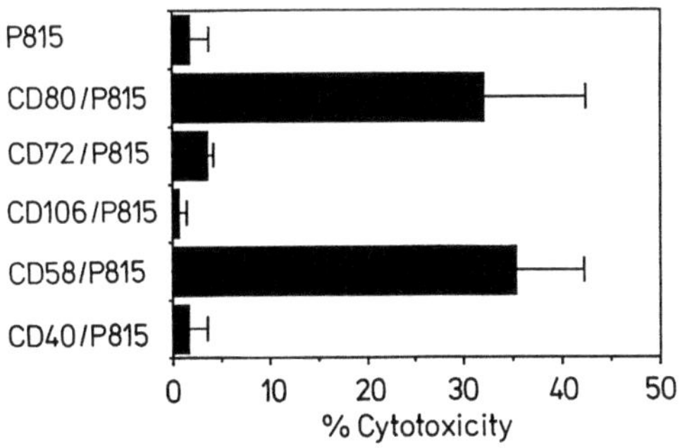

Fig. 1. Costimulation of anti-CD3 induced cytotoxicity. ^{51}Cr-labeled P815 transfectants or untransfected, parental P815 cells were incubated with anti-CD3 monoclonal antibody (mAb) for 30 min, freshly isolated small, resting T cells were added, and the assay was harvested after 4 h. Results are presented as mean % cytotoxicity ± SE for three experiments using different blood donors. E:T ratio was 20:1. We have demonstrated that the lysis of CD80$^+$ P815 or CD58 (LFA-3)$^+$ P815 transfectants is inhibited by anti-CD80 or anti-CD58 mAb, respectively

P815 transfectants in a 4–8 h ^{51}Cr release assay using small, resting T cells as effectors. However, significant lysis was observed against P815 cells transfected with human CD80 or CD58. The ability of CD58 transfectants to costimulate T cells is consistent with prior studies demonstrating facilitated adhesion and signal transduction by the interaction between CD2 and CD58 (Moingeon et al. 1989; Bierer et al. 1988). Additionally, CD2/CD58 interactions have been shown to participate in both antigen-dependent and -independent cell-mediated cytotoxicity (Siliciano et al. 1985; Shaw et al. 1986; Shaw and Luce 1987; Scott et al. 1989; Galocha et al. 1993). However, costimulation via the CD2/CD58 pathway is qualitatively different than CD28-mediated activation. CD2/CD58-dependent activation is inhibited by cyclosporin A (Hess and Bright 1991; van Gool et al. 1993) and in T cells is dependent upon functional expression of a CD3/TCR complex (Breitmeyer et al. 1987; Fox et al. 1986; Bockenstedt et al. 1988; Hahn et al. 1993). In contrast, CD28-dependent activation is resistant to cyclosporin A inhibition (June et al. 1987; Hess and Bright 1991). Moreover, anti-CD28 and anti-CD2 mAbs synergize in inducing T cell proliferation (Yang et al. 1988; van Lier et al. 1988; Pierres et al. 1988), suggesting the existence of distinct signaling pathways for CD2/CD58 and CD28/CD80.

3.2 Effectors of CD28-Dependent T Cell-Mediated Cytotoxicity

In vitro activated CTLs and a subset of presumably in vivo activated T cells lyse untransfected Fc receptor bearing P815 targets in the presence of anti-CD3 mAb (Leeuwenberg et al. 1985; Phillips and Lanier 1986; Segal et al. 1990). In contrast, small, resting T cells do not efficiently lyse P815 targets in the presence of anti-CD3 mAb. We and others have been unable to detect transcripts for serine esterases, perforin tumor necrosis factor-α (TNF-α), or other cytokines in this population of small, resting T cells (Smyth et al. 1992). The fraction of high density

T cells isolated from Percoll gradients used in our experiments does not contain $CD28^{-}8^{+}$ large granular T cells that possess cytolytic activity against P815 without CD28 costimulation (Azuma et al. 1993c). The lytic activity was not due to the presence of in vivo activated T cells contaminating the small, resting T cell population used for these assays. This was excluded by demonstrating that removal of cells expressing HLA-DR, IL-2Rα, CD56 and CD57 did not substantially affect lytic function (Azuma et al. 1993a). Both $CD4^{+}$ and $CD8^{+}$ small, resting T cells killed anti-CD3-coated $CD80^{+}$ P815 target cells.

The ability to generate CTLs from small, resting T cells using CD28 costimulation also has been reported by van Gool et al. (1993) using a different assay system. In these experiments, resting T cells were cultured with a suboptimal concentration of immobilized anti-CD3 and anti-CD28 for 3–4 days, which resulted in the efficient generation of CTLs able to mediate anti-CD3 redirected cytotoxicity against P815 or $Fc\gamma R^{+}$ murine fibroblast L cell targets. CD28-dependent generation of CTL activity is resistant to cyclosporin A and is mediated by both $CD4^{+}$ and $CD8^{+}$ T cells. The ability of $CD4^{+}$ T cells to mediate anti-CD3 redirected cytotoxicity has been confirmed using $CD4^{+}$ helper T cell clones. Therefore, $CD4^{+}$ T cells apparently possess the necessary cellular components for cytolytic function.

In our own studies, we could further distinguish T cell populations able to mediate CTL activity based on their display of cell surface antigens associated with "virgin" or "memory" T cell subsets. When cytotoxicity was measured using a standard 4 h ^{51}Cr release assay, the effector cells were contained predominantly within the memory T cell subset, identified by expression of the CD45RO antigen (Azuma et al. 1993a). In this time period, only minimal cytotoxicity was mediated by the $CD45RO^{-}$ virgin T cell subset. However, kinetic studies revealed that the virgin T cell population was capable of mediating substantial cytolytic activity when measured using a 6 or 8 h assay. We further demonstrated that the cocultivation of anti-CD3 mAb-coated $CD80^{+}$ P815 cells and small, resting T cells results in a rapid induction of the early activation antigens, CD69 and CD25 (IL-2Rα), on both virgin and memory T cells (Azuma et al. 1993a). Prior studies have established that stimulation of $CD45RA^{+}$ T cells with polyclonal mitogens results in a switch to $CD45RO^{+}$ isoform expression, after a period of coexpression of both CD45RO and RA antigens (Akbar et al. 1988; Sanders et al. 1988; Bell and Spartshott 1990; Beverley 1990) Small, resting virgin T cells acquire other antigenic markers and functions with activation, and activated T cells have fewer requirements to initiate effector function upon encountering antigen. It is clear that the memory T cell population can respond more rapidly than virgin T cells to recall antigens.

Both memory and virgin T cells proliferate and produce IL-2 upon cross-linking of the CD3 complex when costimulated by CD80 on antigen-presenting cells (van de Velde et al. 1993). However, when CD80 stimulation is delivered from bystander cells which do not present specific antigens, memory T cells, but not virgin T cells, can be efficiently activated. In other words, the separate delivery of specific antigen and costimulatory signals from two separate cells appears to be less efficient than the presentation by a single cell (LIU and JANEWAY 1992). Consistent with these observations, we do not detect bystander killing of untransfected P815 cells when CTLs are rapidly generated by coculture of small, resting T cells with anti-CD3 mAb and CD80 transfected P815 targets in a 4–8 h cytotoxicity assay. Therefore, in the initial phase of CTL generation, cells expressing CD80 or other potent costimulatory molecules may be lysed preferentially. However, after 24–48 h of stimulation via CD3 and CD28, the fully developed CTLs lyse both CD80-negative and CD80-positive targets equivalently. This is biologically advantageous because CTLs frequently need to eliminate virus-infected cells that do not express CD80. Thus, the CD28 pathway may augment and control the generation of CTLs, but then not be required for effector function by the fully differentiated and activated CTLs.

While our studies of human T cells have used anti-CD3 to demonstrate the role of CD28 costimulation in the generation of CTL, HARDING and ALLISON (1993) have demonstrated that murine splenic T cells can mount an mixed lymphocyte reaction (MLR) against allogeneic P815 cells transfected with CD80, and $CD8^+$ alloantigen-specific CTLs were generated. In contrast, untransfected P815 cells were unable to elicit an allogeneic MLR or generate CTLs P815 cells express high levels of MHC class I, but not class II, and $CD4^+$ T cells were not required for the generation of $CD8^+$ CTLs when $CD80^+$ P815 cells were used to stimulate the allogeneic MLR. Generation of CTL was dependent upon the production of IL-2 from the CTL precursors because anti-IL-2 mAb blocked CTL development. In our own studies using human T cells cocultured with $CD80^+$ P815 and anti-CD3 mAb, IL-2 did not appear to be required for the CTLs generated during the first 4–8 h of coculture because neutralizing mAb against IL-2 failed to prevent the rapid generation of CTLs in this system.

Moreover, the addition of exogenous IL-2 was unable to efficiently generate CTLs when small, resting T cells were cocultured with untransfected P815 and anti-CD3 for 4–8 h. However, in longer culture periods (e.g., 24–48 h) CTLs were generated by coculturing small, resting T cells with anti-CD3 mAb, untransfected P815 and exogenous IL-2. Therefore, while IL-2 may participate in the expansion of CTL from precursors, this cytokine alone is less efficient than CD28-dependent costimulation in the rapid development of CTL, a process which may require a direct activation signal via CD28 or possibly other cytokines for optimal activity.

3.3 Mechanisms Involved in the Generation and Effector Function of CTLs Generated by CD28-Dependent Activation

The mechanism responsible for CD28-dependent cytotoxicity against P815 is still unknown. We excluded the effect of cytolytic cytokines like interferon-γ (INF-γ) and TNF-α because P815 cells are resistant to high concentrations of human recombinant IFN-γ (rIFN-γ) and rTNF-α. Moreover, soluble cytolytic factors are unlikely to account for cytotoxicity in these anti-CD3 redirected cytotoxicity assays because cell-cell contact between the T cells and targets is necessary.

Additionally, no bystander lysis of third party cells was observed (our unpublished observations).

CTLs and NK cells have intracytoplasmic granules that contain factors such as perforin which can destroy the integrity of most cell membranes (PODACK et al. 1991). In addition, it appears that serine proteases in the granules may initiate programmed target cell death when they gain access into target cells possibly via a perforin pore (SHI et al. 1992). While small, resting T cells do not possess perforin or serine esterases, it is possible that the activation of small, resting T cells through CD3 and CD28 results in transcription of perforin and serine esterases (SMYTH et al. 1992; NAKATA et al. 1990, 1992). Studies with metabolic inhibitors and the kinetics of CD28-dependent generation of CTLs certainly are compatible with the possibility that transcription and production of lytic factors could be involved in cytotoxicity, although we have been unable to directly implicate such factors in the process.

Cell-mediated cytotoxicity also involves contact-dependent signals transmitted from the killer cells to the target cells. For example, membrane-bound TNF-α permits activated T cells to lyse TNF-α sensitive target cells as a consequence of direct cell-cell interactions (PEREZ et al. 1990). Likewise, interactions between Fas ligand on activated T cells and Fas on certain sensitive target cells can induce target cell death via apoptosis (ROUVIER et al. 1993; HANABUCHI et al. 1994). Golstein and colleagues have suggested that most of the calcium-independent T cell-mediated cytotoxicity may operate via a Fas-based pathway. The Fas and Fas ligand cytolytic pathway can be mediated by both CD4$^+$ and CD8$^+$ T lymphocytes and clones. In our studies measuring anti-CD3 redirected lysis of CD80-transfected P815 targets, the Fas/Fas ligand pathway probably is not involved in the lytic mechanism. We have been unable to detect Fas on the surface of P815 using anti-mouse Fas mAb and coculture of P815 with anti-mouse Fas mAb does not kill this cell line (unpublished observations).

3.4 Differential Requirements for Costimulatory Molecules Depending on the State of T Cell Activation or Differentiation

The requirements for costimulatory molecules in the generation and effector function of CTLs are influenced by the differentiation and activation status of the T cell. The cytolytic function of in vitro activated CTL clones is resistant to inhibitors of RNA and protein synthesis because the factors necessary for cytotoxicity presumably are present in these mature effector cells (JU 1991; SMYTH et al. 1992). In contrast, we have observed that the generation of CTLs from virgin T cells within the resting T cell population is completely inhibited in the presence of these metabolic inhibitors (AZUMA et al. 1993a). Thus, virgin T cells presumably do not possess the requisite components for cell-mediated cytotoxicity, but these factors are induced de novo during the culture with anti-CD3 and CD80$^+$ P815 transfectants. The lytic activity of memory T cells residing within the small, resting T cell population against anti-CD3-coated CD80$^+$ P815 target cells is only partially blocked by actinomycin D and cyclohexamide. These observations suggest that lysis of CD80+ P815 target cells by small, resting human T cells is mediated by both preexisting cytotoxic effector cells within the memory subset that are rapidly reactivated by costimulation via CD28 and by the de novo induction of CTL recruited from both the virgin and memory T cell populations.

Although the CD28–CD80 pathway is required for induction of anti-CD3 redirected cytotoxicity mediated by small, resting T cells, this interaction alone is insufficient to initiate efficient target cell lysis. We have demonstrated that a functional cooperation exists between LFA-1/CD54 and CD28/CD80 by experiments using murine L cell transfectants cotransfected with both CD54 and CD80 (Azuma et al. 1993a). CD54 alone is insufficient to initiate a cytolytic response mediated by small, resting T cells. However, in vitro activated CTL clones are able to mediate efficient anti-CD3 redirected cytotoxicity against P815 in the absence of CD28 costimulation by a mechanism that is LFA-1 dependent (Azuma et al. 1992a). Moreover, the presence of CD80 on target cells does not substantially affect the anti-CD3 redirected lysis mediated by in vitro activated CTL clones. Thus, CD28 costimulation is not essential for the effector phase of CTL function, provided that the CTL is optimally activated. This conclusion is also supported by studies using human antigen (influenza peptide)-specific CD8+ CTL clones (de Waal Malefyt et al. 1993). In these experiments, CTL were unable to lyse murine L cells transfected with HLA-A2 and loaded with the appropriate influenza peptide. However, cotransfection of the L cells with HLA-A2 and CD54 or CD58 enabled these CTLs to efficiently kill the targets after peptide loading. In contrast, cotransfection of L cells with CD80 and HLA-A2 did not render these targets susceptible to antigen-specific lysis. Therefore, CD28 costimulation appears most critical in the generation of CTLs, not in the final effector phase when fully activated CTLs are operational.

Presently, it is unclear whether there are different requirements for T cell proliferation and cytokine production vs the generation of CTLs. Many of the interactions that promote CTL generation or effector function also have been implicated in T cell proliferation. For example, in many experimental systems LFA-1 has been shown to be an important accessory molecule in CTL effector function. Damle et al. have claimed that ICAMs are able to costimulate T cell proliferation, based on the ability of plastic-immobilized ICAM-Ig fusion proteins to promote mitogen-induced T cell proliferation (Damle et al. 1992a,b). However, in our own studies, human CD54-transfected murine L cells or human CD54-transfected P815 cells are completely unable to costimulate T cell proliferative responses induced by anti-CD3 (Cayabyab et al. 1994 and our unpublished observations). Additionally, anti-LFA-1 mAbs do not inhibit proliferative responses induced with a combination of anti-CD3 and CD80-transfected FcγR+ L cells or by phorbor ester and anti-CD28 mAb (our unpublished observations). Green et al. have suggested that LFA-1 functions not as a costimulatory molecule, but serves primarily to modulate the signal delivered through TCR (Green et al. 1994). Further studies are required to precisely determine the relative contribution of these various receptor/ligand interactions to T cell proliferation and cytotoxicity.

In conclusion, CD28 costimulation participates in the primary generation of CTL, but after activation CD28–CD80 interactions are not essential for effector function. The relatively restricted distribution of the CD28 ligands on antigen-presenting cells may serve to control the initial generation of CTLs in vivo. However, once CTLs are generated they may then use the ICAMs and CD58 molecules expressed on many cell types for binding and triggering cytotoxicity. This may be biologically advantageous, permitting elimination of virus-infected cells of any tissue origin. We summarize the requirements for CD28 costimulation in T cell responses in Table 1.

In our experiment system, we have not clearly distinguished whether CD28 costimulation participates in the effector phase of CTL activity or only in the induction or reactivation of CTLs. Nonetheless, the finding that the CD28–CD80 interaction is not essential for cytolytic function mediated by in vitro activated CTLs does not exclude possible participation in cytotoxicity under some circumstances. Unless the CTLs are optimally activated, the CD28–CD80/CD86 interaction may enhance cytotoxicity, by contributing to effector/target cell binding or by augmenting a suboptimal signal generated through the TCR (Azuma et al. 1992a; van Gool et al. 1993).

We previously reported that the CD28–CD80 interaction is involved in MHC-unrestricted, TCR-independent cytotoxicity mediated by an NK leukemia cell line, YT2C2 (Azuma et al. 1992b). YT2C2 was derived from a human thymic lymphoma and expresses several characteristics of immature or fetal NK cells. However, unlike normal adult human NK cells which do not express CD28, YT2C2 cells express CD28 and kill EBV-transformed B cell lines expressing CD80 and CD86 and murine P815 cells transfected with CD80. YT2C2-mediated cytotoxicity is blocked by anti-CD28 mAb or by a combination of anti-CD80 and anti-CD86 mAbs (Azuma et al. 1992b, 1993b). However, similar to the situation with T cells, YT2C2-mediated lysis requires cooperation between the CD28–CD80/CD86 and the LFA-1/ICAM pathways. Cytotoxicity mediated by YT2C2 is resistant to DNA, RNA and protein synthesis inhibitors, indicating that YT2C2 possesses the necessary components required for cytolytic function. Presently, we have no evidence that

Table 1. Requirements of CD28 costimulation

Requirements	Effector T cells		
	Small, resting T cells		Large T cells
	"Virgin" $CD45RA^+RO^-$	"Memory" $CD45RA^{\pm}RO^+$	"Primed" $CD45RA^-RO^+$
CD28 costimulation			
Response rate	Slow	Immediate	Rapid
Proliferation	++	++	+
Cell-mediated cytotoxicity	++	++	–/±
Cooperation with other signals			
Proliferation	Independent	Independent	Independent
Cell-mediated cytotoxicity	Dependent on LFA-1	Dependent on LFA-1	?

either CD28 or LFA-1 mediates signal transduction in YT2C2. It is possible that the interaction of CD28/CD80–CD86 and LFA-1/CD54 just strengthens the conjugate formation between effectors and target cells. Many studies have demonstrated the involvement of several receptor-ligand interactions (e.g., LFA-1/CD54, CD2/CD58) in MHC-unrestricted cytotoxicity mediated by NK cells and certain T cells (SHAW and LUCE 1987; NAKAMURA et al. 1990; HERSEY and BOLHUIS 1987; SILICIANO et al. 1985; SCOTT et al. 1989; GALANDRIRNI et al. 1992). However, in the context of a cytolytic process it is difficult to clearly distinguish between binding and signal transduction through these receptor-ligand pairs. With respect to MHC-unrestricted cytotoxicity, it is possible interactions between an appropriate array of adhesion molecules with their ligands may in fact deliver signals that are sufficient to trigger the lytic response. Further studies are required to test this hypothesis.

4 Therapeutic Implications of CD28 Costimulation

The observation that CD28 costimulation can potentiate a primary immune response and rapidly generate cytolytic function in small, resting T cells suggests several practical applications in cancer immunotherapy. An effective T cell response against tumor cells requires at least three components. First, *antigen*: tumor-associated peptide antigens that are intracellularly processed and efficiently presented by MHC class I molecules on the cell surface. As a consequence of mutations in normal genes during the transformation process, many tumors may express tumor-associated antigens (TAAs) that can be recognized by T cells in a syngeneic host (PARDOLL 1993). Second, MHC class I or class II molecules: peptide antigens are recognized only when appropriately presented by MHC molecules. While tumor cells frequently alter the level of MHC expression (GARRIDO et al. 1993), nevertheless most tumor cells do express MHC class I or in some cases MHC class II molecules. Additionally, IFN-γ can be used to enhance expression of MHC molecules on transformed cells. Third, *costimulatory molecules*: CD80, ICAMs and possibly other membrane molecules that cooperate with the TCR to generate an efficient immune response. Even when MHC molecules and TAA peptides are present, host T cells may fail to mount an immune response against tumor cells. In some cases, this may be caused by inadequate expression of appropriate costimulatory molecules on the tumor cells. If so, this provides the possibility to provide costimulation as a strategy to elicit tumor-specific immunity.

TOWNSEND and ALLISON (1993) reported that transfer of the CD80 cDNA into murine K1735 melanoma cells caused them to be rejected by immunocompetent syngeneic C3H mice. Moreover, the immunized mice not only rejected the $CD80^+$ melanoma cells but also were resistant to subsequent challenge with CD80-negative K1735 melanoma cells. Growth of the $CD80^+$ K1735 tumor was greatly enhanced by depletion of $CD8^+$ T cells but was only slightly enhanced by $CD4^+$ T cell depletion, suggesting that rejection was mediated predominantly by

CD8⁺ T cells. Presumably in this experimental system, CD8⁺ CTL were directly responsible for tumor elimination. Similar results using the K1735 melanoma cell line were reported by Linsley and colleagues (CHEN et al. 1992). In these experiments, the melanoma cell line was cotransfected with CD80 and the E7 antigen from human papilloma virus (HPV) 16 to provide a strong antigen for T cell recognition. Possibly the K1735 melanoma cell line used by Townsend and Allison had acquired a genetic mutation that provided a sufficiently immunogenic endogenous tumor-associated antigen, therefore eliminating the need to introduce a viral antigen to elicit an immune response.

Recently, we have obtained results demonstrating that transfection of a murine Meth A sarcoma cell line with CD80 cDNA can elicit tumor-specific immunity. In these studies, BALB/c mice rejected the growth of syngeneic $CD80^+$ Meth A sarcoma cells, but not $CD80^-$ Meth A tumors. When rechallenged after immunization with $CD80^+$ Meth A sarcoma cells, immunocompetent mice rejected $CD80^-$ Meth A tumor cells, but not an irrelevant syngeneic renal carcinoma cell, Renca. These results suggested that a tumor-specific immune response against Meth A sarcoma was elicited. This conclusion was supported further by the ability of splenic T cells from $CD80^+$ Meth A immunized mice to mediate an MLR against both the $CD80^+$ and $CD80^-$ Meth A tumor cells in vitro, (unpublished observations).

Cotransfection of CD80 and a truncated MHC class II molecule into murine Sal sarcoma cells also resulted in efficient enhancement of immunogenicity (BASKAR et al. 1993). In this case, $CD4^+$ T cells were required for tumor rejection. An optimal response to tumor antigen may require both $CD8^+$ CTL and $CD4^+$ T helper cells. However, since most tumor cells lack MHC class II molecules this implies that in these circumstances $CD4^+$ T cells must play an indirect role in the generation of an immune response, possibly as a consequence of uptake and processing of secreted or shed tumor-associated peptides by antigen-presenting cells.

On the basis of these concepts, CD28 costimulation provides an attractive new approach to enhance anti-tumor immunity. Recently, CHEN et al. summarized the results from studies using 18 murine tumor cell lines, including melanomas, sarcomas, lymphomas, carcinomas and a mastocytoma, that either expressed or were transduced with CD80 (CHEN et al. 1994). Four of six B cell lymphomas expressing CD80 and four of eight tumors transduced with CD80 failed to induce a T cell immune response. They concluded that these failures may be due to inadequate MHC molecules or the lack of immunogenic tumor-accociated peptides. Therefore, if this strategy is to be successful in the treatment of human cancer, it will be necessary to select appropriate tumors that express adequate levels of MHC molecules and potentially possess immunogenic tumor-associated antigens.

An understanding of the molecules and mechanisms involved in CD28 costimulation may be important not only in the context of eliciting an anti-tumor response, but also may provide new approaches for the management of autoimmune disease and graft rejection. Already CTLA4-Ig fusion proteins have been shown to prevent heart allograft rejection in experimental model systems (Turka et al. 1992) and there is a suggestion that inhibiting CD28 costimulation may be beneficial in the treatment of autoimmune diabetes in nonobese diabetes mice.

Acknowledgement. DNAX Research Institute is supported by Schering Plough Cooperation.

References

Akbar AN, Terry L, Timms A, Beverley PCL, Janossy G (1988) Loss of CD45R and gain of UCHL1 reactivity is a feature of primed T cells. J Immunol 140: 2171–2178

Azuma M, Cayabyab M, Buck D, Phillips JH, Lanier LL (1992a) CD28 interaction with B7 co-stimulates primary allogeneic proliferative responses and cytotoxicity mediated by small, resting T lymphocytes. J Exp Med 175: 353–360

Azuma M, Cayabyab M, Buck D, Phillips JH, Lanier LL (1992b) Involvement of CD28 in major histocompatibility complex-unrestricted cytotoxicity mediated by a human NK leukemia cell line. J Immunol 149: 1115–1123

Azuma M, Cayabyab M, Phillips JH, Lanier LL (1993a) Requirements for CD28-dependent T cell-mediated cytotoxicity. J Immunol 150: 2091–2101

Azuma M, Ito D, Yagita H, Okumura K, Phillips JH, Lanier LL, Somoza C (1993b) B70 antigen is a second ligand for CTLA-4 and CD28. Nature 366: 76–79

Azuma M, Phillips JH, Lanier LL (1993c) CD28-T lymphocytes: antigenic and functional properties. J Immunol 150: 1147–1159

Baskar S, Ostrand-Rosenberg S, Nabavi N, Nadler LM, Freeman GJ, Glimcher LH (1993) Constitutive expression of B7 restores immunogenicity of tumor cells expressing truncated major histocompatibility complex class II molecules. Proc Natl Acad Sci USA 90: 5687–5690

Bell EB, Sparshott SM (1990) Interconversion of CD45R subsets of CD4 T cells in vivo. Nature 348: 163–166

Beverley PCL (1990) Is T-cell memory maintained by crossreactive stimulation? Immunol Today 11: 203–205

Bierer B, Barbosa J, Herrmann S, Burakoff SJ (1988) Interaction of CD2 with its ligand, LFA-3, in human T cell proliferation. J Immunol 140: 3358–3363

Bockenstedt LK, Goldsmith MA, Dustin M, Olive D, Springer TA, Weiss A (1988) The CD2 ligand LFA-3 activates T cells but depends on the expression and function of the antigen receptor. J Immunol 141: 1904–1911

Breitmeyer JB, Daley JF, Levine HB, Schlossman SF (1987) The T11 (CD2) molecule is functionally linked to the T3/Ti T cell receptor in the majority of T cells. J Immunol 139: 2899–2905

Brunet J-F, Denizot F, Luciani M-F, Roux-Dossetto M, Suzan M, Mattei M-G, Golstein P (1987) A new member of the immunoglobulin superfamily-CTLA-4. Nature 328: 267–270

Caux C, Vanbervliet B, Massacrier C, Azuma M, Okumura K, Lanier L, Barchereau J (1995) B70/B7-2 is identical to CD86 and is the major functional ligand for CD28 expressed on human dendritic cells. J Exp Med 180: 1841–1847

Cayabyab M, Phillips JH, Lanier LL (1994) CD40 preferentially costimulates activation of CD4+ T lymphocytes. J Immunol 152:1523–1531

Cerdan C, Martin Y, Courcoul M, Brailly H, Mawas C, Birg F, Olive D (1992) Prolonged IL-2 receptor alpha/CD25 expression after T cell activation via the adhesion molecules CD2 and CD28: demonstration of combined transcriptional and post-transcriptional regulation. J Immunol 149: 2255–2261

Chen L, Ashe S, Brady WA, Hellström I, Hellström KE, Ledbetter JA, McGowan P, Linsley PS (1992) Costimulation of antitumor immunity by the B7 counterreceptor for the T lymphocyte molecules CD28 and CTLA-4. Cell 71: 1093–1102

Chen L, McGowan P, Ashe S, Johnson J, Li Y, Hellström I, Hellström KE (1994) Tumor Immunogenicity determinants the effect of B7 costimulation on T cell-mediated tumor immunity. J Exp Med 179: 523–532

Damle NK, Doyle LV, Grosmaire LS, Ledbetter JA (1988) Differential regulatory signals delivered by antibody binding to the CD28 (Tp44) molecule during the activation of human T lymphocytes. J Immunol 140: 1753–1761

Damle NK, Klussman K, Linsley PS, Aruffo A (1992a) Differential costimulatory effects of adhesion molecules B7, ICAM-1, LFA-3, and VCAM-1 on resting and antigen-primed CD4+ T lymphocytes. J Immunol 148:1985–1992

Damle NK, Klussman K, Linsley PS, Aruffo A, Ledbetter JA (1992b) Differential regulatory effects of intercellular adhesion molecule-1 on costimulation by the CD28 counter-receptor B7. J Immunol 149: 2541–2548

Dariavach P, Mattei M-G, Golstein P, Lefranc M-P (1988) Human Ig superfamily CTLA-4 gene: chromosomal localization and identify of protein sequence between murine and human CTLA-4 cytoplasmic domains. Eur J Immunol 18: 1901–1905

de Jong R, Stokkers P, Lamme E, Kool JM, Borst F, Brouwer M, Miedema F, van Lier RAW (1991) Regulation of T-cell differentiation by CD2 and CD28 accessory molecules. Immunology 74:175–182

de Waal Malefyt R, Verma S, Bejarano MR, Hill M, Spits H (1993) CD2/LFA-3 or LFA-1/ICAM-1 but not CD28/B7 interaction can augment cytotoxicity by virus- specific CD8+ cytotoxic T lymphocytes. Eur J Immunol 23: 418–424

Engel P, Gribben JG, Freeman GJ, Zhou L-J, Nozawa Y, Abe M, Nadler LM, Wakasa H, Tedder TF (1994) The B7-2 (BDO) Costimulatory molecule expressed by monocytes and activated B lymphocytes is the differentiation antigen Blood 84: 1402–1407

Fox DA, Schlossman SF, Reinherz EL (1986) Regulation of the alternative pathway of T cell activation by anti-CD3 monoclonal antibody. J Immunol 136: 1945–1950

Fraser JD, Irving BA, Crabtree GR, Weiss A (1991) Regulation of interleukin-2 gene enhancer activity by the T cell accessory molecule CD28. Science 251: 313–316

Freeman GJ, Freedman AS, Segil JM, Lee G, Whitman JF, Nadler LM (1989) B7, a new member of the Ig superfamily with unique expression on activated and neoplastic B cells. J Immunol 143: 2714–2722

Freeman GJ, Gray GS, Gimmi CD, Lombard DB, Zhou L-J, White M, Fingeroth JD, Gribben JG, Nadler LM (1991) Structure, expression, and T cell costimulatory activity of the murine homologue of the human B lymphocyte activation antigen B7. J Exp Med 174: 625–631

Freeman GJ, Borriello F, Hodes RJ, Reiser H, Gribben JG, Ng JW, Kim J, Goldberg JM, Hathcock K, Laszlo G, Lombard LA, Wang S, Gray GS, Nadler LM, Sharpe AH (1993a) Murine B7-2, an alternative CTLA4 counter-receptor that costimulates T cell proliferation and interleukin-2 production. J Exp Med 178: 2185–2192

Freeman GJ, Gribben JG, Boussiotis VA, Ng JW, Restivo VA Jr, Lombard LS, Gray GS, Nadler LM (1993b) Cloning of B7-2: A CTLA-4 counter-receptor that costimulates human T cell proliferation. Science 262: 909–911

Freeman GL, Borriello F, Hodes JR, Reiser H, Hathcock KS, Laszlo G, McKnight AJ, Kim J, Du L, Lombard DB, Gray GS, Nadler LM, Sharpe AH (1993c) Uncovering of functional alternative CTLA-4 counter-receptor in B7-deficient mice. Science 262: 907–909

Galandrirni R, Albi N, Zarcone D, Grossi CE, Velardi A (1992) Adhesion molecule-mediated signals regulate major histocompatibility complex-unrestricted and CD3/T cell receptor-triggered cytotoxicity. Eur J Immunol 22: 2047–2053

Galocha B, Lopez D, de Castro J L (1993) Clonal heterogeneity in LFA-3 and ICAM-1 requirement for lysis by alloreactive T lymphocytes. J Immunol 150: 1653–1662

Garrido F, Cabrera T, Concha A, Glew S, Ruiz-Cebello F, Stern PL (1993) Natural history of HLA expression during tumour development. Immunol Today 14: 491–499

Green JM, Zheng X-G, Shimizu Y, Thompson CB, Turka LA (1994) T cell receptor stimulation, but not CD28 costimulation, is dependent on LFA-1 mediated events. Eur J Immunol 24: 265–272

Hahn WC, Burakoff SJ, Bierer BE (1993) Signal transduction pathways involved in T cell receptor-induced regulation of CD2 avidity for CD58. J Immunol 150: 2607–2619

Hanabuchi S, Koyanagi M, Kawasaki A, Shinohara N, Matsuzawa A, Nishimura Y, Kobayashi Y, Yonehara S, Yagita H, Okumura K (1994) Fas and its ligand in a general mechanism of T cell-mediated cytotoxicity. Proc Natl Acad Sci USA (in press)

Hara T, Fu SM, Hansen JA (1985) Human T cell activation II. A new activation pathway used by a major T cell population via a disulfide-bonded dimer of a 44 kilodalton polypeptide (9.3 antigen). J Exp Med 161: 1513–1524

Harding FA, Allison JP (1993) CD28-B7 interaction allow the induction of CD8+ cytotoxic T lymphocytes in the absence of exogenenous help. J Exp Med 177:1791–1796

Harding FA, McArthur JG, Gross JA, Raulet DH, Allison JP (1992) CD28 mediated signalling costimulates murine T cells and prevents induction of anergy in T cell clones. Nature 356: 607–609

Harper K, Balzano C, Rouvier E, Mattei M-G, Luciani M-F, Golstein P (1991) CTLA-4 and CD28 activated lymphocyte molecules are closely related in both mouse and human as to sequence, message expression, gene structure, and chromosomal location. J Immunol 147: 1037–1044

Hathcock KS, Laszlo G, Dickler HB, Bradshaw J, Linsley P, Hodes RJ (1993) Identification of an alternative CTLA-4 ligand costimulatory for T cell activation. Science 262: 905–907
Hersey P, Bolhuis R (1987) 'Nonspecific' MHC-unrestricted killer cells and their receptors. Immunol Today 8: 233–239
Hermann P, Blanchard D, de Saint-Vis B, Fossiez F, Gaillard C, Vanbervliet B, Briere F, Banchechereau J, Galizzi J-P (1993) Expression of a 32-kDa ligand for the CD40 antigen on activated human T lymphocytes. Eur J Immunol 23:961–964
Hess AD, Bright EC (1991) Cyclosporine inhibits T-cell activation at two distinct levels: role of the CD28 activation pathway. Transplant Proc 23: 961–966
Ju S-T (1991) Distinct pathways of CD4 and CD8 cells induce rapid target DNA fragmentation. J Immunol 146: 812–818
June CH, Ledbetter JA, Gillespie MM, Linsten T, Thompson CB (1987) T-cell proliferation involving the CD28 pathway is associated with cyclosporine- resistant interleukin 2 gene expression. Mol Cell Biol 7: 4472–4481
June CH, Ledbetter JA, Linsley PS, Thompson CB (1990) Role of the CD28 receptor in T-cell activation. Immunol Today 11: 211–216
Jung G, Freimann U, von Marschall Z, Reisfield RA, Wilmanns W (1991) Target cell-induced T cell activation with bi- and trispecific antibody fragments. Eur J Immunol 21: 2431–2435
Lane P, Traunecker A, Hubele S, Inui S, Lanzavecchia A, Gray D (1992) Activated human T cells express a ligand for the human B cell- associated antigen CD40 which participates in T cell-dependent activation of B lymphocytes. Eur J Immunol 22: 2573–2578
Lanier L, O'Fallon S, Somoza C, Phillips SH, Linsley PS, Okumura K, Ito D, Azuma M (1995) CD80(B7) and CD86(B70) provides similar costimulatory signals for T cell proliferation, cytokine production and generation of cytotoxic T lymphocytes. J Immunol (in press)
Ledbetter JA, Imboden JB, Schieven GL, Grosmaire LS, Rabinovitch PS, Lindsten T, Thompson CB, June CH (1990) CD28 ligation in T-cell activation: evidence for two signal transduction pathways. Blood 75: 1531–1539
Leeuwenberg JFM, Spits H, Tax WJM, Capel JA (1985) Induction of nonspecific cytotoxicity by monoclonal anti-T3 antibodies. J Immunol 134: 3770–3775
Lenschow DJ, Zeng Y, Thistlethwaite JR, Montag A, Brandy W, Gibson MG, Linsley PS, Bluestone JA (1992) Long-term survival of xenogeneic pancreatic islet grafts induced by CTLA41g. Science 257: 789–792
Lindsten T, June CH, Ledbetter JA, Stella G, Thompson CB (1989) Regulation of lymphokine messenger RNA stability by a surface-mediated T cell activation pathway. Science 244: 339–343
Lindsten T, Lee KP, Harris ES, Petryniak B, Crayghead N, Reynolds PJ, Lombard DB, Freeman GJ, Nadler LM, Gray GS, Thompson CB, June CH (1993) Characterization of CTLA-4 structure and expression on human T cells. J Immunol 151: 3489–3499
Linsley PS, Brady W, Grosmaire L, Aruffo A, Damle NK, Ledbetter JA (1991a) Binding of the B cell activation antigen B7 to CD28 costimulates T cell proliferation and interleukin 2 mRNA accumulation. J Exp Med 173: 721–730
Linsley PS, Brady W, Urnes M, Grosmaire LS, Damle NK, Ledbetter JA (1991b) CTLA-4 is a second receptor for the B cell activation antigen B7. J Exp Med 174: 561–569
Linsley PS, Greene JL, Tan P, Gradshaw J, Ledbetter JA, Anasetti C, Damle NK (1992) Coexpression and functional cooperation of CTLA-4 and CD28 on activated T lymphocytes. J Exp Med 176: 1595–1604
Liu Y, Janeway Jr CA (1992) Cells that present both specific ligand and costimulatory activity are the most efficient inducers of clonal expansion of normal CD4 T cells. Proc Natl Acad Sci USA 89: 3845–3849
Martin PJ, Ledbetter JA, Morishita Y, June CH, Beatty PG, Hansen JA (1986) A 44 kilodalton cell surface homodimer regulates interleukin 2 production by activated human T lymphocytes. J Immunol 136: 3282–3287
Moingeon P, Chang H-C, Wallner BP, Stebbins C, Frey AZ, Reinherz EL (1989) CD2-modiated adhesion facilitates T lymphocyte antigen recognition function. Nature 339: 312–314
Nakamura T, Takahashi K, Fukazawa T, Koyanagi M, Yokoyama A, Kato H, Yagita H, Okumura K (1990) Relative contribution of CD2 and LFA-1 to murine T and natural killer cell functions. J Immunol 145: 3628–3634
Nakata M, Smyth MJ, Norihisa Y, AK, Shinkai Y, Okumura K, Yagita H (1990) Constitutive expression of pore-forming protein in peripheral blood γ/δ T cells: implications for their cytotoxic role in vivo. J Exp Med 172: 1877–1880
Nakata M, Kawasaki A, Azuma M, Tsuji K, Matuda H, Shinkai Y, Yagita H, Okumura K (1992) Expression

of perforin and cytolytic potential of human peripheral blood lymphocyte subpopulations. Int Immunol 4: 1049–1054
Pardoll DM (1993) Cancer vaccines. Immunol Today 14: 310–316
Perez C, Albert I, DeFay K, Zachariades N, Gooding L, Kriegler M (1990) A nonsecretable cell surface mutant of tumor necrosis factor (TNF) kills by cell-cell contact. Cell 63: 251–258
Phillips JH, Lanier LL (1986) Lectin-dependent and anti-CD3 induced cytotoxicity are preferentially mediated by peripheral blood cytotoxic T lymphocytes expressing Leu-7 antigen. J Immunol 136: 1579–1585
Pierres A, Lopez M, Cerdan C, Nunes J, Olive D, Mawas C (1988) Triggering CD28 molecules synergize with CD2 (T11.1 and T11.2)-mediated T cell activation. Eur J Immunol 18: 685–690
Podack ER, Hengartner H, Lichtenheld MG (1991) A central role of perforin in cytolysis? Annu Rev Immunol 9:129–157
Rouvier E, Luciani M-F, Golstein P (1993) Fas involvement in Ca^{2+}-independent T cell-mediated cytotoxicity. J Exp Med 177: 195–200
Sanders ME, Makgoba MW, Shaw S (1988) Human naive and memory T cells: reinterpretation of helper-inducer and suppressor-inducer subsets. Immunol Today 9: 195–198
Scott CF, Boloender S, GDM, Holldack J, Lambert JM, Venkatesh YP, Morimoto C, Ritz J, Schlossman SF (1989) Activation of human cytolytic cells through CD2/T11: comparison of the requirements for the induction and direction of lysis of tumor targets by T cells and NK cells. J Immunol 142: 4105–4112
Segal DM, Garrido MA, Qian J-H, Mezzananica D, Andrew S, Perez P, Kurucz I, Valdayo MJ, Titus JA, Winkler DF, Wunderlich JR (1990) Effectors of targeted cellular cytotoxicity. Mol Immunol 27: 1339–1342
Shaw S, Luce GEG (1987) The lymphocyte function-associated antigen (LFA)-1 and CD2/LFA-3 pathways of antigen-independent human T cell adhesion. J Immunol 139:1037–1045
Shaw S, Luce GEG, Quinones R, Gress RE, Springer TA, Sanders ME (1986) Two antigen-independent adhesion pathways used by human cytotoxic T-cell clones. Nature 323: 262–264
Shi L, Kraut RP, Aebersold R, Greenberg AH (1992) A natural killer cell granule protein that induces DNA fragmentation and apoptosis. J Exp Med 175: 553–566
Siliciano RF, Pratt JC, Schmidt RE, Ritz J, Reinherz EL (1985) Activation of cytolytic T lymphocyte and natural killer cell function through the T11 sheep erythrocyte binding protein. Nature 317: 428–430
Smyth MJ, Norihisa Y, Ortaldo JR (1992) Multiple cytolytic mechanisms displayed by activated human peripheral blood T cell subsets. J Immunol 148: 55–62
Sommer F, Rollinghoff M, Lohoff M (1993) Signaling via CD28 costimulates lymphokine production, but does not reverse unresponsiveness to interleukin-2 in anti-CD3 triggered Th1 cells. Eur J Immunol 23: 2498–2502
Townsend SE, Allison JP (1993) Tumor rejection after direct costimulation of $CD8^+$ T cells by B7-transfected melanoma cells. Science 259: 368–370
Turka LA, Linsley PS, Lin H, Brady W, Leiden JM, Wei R-Q, Gibson ML, Zheng X-G, Myrdal S, Gordon D, Bailey T, Bolling SF, Thompson CB (1992) T-cell activation by the CD28 ligand B7 is required for cardiac allograft rejection in vivo. 89: 11102–11105
van de Velde H, Lorre K, Bakkus M, Thielemans K, Ceuppens JL, de Boer M (1993) $CD45RO^+$ memory T cells but not $CD45RA^+$ naive T cells can be efficiently activated by remote co-stimulation with B7. Int Immunol 11: 1483–1487
van Gool SW, de Boer M, Ceuppens JL (1993) CD28 ligation by monoclonal antibodies or B7/BB1 provides an accessory signal for the cyclosporin A-resistant generation of cytotoxic T cell activity. J Immunol 150: 3254–3263
van Lier RA, Brouwer M, Aarden LA (1988) Signals involved in T cell activation. T cell proliferation induced through the synergistic action of anti-CD28 and anti-CD2 monoclonal antibodies. Eur J Immunol 18: 167–172
van Lier RA, Brouwer M, Groot ED, Kramer I, Aarden LA, Verhoeven AJ (1991) T cell receptor/CD3 and CD28 use distinct intracellular signaling pathways. Eur J Immunol 21: 1775–1778
Yamada H, Martin PJ, Braun MP, Beatty PG, Sadamoto K, Hansen JA (1985) Monoclonal antibody 9.3 and anti-CD11 antibodies define reciprocal subsets of lymphocytes. Eur J Immunol 15: 1164–1168
Yang SY, Denning SM, Mizuno S, Dupont B, Haynes BF (1988) A novel activation pathway for mature thymocytes: costimulation of CD2 (Tp50) and CD28 (Tp44) induces autocrine interleukin 2/ interleukin 2 receptor-mediated cell proliferation. J Exp Med 168: 1457–1468
Yokochi T, Holly RD, Clark EA (1982) B lymphoblast antigen (BB-1) expressed on Epstein-Barr virus-activated B cell blasts, B lymphoblastoid cell lines, and Burkitt's lymphomas. J Immunol 128: 823–827

Degranulating Cytotoxic Lymphocytes Inflict Multiple Damage Pathways on Target Cells

P.A. HENKART, M.S. WILLIAMS, and H. NAKAJIMA

1 The Granule Exocytosis Model

1.1 Physiology of Lymphocyte Cytotoxicity

In this review, we will describe our recent studies on lymphocyte cytotoxicity using a molecular approach to the understanding of cytotoxic lymphocyte function. However, before describing this work, it is important to place these studies in a broader perspective, as providing a biochemical basis to the general physiological picture of lymphocyte cytotoxic mechanisms which had emerged previously. After the development of the ^{51}Cr release assay, which allowed

Experimental Immunology Branch, Building 10, Room 4B17, NCI–National Institutes of Health, Bethesda, MD 20892, USA convenient quantitation of cell death in vitro, several excellent laboratories studying alloreactive cytotoxic T lymphocytes (CTLs) established a number of important properties of what might be called the physiology of CTL-mediated cytotoxicity (GREEN and HENNEY 1981; GOLSTEIN and SMITH 1977; MARTZ 1977). Three distinct phases of the cytotoxic process were distinguished: (1) adhesion; (2) lethal hit; and (3) target cell disintegration, or killer cell-independent lysis. The first two of these phases occurred within minutes at 37° C), and required the divalent cations Mg^{+2}, Ca^{+2}, respectively. They could be blocked by various inhibitors of energy production and cytoskeletal function. The target disintegration step, which occurred after the CTL had delivered the "kiss of death" to the target, occurred over the course of hours and was difficult to block by drugs. This multihour "death agony" of the target cell is relevant to considerations of the pathways described below. The above three phases of the cytotoxic process were later described in NK cell-mediated cytotoxicity, leading to the implication that although there were two distinct lineages of cytotoxic lymphocytes utilizing different receptors for target cell recognition, the basic mechanism of lethal hit delivery was likely to be the same.

One of the hallmark properties of CTL killing is the highly selective nature of its lethal effects. These were shown by "innocent bystander" experiments in which bystander cells not bearing target antigen were mixed with antigenic target cells and CTLs added. The results showed that during the process of killing target cells there was little or no killing of the copelleted bystander cells. Thus, the lethal hit was highly polarized and these experiments were widely interpreted as ruling out a secretory mechanism for lethal hit delivery.

1.2 Secretory Granules in Lymphocytes

An early suggestion that a secretory process accompanied CTL killing was provided by Zagury and colleagues, who reported a histochemical study of in vivo derived CTLs bound to target cells. They observed acid phosphatase staining CTL granules in a polarized distribution, preferentially localized towards the bound target (THIERNESSE et al. 1977). With time this histochemical stain was found in the extracellular space between the killer and target, clearly implying a secretory process. At the time a secretory process involving lysosomes did not seem plausible, but subsequent studies of both mast cells and neutrophils as well as lymphocytes have shown that all of these cells have secretory granules which also contain lysosomal enzymes.

Classically lymphocytes are white blood cells without prominent granules. After years of work trying to define the "natural killing" phenomenon, a major advance came with the recognition that the natural killer (NK) effector cells responsible for this activity are a physically and morphologically distinct normal lymphocyte subpopulation containing azurophilic cytoplasmic granules (TIMONEN et al. 1981). The cells were termed large granular lymphocytes (LGLs), and their granules were shown by histochemical techniques to contain lysosomal enzymes (GROSSI et al. 1982).

After the technology of lymphocyte growth and cloning in vitro using interleukin-2 (IL-2) was developed, it became far easier to study CTLs as cells, and it was widely recognized that CTLs, like LGLs, contain cytoplasmic granules (DENNERT and PODACK 1983). Their structure, development and cell biology are described in other chapters in this volume. It should be noted that cloned $CD4^+$ "helper" T cells, which recognize soluble antigenic peptides bound to MHC class II molecules, have also been shown to contain granules, and to degranulate when the T cell receptor (TCR) is cross-linked (TAPLITS et al. 1988). The physiological relevance of this secretory process is still unclear.

There has been some controversy in the literature as to whether or not in vivo CTLs have granules. It is clear that IL-2 can up-regulate expression of CTL granule components 10- to 100-fold (BERKE et al. 1993; SMYTH et al. 1990), and it is likely that CTLs which arise in vivo contain considerably smaller amounts of these molecules than clones maintained in vitro. Nevertheless, sensitive techniques have clearly shown that in vivo CTLs express cytolysin/perforin and granzyme A (NAGLER-ANDERSON et al. 1989). It would appear that failure to detect granules morphologically or granule components biochemically in these CTLs can best be explained by limitations in the detection methods being used, but this issue is still controversial (HELGASON et al. 1992).

1.3 The Granule Exocytosis Model

As shown in Fig. 1, the essential feature of this proposal is that a polarized and highly regulated secretion, similar to that in nerve cells, could explain most of the known physiology of lymphocyte-mediated cytotoxicity. A "temporary synapse" is envisioned to exist between killer cells and their bound targets, with degranulation releasing the granule mediators into this restricted volume. The model proposes that the lethal moieties are presynthesized and stored in secretory granules in the effector cells.

Several lines of evidence show that degranulation occurs in lymphocytes in response to triggering by the receptors recognizing target cells. Microscopic studies, both at the light and electron microscopy (EM) levels, have documented such secretion (YANNELLI et al. 1986; HENKART and HENKART 1982; DENNERT and PODACK 1983). Measurments of granzyme, proteoglycan, and lysosomal enzyme secretion have verified that degranulation accompanies target recognition in both CTL and NK cells (PASTERNACK et al. 1986; SCHMIDT et al. 1985). Deposition of cytolysin pore structures on target membranes (DOURMASHKIN et al. 1980; DENNERT and PODACK 1983) can only occur as a result of effector cell granule exocoytosis.

From the point of view of cell biology, degranulation corresponds to the "regulated" pathway of protein secretion; it is the secretion event itself (vesicle or granule exocytosis) which is regulated, typically by engagement of a membrane receptor. Newly synthesized proteins which are secreted by this pathway have to be sorted from others after being processed by the Golgi. In contrast, those which are immediately secreted without storage utilize the "constitutive"

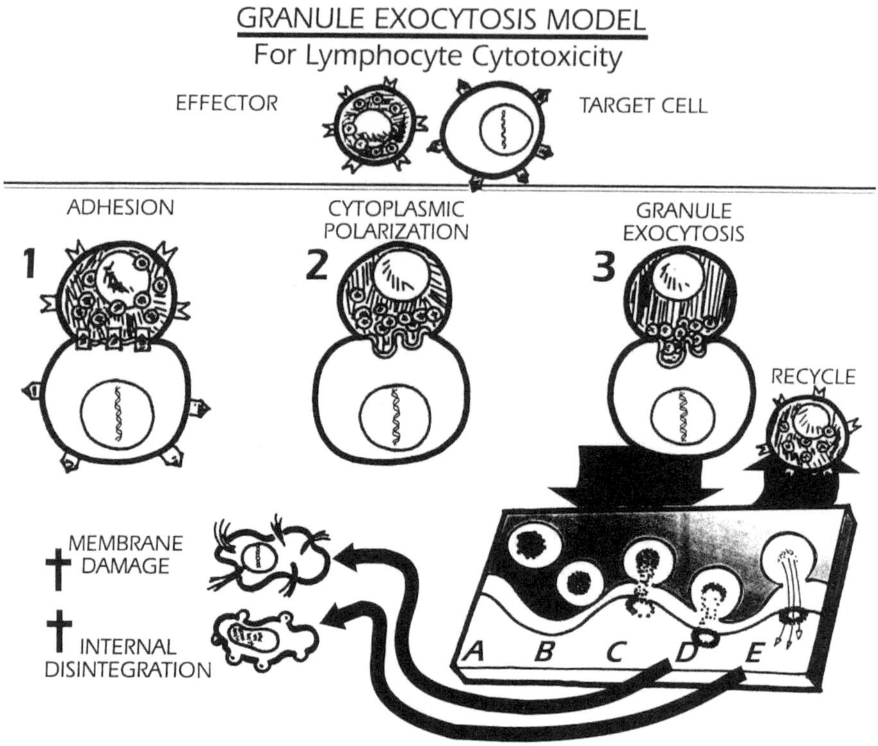

Fig. 1. The granule exocytosis model. In stage *1* the cytotoxic lymphocyte recognizes and binds to the target cell by a combination of specific receptor-ligand interactions and adhesion molecules. Their engagement triggers stage *2*, a polarization of the cytoplasm of the cytotoxic lymphocyte. This polarization involves most of the cytoplasmic organelles, particularly the secretory granules, which move towards the bound target. The combination of polarization and adhesion creates a synapse-like region between the two cells. Stage *3* is the critical secretion step, in which the secretory granule (*A*) membranes fuse with the plasma membranes (*B*), releasing the granule contents into the synapse-like junctional volume between the two cells (*C*). The cytolysin inserts into the target membrane and polymerizes, creating membrane pores (*D*) which can cause target cell lysis by the membrane damage pathway. The pores also allow the penetration into the target cytoplasm of other molecules which were released from the effector cell's secretory granules (*E*). As described in this review, granzymes have been identified as major mediators of the internal disintegration pathway of target cell damage. (From HENKART et al. 1992)

secretory process, in which small vesicles are shuttled directly to the plasma membrane where they undergo exocytosis immediately (KELLY 1985). In this case the secretory process itself is not regulated, although the expression of the secreted proteins commonly is. Such proteins appear to include lymphokines, such as interferon-γ (IFN-γ), which are produced by both the LGLs and the CTLs in response to target recognition. Thus it is likely that cells like CTLs secrete proteins by both secretory pathways, and while secretion is controlled by the TCR in both cases, different means are used and different kinetics result.

In a more general view, it is satisfying to note that the granule exocytosis model proposes that cytotoxicity is like other known lymphocyte effector functions in that all operate principally by secretion: Immunoglobulins secreted by the constitutive secretory pathway in plasma cells and B cells, lymphokines secreted by the constitutive pathway for T and NK cells, and granule proteins secreted by the regulated pathway in cytotoxic cells.

The various lines of evidence showing that degranulation of cytotoxic lymphocytes describes a major pathway of lymphocyte cytotoxicity have been reviewed elsewhere (TSCHOPP and NABHOLZ 1990; HENKART et al. 1992; PODACK et al. 1991). In addition to the evidence for a degranulation accompanying cytotoxicity cited above, there is the following: (1) expression of cytolysin/preforin mRNA and protein is limited to lymphocytes known to be cytotoxic, and when protein expression is examined using antibody localization techniques it is found in granules (PODACK et al. 1991); (2) the cytolysin pore structures are deposited on target membranes after killing by LGLs and CTLs (DOURMASHKIN et al. 1980; DENNERT and PODACK 1983); (3) inhibition of the induction of cytotoxicity by an antisense oligonucleotide against cytolysin/perforin (ACHA-ORBEA et al. 1990); (4) loss of cytotoxic activity in CTL from in perforin "knock-out" mice (KAGI et al. 1994); (5) the rat basophilic leukemia (RBL) transfection experiments described below.

2 The Rat Basophilic Leukemia Cell Transfectant Approach

In the last several years, we have utilized a novel strategy to probe the properties of several granule components with respect to the granule exocytosis model. This approach is based on the philosophy that building up a functional system by adding known components together is a stringent test of one's knowledge of that system. It complements the opposite and more common approach of perturbing a functional system by knocking out one of its functional components. In this case we have started with a noncytotoxic cell which possesses a regulated secretory pathway, expressed lymphocyte granule components by transfection with the appropriate cDNAs, and then assessed the ability of the resulting transfectants to kill target cells. The parental cells for these studies are RBLs, a rat mast cell tumor line which has been used extensively as a model system for mast cell degranulation triggered by the IgE Fc receptor (FEWTRELL and METZGER 1981). We reasoned that target cell recognition could be achieved by using a monoclonal IgE antihapten after covalently haptenating the target surface proteins. Initial experiments showed that labeled histamine was released from RBL granules after contact with such IgE coated target cells.

Transfection of RBL cells proved much easier than cloned T lymphocyte lines, which seem to be resistant to standard transfection approaches. A critical issue with this whole approach is the proper targeting to rat mast cell secretory granules of the expressed murine lymphocyte granule proteins. The criteria by which proteins are sorted to secretory granules from the ER are not fully understood and we feared that this problem could cripple this approach. Nevertheless, we have found that in all three cases in which we have done this have resulted in proper granule targeting in RBLs, perhaps reflecting the homology between mast cells and lymphocytes and between mouse and rat.

3 The Membrane Damage Pathway of Target Injury

The original granule exocytosis model proposed that the granule mediator that caused lethal target damage was cytolysin/perforin. This was identified after several lines of evidence had suggested that membrane damage was an essential component of target cell attack by CTLs. These aspects of the granule exocytosis model have been reviewed previously (HENKART and YUE 1988; PODACK et al. 1991; YOUNG and COHN 1986; YAGITA et al. 1992). We will summarize them only briefly here and deal with selected aspects of this pathway.

3.1 Target Cell Membrane Damage: Role of Cytolysin

Early experiments showed that CTLs could kill target cells which were enucleated or lightly fixed (SILICIANO and HENNEY 1978; BUBBERS and HENNEY 1975). Thus, we were attracted to the hypothesis that the CTL lethal hit was directed at the membrane. Accordingly, we set up a model system in which cytotoxic LGLs attacked antibody-coated red blood ghosts which were filled with various sized marker proteins (Fig. 2) (SIMONE and HENKART 1980). The results were a striking suggestion that membrane pores were created by the attacking LGLs and when the membranes were examined by EM after negative stain, pore-like structures were observed (DOURMASHKIN et al. 1980). We then had to conclude that this pore material was probably deposited as a result of secretion by the LGLs and we began a search for this membrane damaging material. Using LGL tumors as a convenient source for biochemical purification, we purified secretory granules from homogenates and then identified a potent lytic agent which we termed cytolysin (MILLARD et al. 1984). This activity caused all types of cells to lyse within minutes and was exquisitely dependent on calcium. Upon purification, it was found to be comprised of a single 65 kDa protein which has the calcium-dependent ability to insert into pure lipid bilayer membranes, aggregate to form pore-like structures, and create functional aqueous pores in membranes (BLUMENTHAL et al. 1984). Similar experiments were done with CTL granules (PODACK and KONIGSBERG 1984).

Cytolysin/perforin's primary structure has been elucidated from cDNA cloning and sequencing, and it has modest homology to the complement "membrane attack" proteins C6, C7, C8, and C9 (YOUNG et al. 1986). A detailed picture of how this protein transforms under the influence of calcium from a water soluble protein to an aggregated amphipathic membrane-inserted pore has still not been developed.

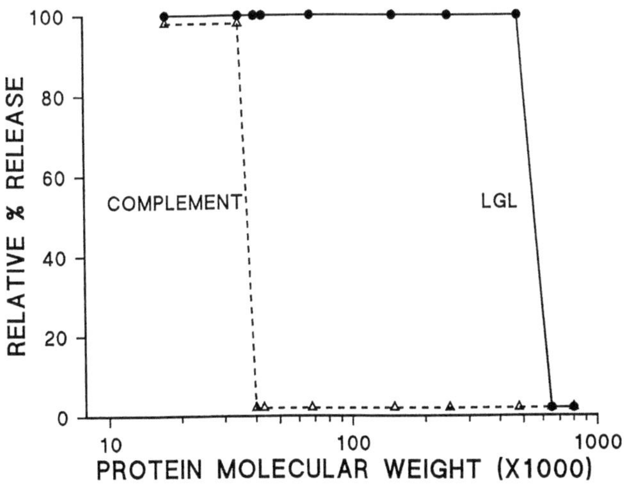

Fig. 2. Size dependence of proteins diffusing out of large granular lymphocyte (LGL)-attacked red cell ghosts. Trinitrophenyl (TNP)-modified resealed ghosts containing rhodamine-modified marker proteins along with carboxyfluorescein were prepared from human red blood cells. These were attached to the bottom of culture wells, and human peripheral blood lymphocytes (PBLs) were added. The anti-TNP-dependent release of protein and carboxyfluorescein markers was measured after 4 h of incubation. The data shown summarize a series of experiments with different size protein markers. The relative release plotted shows the percentage of protein marker released relative to the percentage of carboxy fluorescein. These experiments also showed that complement created smaller pores. (From SIMONE and HENKART 1980)

3.2 Lysis of Nucleated Cells by Pore-Forming Agents

It is often assumed that cell death via membrane damage occurs by colloid osmotic lysis, a model developed based almost entirely on evidence with red cells. In this model water moves into the cell in response to the unbalanced osmotic pressure after small molecules have equilibrated across the damaged membrane. The resulting pressure on the membrane then causes the cell membrane to rupture, releasing all the cell contents. Experiments with complement killing of nucleated cells have suggested that colloid osmotic lysis does not operate in this case as it does with red cells (KIM et al. 1989), and because of the larger maximal pores it seems less likely that cytolysin-damaged cells behave like complement-damaged cells. In any case, there is no good evidence that such colloid osmotic lysis takes place after CTLs attack nucleated target cells. It is plausible that such cells die as a result of directly losing essential metabolites and/or macromolecules through the damaged membrane. Nevertheless, this is difficult way to kill most tumor cells, as their complement-mediated lysis requires more pores of the membrane attack complex to be inserted in the membrane than the single one required for red cell lysis (MORGAN 1989). One explanation is that tumor cells have the ability to repair this type of membrane damage by blebbing off and endocytosing patches of membrane containing pores, and cytolysin damaged tumor cells have been shown to recover their membrane integrity with time (BASHFORD et al. 1988).

One unappreciated problem for the granule exocytosis model with cytolysin as the sole mediator is the rapid lysis of tumor cells when even small amounts of

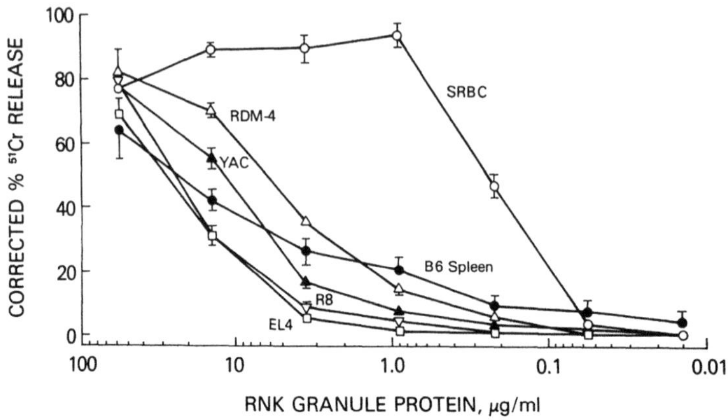

Fig. 3. Lytic effect of rat NK (*RNK*) cell tumor granules on various cells. It can be seen that sheep red blood cells (*SRBC*) are 50- to 200-fold more sensitive to the granule cytolysin than normal splenic lymphocytes (*B6 spleen*) or the various lymphoid tumor cells (*RDM-4, YAC, R8, EL4*). (From HENKART et al. 1984)

cytolysin are applied to a cell suspension. We showed that such lysis is complete within a few minutes at 25° C, and in less than a minute at 37°C (HENKART et al. 1984). It is hard to explain the multihour death agony of lethally injured target cells (see above) if cytolysin is the sole mediator of death. Curiously, this issue has not been actively raised in the prolonged debate over the granule exocytosis model.

In our original description of cytolysin's lytic activity, we noted that red blood cells were about 100x more sensitive to lysis by granule extracts than nucleated cells (Fig. 3). Once redirected lysis techniques were developed, it became clear

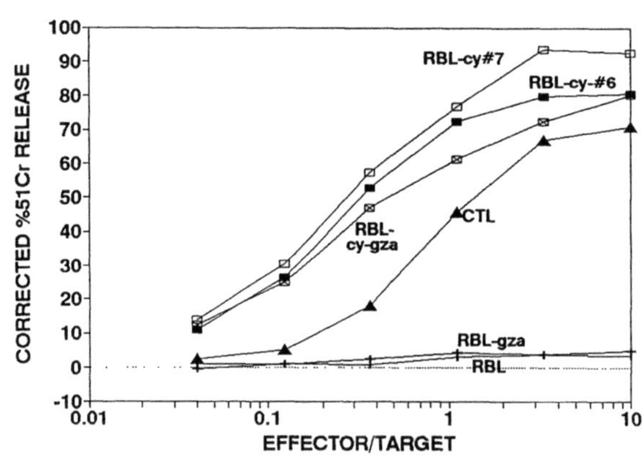

Fig. 4. Cytolysin-transfected rat basophilic leukemia (*RBL*) cells (*RBL-cy*) are potent killers of red cell targets. In this experiment using trinitrophenyl (TNP)-modified human red cell targets, two different clones of RBL-cy transfectants are compared with one RBL-granzyme A (*RBL-gza*) clone, one RBL-cy-gza double transfectant clone, the parental RBL line, and a mouse cloned CTL line. Lysis by the RBL transfectants was dependent on monoclonal IgE anti-dinitrophenyl (DNP), and that by CTL on α-DNPxαCD3 heteroconjugate (redirected lysis). This experiment shows that cytolysin is required for red cell killing, but granzyme A expression does not influence this activity. (From SHIVER and HENKART 1991)

that red cell targets were about equally susceptable to CTLs as tumor cells (Figs. 4, 5). We thought that this could still be compatible with cytolysin as the mediator of the lethal damage if it was assumed that the quantal local delivery predicted by the granule exocytosis model might mean that cytolysin's lytic potency is not limiting. However, this issue was hard to assess prior to the experiments described in the next section.

3.3 Cytolysin-Transfected Rat Basophilic Leukemia Cells

The first and most obvious protein to express in RBLs was cytolysin. We found that mRNA levels close to those of cloned CTLs could be achieved, and the resulting RBL-cy had an IgE-dependent, calcium-dependent cytotoxic activity against red blood cell targets that was comparable to cloned CTL (Fig. 4) (SHIVER and HENKART 1991). While this was a satisfying result confirming the viability of the RBL transfectant approach, the original RBL-cy transfectants showed only marginal cytotoxic activity on tumor targets. A subsequent group of RBL-cy transfectants did show significant cytotoxicity on these targets (Fig. 5), but this was still only a small fraction of the activity of CTL. One possible explanation was that nucleated cells were less efficient at triggering degranulation, but this was ruled out by measuring granzyme A secretion, which was equal. Another possibility which was difficult to assess was that the RBL-target cell junctional regions were not as well sealed as were those between CTLs and their targets, perhaps because the RBL's adhesion molecules were not as potent as those of CTL.

An obvious rationale for the limited activity of RBL-cy on tumor targets was that other important granule components were missing in the RBL-cy

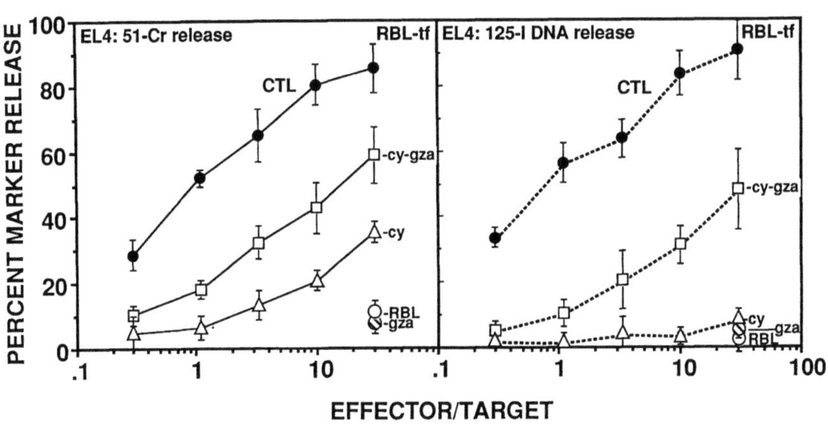

Fig. 5. Tumor cell killing by RBL transfectants (see Fig. 4). The murine thymoma line EL4 was prelabeled with ^{125}I-UdR to label DNA, labeled with ^{51}Cr and trinitrophenyl (TNP) modified, and then used in a standard 4 h cytotoxicity experiment. ^{125}I-DNA released was combined from both initial supernatants and again after 0.05% Triton X100 treatment. Data shown are means of multiple experiments with the following number of independent clones of each type: RBL-cy-gza, 4; RBL-cy, 8; RBL-gza, 2

transfectants, and these were particularly important for killing nucleated cells. Because they were major granule protein components, granzymes seemed like promising prospects, and there were three possible ways in which these proteases could be envisioned to work. One way was that they participated in processing the cytolysin so that it could more efficiently attack membranes, especially those of nucleated cells. A case for granzyme processing of cytolysin has been made based on experiments with specific protease inactivators (HUDIG et al. 1991), but more evidence is required to show the nature of the cytolysin protein processing for this to become convincing. Another possibility was that granzymes digested away external domains of membrane glycoproteins, thereby providing better access for cytolysin to insert in the lipid bilayer. There has been no evidence for this.

A third possible role for granzymes was that they enter target cells and trigger an "internal disintegration" process as proposed by JOHN RUSSELL (1983). An attractive feature of this hypothesis is that it provides an explanation for the difference between the lysis of red cell and tumor targets by RBL-cy. In this regard, a critical issue for cytotoxicity with target cells was whether DNA fragmentation accompanied lysis as it generally does with both CTL and NK effectors (RUSSELL 1983). Confirming experiments with purified cytolysin in solution, RBL-cy lysed targets without detectable DNA breakdown (Fig. 4).

4 The Internal Disintegration Pathway

4.1 The Search for Granule Agents Inducing DNA Breakdown

Our laboratory recognized that the granule exocytosis model as originally formulated failed to explain the apoptotic character of target cell death typically inflicted by cytotoxic lymphocytes. While some interpreted this as a failure of the model, we considered whether another granule component might enter the target cell after exocytosis and trigger the apoptotic changes. The idea that proteins secreted by the effector cell could cross the cytolysin-attacked target cell membrane has met with some resistance. Based on the experiments with resealed red cell ghosts attacked by LGLs (Fig. 2), we would argue this is eminently plausible. Given the high local concentration of granule components in the synapse-like junctional region after exocytosis, target cell penetration would be expected even if the pores are repaired fairly quickly by the target cell.

We considered the hypothesis that a granule component might gain access to the target cell in this way and thus be considered to trigger a target internal disintegration. Experimentally, we tested whether CTL granules contained a component which would cause DNA fragmentation when added to nuclei. Such activity was clearly present, and its purification gave us the unexpected result that it was granzyme A (HAYES et al. 1989). A combination of purified cytolysin and purified granzyme. A was shown to kill cells with accompanying DNA breakdown. In an independent approach to the role of granzymes in cytotoxicity, we examined the cytotoxic activity of CTLs pretreated with the irreversible serine protease inhibitor PMSF and found that target DNA fragmentation was severely compromised under conditions favoring PMSF inactivation of proteases in acidic compartments (HAYES et al. 1989).

Beautiful studies of the action of rat LGL granzymes and sublethal doses of cytolysin on tumor cells have been carried out by A. Greenberg and colleagues. These have revealed considerable molecular detail about the mechanism of granzyme-mediated attack on the nucleus, and are summarized in another chapter in this volume. These studies argue for a lytic target cell response to cytoplasmic granzymes, but, as discussed below, it is presently unclear whether nuclear damage contributes to target cell death in this situation.

4.2 Granzyme-Transfected Rat Basophilic Leukemia Cells

In order to assess the functional role of granzymes we began by expressing granzyme A in RBL both alone and in combination with cytolysin. As expected, this protease by itself had no effect on target cells when delivered by target-triggered granule exocytosis. However, when coexpressed with cytolysin, granzyme A was found to add two properties to the RBL-cy single transfectants. First and most satisfying, DNA fragmentation now accompanied lysis (SHIVER et al. 1992), at least to a considerable degree even if not to the same quantitative level as seen with CTLs. This is seen with a number of different tumor target cells (NAKAJIMA and HENKART 1994). The second important property of RBL-cy-gza is that they are more lytic towards tumor target cells. Careful comparison of multiple RBL-cy and RBL-cy-gza transfectant clones over the course of several experiments led to the clear conclusion that RBL-cy-gza transfectants have a more potent ability to release ^{51}Cr from several different tumor target cells. This had not been our conclusion from the first set of double transfectants studied (SHIVER et al. 1992), but subsequent studies with a greater number of stable transfectant clones led to this conclusion (NAKAJIMA and HENKART 1994). In order to exclude the possibility that the superiority of RBL-cy-gza over RBL-cy was the result of more efficient delivery of cytolysin to the target cell, we compared red cell targets with tumor targets. Lysis of red cell targets, with virtually no cytoplasmic apparatus, are expected to be a measure of the efficiency of cytolysin delivery to target membranes by the degranulation process. It can be seen from Fig. 5 that RBL-cy-gza double transfectants are equal in cytolytic potency to RBL-cy single transfectants in their ability to lyse red cells in the same experiments in which they were superior in their ability to lyse tumor cells. These results show that the internal disintegration pathway which granzymes initiate contributes to target lysis and to DNA fragmentation.

Recently we have studied RBLs transfected with granzyme B in addition to the above granule components, and have made transfectants expressing various combinations. While a final judgment awaits analysis of numerous individual clones and serial transfections, at this point we can make the following conclusions: (1) RBL-gzb transfectants are not detectably cytotoxic; (2) RBL-cy-gzb mediate tumor cell lysis with accompanying DNA fragmentation; (3) RBL-cy-gzb are more lytic to tumor cells but not red cells than RBL-cy; (4) RBL-cy-gza-gzb can show potent cytolytic activity against tumor targets, approaching that of cloned CTLs. In general it appears that granzyme A and granzyme B, with very different specificities, behave in a generally similar way. We have no convincing evidence that these two granzymes act in a synergistic way.

Overall, we can say that the RBL transfection approach has proved highly successful in understanding the role of the various lymphocyte granule proteins in cytotoxicity. It provides us with information which is complementary to "knock-out" constructions, and indeed the recent description of CTLs from granzyme B knock-out mice (HEUSEL et al. 1994) is very compatible with our conclusions from the RBL. It has been instrumental in developing the concept that granzymes participate in target cell damage via an internal disintegration pathway associated with target DNA breakdown. One limitation of the RBL transfection approach which must be borne in mind is that RBL cells contain granules which have some components of their own. These have not been fully analyzed, but may contain some granzymes which are known to be expressed in mast cells (e.g., RMCP2, whose mRNA is expressed in RBL; BENFEY et al. 1987).

4.3 Target Cells Loaded Internally with Protease Inhibitor Resist Lymphocyte Attack

After we became convinced by the RBL transfection experiment that cytotoxic lymphocyte granzymes played a role in killing nucleated target cells we decided to test the hypothesis that the physiological substrate was in the target cytoplasm by introducing macromolecular protease inhibitors into the target cell cytoplasm. In order to accomplish this we used the technique of osmotic lysis of pinosomes (OKADA and RECHSTEINER 1982), an established method of loading macromolecules into cells which have reasonable pinocytic rates. This method is not efficient in that mg/ml concentrations of the protein to be loaded must be used during the loading incubation, so it is not good for proteins available in limited amounts. After trying several different protease inhibitors, we settled on aprotinin as one in which we were able to see significant effects on cytotoxic lymphocyte attack. This 6 kDa basic protein was known from previous studies to inhibit both granzymes A and B and is well known as an inhibitor of most serine proteases. Its low molecular weight may be helpful in delivering more molecules per unit weight than other protease inhibitors.

Figure 6 shows a summary of our experiments with aprotinin-loaded target cells. We took advantage of our RBL transfectants to control for a possibility which is otherwise difficult to assess. If aprotinin-loaded target cells were for some reason more resistant to the membrane attack pathway, the secondary

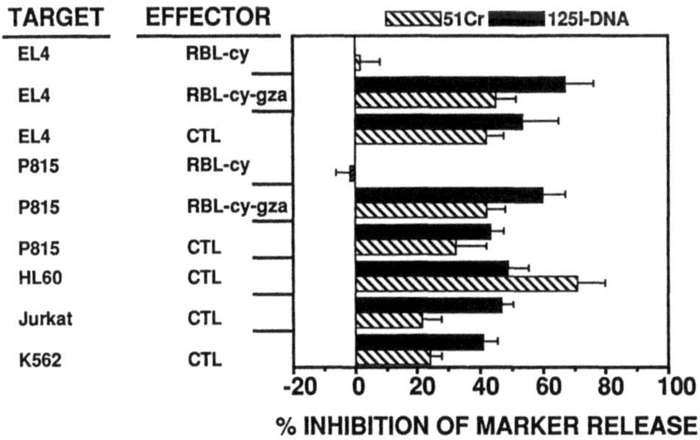

Fig. 6. Effect of target cell intracellular aprotinin loading on CTLs and on rat basophilic leukemia (RBL)-transfectant (*RBL-cy*) cytotoxic activity. Summary data from 4 h cytotoxicity experiments using target cells labeled with both ^{51}Cr and ^{125}I -DNA in which the marker release of aprotinin-loaded targets was compared to bovine serum albumin (BSA)-loaded targets. With RBL-cy effector cells, ^{125}I-DNA release was negligible. Data are means ± SEM of 3–6 experiments. (From NAKAJIMA and HENKART 1994)

damage of the internal disintegration pathway would also be inhibited. Fortunately, RBL-cy were available to control for this. As can be seen, aprotinin-loaded but not BSA-loaded EL-4 target cells show markedly less DNA release and considerably less ^{51}Cr release with CTL effector cells than do unloaded EL-4 targets. RBL-cy-gza effectors behave very similarly to CTL with these three targets, while RBL-cy lyse all three with similar efficiency (NAKAJIMA and HENKART 1994). When multiple experiments with several target cells are expressed as a percentage inhibition of marker release by aprotinin relative to BSA and averaged, the results are quite striking (Fig. 6). It is clear that for both CTL and RBL-cy-gza aprotinin loading inhibits release of DNA and also of ^{51}Cr. These results argue strongly that granzymes are mediators of both target lysis and apoptotic damage, and do so by acting on a substrate inside the target cell.

4.4 Cell Death by Protease Injection

In another attempt to understand the protease-triggered internal disintegration pathway, we decided to examine the effect of protease injection into cells, again using the osmotic lysis of pinosomes technique. Because of the relative inefficiency of this method, we could not use the limited amounts of granzymes available, but we decided to look at the effects of commercially available, well-characterized proteases. We picked trypsin because its specificity is very generally similar to granzyme A; chymotrypsin, as another serine protease with a distinctly different specificity; and also proteinase K, as a non-specific protease. Using peroxidase to quantitate this loading procedure, we estimate that 10^5–10^6 molecules of protease are introduced under the conditions used.

Introduction of these proteases into five cell lines of different origins led to distinct results with the each of the cell lines. In most combinations target lysis,

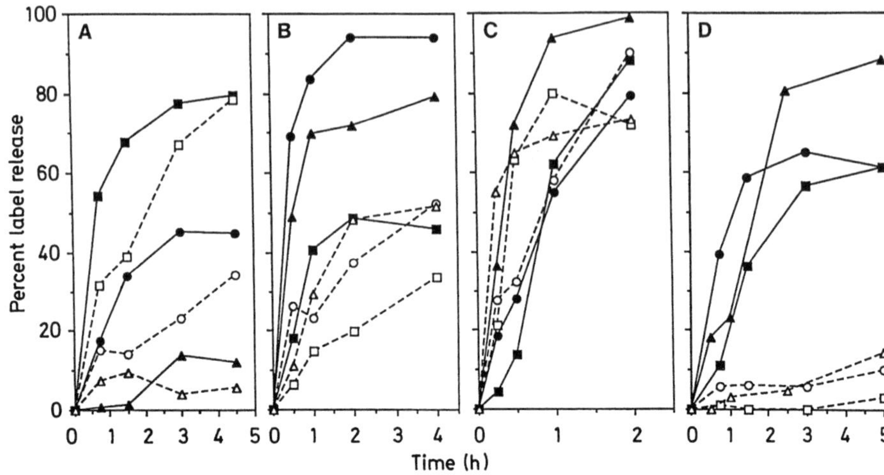

Fig. 7A–D. Cytotoxic effects of cytoplasmic proteolysis. Double labeled cells were injected with proteases at various concentrations via osmotic lysis of pinosomes, and the release of ^{51}Cr (*filled symbols, solid lines*) and ^{125}I-DNA (*open symbols, dashed lines*) was determined over time. **A** Concentration-dependence and kinetics for trypsin loaded into EL-4 lymphoma cells. The extracellular concentrations of trypsin during pinocytic uptake were: 20 mg/ml (*squares*); 5 mg/ml (*circles*); 1 mg/ml (*triangles*). For these cells the lytic potency was trypsin = Proteinase K >> chymotrypsin. **B** Effects of different proteases loaded into P815 mastocytoma cells. Extracellular concentrations of loaded proteases were: chymotrypsin, 20 mg/ml (*squares*) proteinase K, 10 mg/ml (*circles*); trypsin, 20 mg/ml (*triangles*). The lytic potency for these cells was proteinase K > trypsin > chymotrypsin). **C** Effects of different proteases on interleukin-2 (IL-2)-dependent T cell line CTLL. Extracellular protease concentrations loaded were: chymotrypsin, 5 mg/ml (*squares*); proteinase K, 2.5 mg/ml (*circles*), trypsin, 5 mg/ml (*triangles*). The lytic potency of these cells was: proteinase K > trypsin = chymotrypsin. **D** Effects of different proteases on M1 fibroblasts. Extracellular loading concentration of proteases was: chymotrypsin, 5 mg/ml (*squares*); proteinase K, 2.5 mg/ml (*circles*); trypsin, 5 mg/ml (*triangles*). The lytic potency of these cells is: proteinase K > trypsin = chymotrypsin

as evidenced by release of ^{51}Cr, was complete within a couple of hours (Fig. 7). There was a clear dose-response for the proteases, with lower loading levels giving a less complete final level of lysis. However, in some cells, a single protease was not able to yield efficient lysis at any achievable loading concentration.

It was interesting to find that DNA release accompanied cell lysis in most combinations of protease and cell line. In some cases the DNA release was quantitatively comparable to the level of lysis, while in others there was DNA release but it was less than the amount of cell lysis. In other cases there was little or no DNA release. Both the fibroblast line and the erythroleukemia line K562 were generally resistant to DNA damage by injected proteases. This was interesting because when attacked by CTLs, fibroblasts have been found to show minimal DNA damage in spite of good lysis (Sellins and Cohen 1991).

These data exemplify the different susceptibilities to lysis or DNA damage that these cell types show to the different proteases tested. It is clear that for a given target cell type, a particular protease may be relatively ineffective, while another may very efficiently lead to cell death. We do not know the mechanism of this effect—it could be due to the presence of endogenous specific protease inhibitors in some cells but not others, or to differences in the critical substrates which must be cleaved in order to cause cell death. What we find particularly suggestive, however, is the implication that there may be an evolutionary advantage for the existence of multiple granzymes of differing substrate specificity in cytotoxic lymphocytes. This protease diversity would confer on the cytotoxic cell an ability to kill a wider variety of target cells.

As has previously been observed with CTL killing and purified granule proteases, DNA release upon protease injection correlated well with other apoptotic death read-outs, specifically DNA fragmentation into ladders and nuclear morphology changes assessed by the DNA binding dye Hoechst 33342. Since the death caused by protease injection was generally apoptotic, we tested to see if inhibitors of programmed cell death also inhibited this death. Inhibitors of protein or RNA synthesis had no effect on this cell death and neither did the chelation of divalent cations by EDTA or EGTA. In addition, the commonly used inhibitors of apoptotic cell death, aurintricarboxylic acid (ATA) and Zn^{+2}, also had no effect of the loss of cell viability induced by protease injection.

Our conclusions from these studies are that proteases of various specificities are potent cytotoxic agents when introduced into the cytoplasm of various cells. This type of cell death was typically but not invariably apoptotic and seems to mimic the injection of granzymes by CTLs. Strikingly, the kinetics of cell death by injected proteases is very compatible with the killer cell-independent lysis phase of target death after the CTL lethal hit.

5 What Does Target DNA Fragmentation Imply About the Molecular Death Pathway?

One interpretation of the target DNA breakdown accompanying lymphocyte-mediated cytotoxicity is that this implies that an apoptotic death program designed for physiological cell death has been triggered by the effector cell. This interpretation is based on a growing body of evidence that cell death is part of normal physiology in many biological systems, especially the immune system. In many cases such programmed cell death has been accompanied by distinctive apoptotic nuclear morphology changes and DNA fragmentation (Wyllie et al. 1980; Arends et al. 1990). The implications of these findings for the mechanism of target cell death inflicted by cytotoxic lymphocytes depend on two related issues, discussed below.

We are not comfortable assuming that because most programmed cell deaths in vertebrates are apoptotic, all apoptotic deaths must be the result of a common biochemical pathway. There has been a tendency to assume that DNA breakdown is a part of this apoptotic death pathway, and indeed it has become an accepted component of the definition of apoptotic death. Clearly cells whose

DNA has fragmented cannot live for long, but does this mean that the DNA breakdown is part of the death pathway? The correlation between apoptotic DNA breakdown and programmed cell death does not really speak to this issue, because it does not establish whether the DNA breakdown is a cause of death or merely associated with it. We would argue that, especially when death is rapid, DNA breakdown may well be a consequence of the cell death pathway but not be a part of it, i.e., an epiphenomenon. Although it has been proposed that DNA breakdown is a step in the apoptotic death program for corticosteroid-induced death of thymocytes (McConkey et al. 1992), and in general schemes for apoptotic death (Collins and Rivas 1993), evidence showing DNA breakdown is part of the death pathway is weak. The single most persuasive argument is based on experiments with the nonspecific inhibitor ATA, in which there is a time and dose correlation between blocking DNA fragmentation and cell death (Mogil et al. 1994). In any case, for most cases of programmed cell death the role of DNA fragmentation has not been clearly resolved, and for two cases of growth factor withdrawl is clearly not required (Jacobson et al. 1994). For CTLs, it is clear that DNA fragmentation is not necessary for target cytotoxicity because it has been shown that enucleated cytoplasts show the same sensitivity as cells with nuclei (Siliciano and Henney 1978). Although one could argue that this is due to the membrane damage pathway, this appears not to be the dominant player in most cases, and we would speculate that it is unlikely that the internal disintegration cell death pathway directly involves DNA fragmentation.

A related issue is whether there are any common biochemical steps in apoptotic death pathways operating in different biological systems. Extending our skeptical position on DNA breakdown, we would argue that there are no such common biochemical steps now known. Whether or not the internal disintegration pathway we have described for lethal damage inflicted by CTLs has anything in common with apoptotic death triggered as a part of physiological programmed cell death remains to be seen. We are obviously intrigued by the evidence for protease involvement in some such systems (Sarin et al. 1993; Yuan et al. 1993; Miura et al. 1993), but feel strongly that more steps in these death pathways need to be worked out before making any conclusions. We would extend this argument to the apoptotic deaths triggered by agents such as toxins, chemotherapeutic drugs, radiation, and oxidants. Our future efforts will be directed at defining as completely as possible the granzyme-triggered internal disintegration death pathway, with the hope that we can then further ask which steps may be shared with other death pathways.

6 Why Might Cytotoxic Lymphocytes Have Multiple Death Pathways with Multiple Effector Molecules?

If there is a theme to this volume it is that lymphocytes have a variety of means by which they can cause the death of other cells, and some other chapters describe death pathways in which lymphocytes can trigger the death of target cells bearing particular receptors on their surface. It seems reasonable that, under specific physiological circumstances, any one of these pathways could dominate target cell damage. However, under the in vitro conditions in which relatively rapid target cell death is commonly studied, it appears that the granule exocytosis mechanism dominates in mediating target cell death, at least for non-Fas expressing targets. Because this granule exocytosis mechanism is also likely to be dominant in vivo (Kagi et al. 1994), it is not surprising that it utilizes several different effector molecules, which may be more selectively effective at damaging different types of target cells. By having a multiplicity of effector molecules, CTLs maximize their chances of killing the variety of cell types which the immune system needs to kill in order to defend the body from viruses or intracellular bacteria. This multiplicity of effector molecules also makes it much more difficult for these microorganisms to evolve resistance pathways.

While we have focused on the molecular mechanism of lymphocyte-mediated cytotoxicity, it is well to consider that the cells having this activity are also endowed with other important pathways of defense against virally infected cells. Using the vesicle-mediated "constitutive" pathway of protein secretion, these cells release potent antiviral cytokines like IFN-γ (Guidotti et al. 1994) and perhaps others yet to be characterized (Mackewicz et al. 1994). Thus the idea that there are multiple mediators within the granule exocytosis pathway seems reasonable given the overall multiplicity of antipathogen defense mechanisms evolved by the immune system. We believe that our current picture of lymphocyte cytotoxicity is entirely expected from the proposed role of these cells as major components of the efferent arm of the immune system.

References

Acha-Orbea H, Scarpellino L, Hertig S, Dupuis M, Tschopp J (1990) Inhibition of lymphocyte mediated cytotoxicity by perforin antisense oligonucleotides. EMBO J 9: 3815–3819

Arends MJ, Morris RG, Wyllie AH (1990) Apoptosis: the role of the endonuclease. Am J Pathol 136: 593–608

Bashford CL, Menestrina G, Henkart PA, Pasternak CA (1988) Cell damage by cytolysin: spontaneous recovery and reversible inhibition by divalent cations. J Immunol 141: 3965–3974

Benfey PN, Yin FH, Leder P (1987) Cloning of the mast cell protease, RMCP II. Evidence for cell-specific expression and a multi-gene family. J Biol Chem 262: 5377–5384

Berke G, Rosen D, Ronen D (1993) Mechanism of lymphocyte-mediated cytolysis: functional cytolytic T cells lacking perforin and granzymes. Immunology 78: 105–112

Blumenthal R, Millard PJ, Henkart MP, Reynolds CW, Henkart PA (1984) Liposomes as targets for granule cytolysin from cytotoxic LGL tumors. Proc Natl Acad Sci USA 81: 5551–5555

Bubbers JE, Henney CS (1975) Studies on the synthetic capacity and antigenic expression of glutaralehyde-fixed target cells. J Immunol 114: 1126–1131

Collins MKL, Rivas AL (1993) The control of apoptosis in mammalian cells. Trends Biochem Sci 18: 307–309
Dennert G, Podack ER (1983) Cytolysis by H-2 specific T killer cells. Assembly of tubular complexes on target membranes. J Exp Med 157: 1483–1495
Dourmashkin RR, Deteix P, Simone CB, Henkart PA (1980) Electron microscopic demonstration of lesions on target cell membranes associated with antibody-dependent cytotoxicity. Clin Exp Immunol 43: 554–560
Fewtrell C, Metzger H (1981) Stimulus-secretion coupling in rat basophilic leukemia cells. KROC Found Ser 14: 295–314
Golstein P, Smith ET (1977) Mechanism of T-cell-mediated cytolysis: the lethal hit stage. Contemp Top Immunobiol 7: 273–300
Green WR, Henney CS (1981) The mechanism of T cell-mediated cytotolysis. CRC Crit Rev Immunol 1: 259–286
Grossi CE, Cadoni A, Zicca A, Leprini A, Ferrarini M (1982) Large granular lymphocytes in human peripheral blood: ultrastructural and cytochemical characterization of the granules. Blood 59: 277–283
Guidotti LG, Ando K, Hobbs MV et al. (1994) Cytotoxic T lymphocytes inhibit hepatitis B virus gene expression by a noncytolytic mechanism in transgenic mice. Proc Natl Acad Sci USA 91: 3764–3768
Hayes MP, Berrebi GA, Henkart PA (1989) Induction of target cell DNA release by the cytotoxic T lymphocyte granule protease granzyme A. J Exp Med 170: 933–946
Helgason CD, Prendergast JA, Berke G, Bleackley RC (1992) Peritoneal exudate lymphocyte and mixed lymphocyte culturee hybridomas are cytolytic in the absence of cytotoxic cell proteinases and perforin. Eur J Immunol 22: 3187–3190
Henkart MP, Henkart PA (1982) Lymphocyte mediated cytolysis as a secretory phenomenon. Adv Exp Med Biol 146: 227–242
Henkart P, Yue CC (1988) The role of cytoplasmic granules in lymphocyte cytotoxicity. Prog Allergy 40: 44–81
Henkart PA, Millard PJ, Reynolds CW, Henkart MP (1984) Cytolytic activity of purified cytoplasmic granules from cytotoxic rat LGL tumors. J Exp Med 160: 75–93
Henkart PA, Hayes MP, Shiver JW (1992) The granule exocytosis model for lymphocyte cytotoxicity and its relevance to target cell DNA breakdown. In: Sitkoushy MV, Henkart PA (eds) Cytotoxic cells: recognition, effector function, generation and methods. Brikhauser, Boston, pp 153–165
Heusel JW, Wesselschmidt RL, Shresta S, Russell JH, Ley TJ (1994) Cytotoxic lymphocytes require granzyme B for the rapid induction of DNA fragmentation and apoptosis in allogeneic target cells. Cell 76: 977–987
Hudig D, Allison NJ, Pickett TM, Winkler U, Kam C-M, Powers JC (1991) The function of lymphocyte proteases: inhibition and restoration of granule-mediated lysis with isocoumarin serine protease inhibitors. J Immunol 147: 1360–1368
Jacobson MD, Burne JF, Raff MC (1994) Programmed cell death and Bcl-2 protection in the absence of a nucleus. EMBO J 13: 1899–1910
Kagi D, Ledermann B, Burki K et al. (1994) Cytotoxicity mediated by T cells and natural killer cells is greatly impaired in perforin-deficient mice. Nature 369: 31–37
Kelly RB (1985) Pathways of protein secretion in eukaryotes. Science 230: 25–32
Kim SH, Carney DF, Papadimitriou JC, Shin ML (1989) Effect of osmotic protection on nucleated cell killing by C5b-9: cell death is not affected by the prevention of cell swelling. Mol Immunol 26: 323–331
Mackewicz CE, Ortega H, Levy JA (1994) Effect of cytokines on HIV replication in $CD4^+$ lymphocytes: lack of identity with the $CD8^+$ cell antiviral factor. Cell Immunol 153: 329–343
Martz E (1977) Mechanism of specific tumor cell lysis by alloimmune T lymphocytes: resolution and characterization of discrete steps in the cellular interaction. Contemp Top Immunobiol 7: 301–361
McConkey DJ, Jondal M, Orrenius S (1992) Cellular signalling in thymocyte apoptosis. Semin Immunol 4: 371–377
Millard PJ, Henkart MP, Reynolds CW, Henkart PA (1984) Purification and properties of cytoplasmic granules from cytotoxic rat LGL tumors. J Immunol 132: 3197–3204
Miura M, Zhu H, Rotello R, Hartwieg EA, Yuan J (1993) Induction of apoptosis in fibroblasts by IL-1B-converting enzyme, a mammalian homolog of the C. elegans cell death gene ced-3. Cell 75: 653–660
Mogil RJ, Shi Y, Bissonnette RP, Bromley P, Yamaguchi I, Green DR (1994) Role of DNA fragmentation in T cell activation-induced apoptosiss in vitro and in vivo. J Immunol 152: 1674–1683

Morgan BP (1989) Complement membrane attack on nucleated cells: resistance, recovery, and non-lethal effects. Biochem J 264: 1–14

Nagler-Anderson C, Lichtenheld M, Eisen HN, Podack ER (1989) Perforin mRNA in primary peritoneal exudate cytotoxic TT lymphocytes. J Immunol 143: 3440–3443

Nakajima H, Henkart PA (1994) Cytotoxic lymphocyte granzymes trigger a target cell internal disintegration pathway leading to cytolysis and DNA breakdown. J Immunol 152: 1057–1063

Okada CY, Rechsteiner M (1982) Introduction of macromolecules in to cultured mammalian cells by the osmotic lysis of pinocytic vesicles. Cell 29: 33–41

Pasternack MS, Verret CR, Liu MA, Eisen HN (1986) Serine esterase in cytolytic T lymphocytes. Nature 322(6081): 740–743

Podack ER, Konigsberg PJ (1984) Cytolytic T cell granules. Isolation, biochemical and functional characterization. J Exp Med 160: 695–710

Podack ER, Hengartner H, Lichtenheld MG (1991) A central role of perforin in cytolysis? Annu Rev Immunol 9: 129–157

Russell JH (1983) Internal disintegration model of cytotoxic lymphocyte-induced target damage. Immunol Rev 72: 97–118

Sarin A, Adams DH, Henkart PA (1993) Protease inhibitors selectively block T-cell receptor-triggered programmed cell death in a murine T cell hybridoma and activated peripheral T cells. J Exp Med 178: 1693–1700

Schmidt RE, MacDermott RP, Bartley G et al. (1985) Specific release of proteoglycans from human natural killer cells during target lysis. Nature 318(6043): 289–291

Sellins KS, Cohen JJ (1991) Cytotoxic T lymphocytes induce different types of DNA damage in target cells of different origins. J Immunol 147: 795–803

Shiver JW, Henkart PA (1991) A Noncytotoxic mast cell tumor line exhibits potent IgE-dependent cytotoxicity after transfection with the cytolysin/perforin gene. Cell 62: 1174–1181

Shiver JW, Su L, Henkart PA (1992) Cytotoxicity with target DNA breakdown by rat basophilic leukemia cells expressing both cytolysin and granzyme A. Cell 71: 315–322

Siliciano RF, Henney CS (1978) Studies on the mechanism of lymphocyte-mediated cytolysis. X. Enucleated cells as targets for cytotoxic attack. J Immunol 121: 186–191

Simone CB, Henkart P (1980) Permeability changes induced in erythrocyte ghost targets by antibody-dependent cytotoxic effector cells: evidence for membrane pores. J Immunol 124: 954–963

Smyth MJ, Ortaldo JR, Shinkai Y et al. (1990) Interleukin 2 induction of pore-forming protein gene expression in human peripheral blood $CD8^+$ T cells. J Exp Med 171: 1269–1281

Taplits MS, Henkart PA, Hodes RJ (1988) T helper cell cytoplasmic granules: exocytosis in response to activation via the T cell receptor. J Immunol 141: 1–9

Thiernesse N, David A, Bernard J, Jeanesson P, Zagury D (1977) Activitie phosphatasique acide de la cellule T cytolytique au cours du processus de cytolyse. CR Acad Sci (Paris) [D]285: 713–715

Timonen T, Ortaldo JR, Herberman RB (1981) Characteristics of human large granular lymphocytes and relationship to natural killer and K cells. J Exp Med 153: 569–582

Tschopp J, Nabholz M (1990) Perforin-mediated target cell lysis by cytolytic T lymphocytes. Annu Rev Immunol 8: 279–302

Wyllie AH, Kerr JFR, Currie AR (1980) Cell death: the significance of apoptosis. Int Rev Cytol 68: 251–306

Yagita H, Nakata M, Kawasaki A, Shinkai Y, Okumura K (1992) Role of perforin in lymphocyte-mediated cytolysis. Adv Immunol 51: 215–242

Yannelli JR, Sullivan JA, Mandell GL, Engelhard VH (1986) Reorientation and fusion of cytotoxic T lymphocyte granules after interaction with target cells as determined by high resolution cinemicrography. J Immunol 136: 377–382

Young JD, Cohn ZA (1986) Cell-mediated killing: a common mechanism? Cell 46: 641–642

Young JD, Cohn ZA, Podack ER (1986) The ninth component of complement and the pore-forming protein (perforin 1) from cytotoxic T cells: structural, immunological, and functional similarities. Science 233(4760): 184–190

Yuan J, Shaham S, Ledoux S, Ellis HM, Horvitz HR (1993) The C. elegans cell death gene ced-3 encodes a protein similar to mammalian interleukin-1B-converting enzyme. Cell 75: 641–652

Granzymes and Apoptosis: Targeting the Cell Cycle

A.H. GREENBERG and D.W. LITCHFIELD

1 Introduction

The ability of cytotoxic T lymphocytes (CTLs) and natural killer (NK) cells to kill susceptible targets has been known for over 20 years. This observation led to an intensive if not somewhat prolonged search for the molecular mediators of cell death. With the knowledge that is now available to us, and which will be discussed here and in the various chapters of this volume, it is clear that killing involves at least two different pathways. The first is initiated by the content of the cytoplasmic granules which are exocytosed towards the target cell (HENKART 1985; GOLSTEIN et al. 1991). The second results from activation of the Fas receptor on the target by its ligand on the effector cell (ROUVIER et al. 1993; ITOH et al. 1991; SUDA et al. 1993; VIGNAUX and GOLSTEIN 1994). If it were not sufficiently complex that a cell uses two killing pathways, it also appears that cell death induced by granule exocytosis may utilize several molecular mediators: the pore forming

Manitoba Institute of Cell Biology, University of Manitoba, 100 Olivia Street, Winnipeg, Manitoba, R3E OV9 Canada protein perforin (Henkart et al. 1984; Podack et al. 1985, 1991), a group of CTL/NK-specific granule serine proteases known as granzymes (Hayes et al. 1989; Shi et al. 1992 a, b, 1994; Shiver et al. 1992) and perhaps a poly(A)-binding protein TIA-1 (Tian et al. 1991; Kawakami et al. 1992). Thus, it is not surprising that the mechanisms of CTL killing have revealed themselves only slowly.

The current framework in which cell death is viewed is heavily influenced by the idea that death occurs either by an external agent acting on the cell membrane, referred to as necrosis, or by signals that trigger the cell to initiate its own death, which has become known as apoptosis or in the developmental literature as programmed cell death (PCD). This classification originated with the descriptions of Wyllie and coworkers (Wyllie et al. 1980; Wyllie 1987). It has been particularly useful in focussing attention on the regulated internal cell death pathway and has led to an explosion of knowledge in the last several years both in mammalian and nonmammalian systems (see Ellis et al. 1991; Golstein et al. 1991; Cohen et al. 1992; Raff 1992; Williams and Smith 1993). The physical features of apoptosis which have been extensively reviewed elsewhere (Wyllie et al. 1980; Wyllie 1987) are characterized by nuclear chromatin condensation, internucleosomal DNA fragmentation and plasma membrane blebbing. This is strikingly contrasted with necrosis in which the cell responds to membrane damage by osmotic changes, swelling of mitochondria and endoplasmic reticulum, but with little effect on the nucleus. Most agents that kill cells induce death by either apoptosis or necrosis. An exception appears to be CTL/NK cell-mediated cytotoxicity which results in the same target cell exhibiting morphological features of both necrosis and apoptosis (Zychlinsky et al. 1991). Target cells become rapidly permeable to ions and show a particularly sharp rise in intracellular Ca^{2+} (Poenie et al. 1987) followed by organelle and plasma membrane disruption. Within minutes nuclear changes and plasma membrane blebbing become evident and are typical of those seen in many other tissues undergoing apoptosis or PCD (Golstein et al. 1991; Cohen et al. 1992). The morphological features of apoptosis are variable (Ucker et al. 1992; Oberhammer et al. 1993; Schwartz et al. 1993; Schwartz and Osborne 1993; Zakeri et al. 1993).

The question overlying the field at the moment is whether or not there is a common biochemical pathway leading to an irreversible committment to cell death that ultimately manifests itself morphologically as what we are calling apoptosis or PCD. If there is, then on this pathway there is a biochemical event that is necessary and sufficient to induce apoptosis/PCD. The consequent morphological features will depend on the limitations of the cell in question. For example, some cells induce chromatin condensation in the absence of a nucleosomal ladder (Ucker et al. 1992; Cohen et al. 1992; Brown et al. 1993; Tomei et al. 1993; Oberhammer et al. 1994) and DNA ladders have been produced in the absence of chromatin condensation (Sun et al. 1994). The alternate and more complex view is that there are several pathways than can lead to apoptosis and PCD. In this model some pathways may lead to the typical morphology of apoptosis but others may have a different morphological end point. There is no strong evidence to exclude either view at this time, and in our estimation this will not be resolved until there is a thorough understanding of the molecular mechanisms of apoptotic cell death. In the following pages, we will examine the evidence for the identity of the cytotoxic cell proteins that mediate apoptosis, and what biochemical pathways in the target cells are recruited by these proteins to initiate nuclear disruption.

2 Induction of Apoptosis and Necrosis by Granzymes

2.1 Plasma Membrane Damage and Necrosis

In 1984, the discovery of perforin or cytolysin by HENKART et al. (1984) and PODACK et al. (1985) in the granules of CTLs and NK cells gave the first rational explanation for the ability of cytotoxic effectors to damage target cell membranes. Following the purification of perforin it became possible to reproduce the membrane lesions and the mitochondrial swelling apparently caused by perforin forming a membrane pore that leads to nonspecific ion flux (PODACK et al. 1991). It was proposed that movement of water with ions into the cell induced osmotic damage (PODACK et al. 1991). Perforin also reproduced the Ca^{2+} influx observed with CTLs and isolated granules (YOUNG et al. 1986; POENIE et al. 1987; ALLBRITTON et al. 1988). If the Ca^{2+} is not rapidly removed by intracellular Ca^{2+} extrusion mechanisms such as the Na^+/Ca^{2+} exchanger (KRAUT et al. 1990, 1992), high Ca^{2+} levels can lead to cell death (reviewed in ORRENIUS et al. 1989). Further evidence of the importance of perforin in inducing membrane damage came subsequently from the transfection of the perforin gene into mast cells and expression in cytoplasmic granules (SHIVER and HENKART 1991). Following cross-linking of the mast cells and target with IgE antibody, which presumably induced exocytosis of perforin, membrane damage was induced on the target cell. Antisense oligonucleotides can also suppress perforin synthesis and membrane lysis (OCHA-ORBEA et al. 1990). More recently, perforin knock-out mice have been constructed in which the absence of perforin expression in CTLs and NK cells resulted in effector cells which are unable to induce cell death as measured by ^{51}Cr release in many target cells (KÄGI et al. 1994). Thus, it has been convincingly demonstrated that perforin can induce membrane damage and that its induction by CTLs is dependent on perforin expression. However, these experiments do not exclude the participation of other CTL molecules and do not directly demonstrate which of these molecules CTLs use to induce membrane damage.

It can be said with some certainty that perforin is not the only molecule capable of mediating CTL-induced plasma membrane damage. Experiments on the cell line d11S, which kill only by the Fas pathway, from the studies of Golstein and colleagues indicate that activation of Fas results in death associated with the release of ^{51}Cr-labeled cytoplasmic proteins (ROUVIER et al. 1993; VIGNAUX and

GOLSTEIN 1994). How stimulation of Fas by its ligand can damage the cell membrane is not known but it is likely an indirect effect. Another pathway was suggested following experiments from our laboratory in which granule serine proteases were shown to induce membrane damage (SHI et al. 1992a). By titrating perforin to sublytic levels, increasing amounts of fragmentin-2, the granzyme B orthologue of rats, produce a profound and rapid increase in ^{51}Cr release. Similar data have been published using mast cell lines transfected with both perforin and granzyme A in which the expression of the granzyme greatly enhances the ability of the cells to induce ^{51}Cr release over cells expressing only perforin (SHIVER et al. 1992). It is not known if all or even some CTLs synthesize and release sufficient perforin to directly damage the cell membrane. Transfection of mast cells to high levels of perforin expression can produce cells that kill by membrane damage alone but they are still not as efficient as CTLs (SHIVER et al. 1993). Thus it is possible that CTL and NK cells may be able to induce membrane disruption through perforin and/or they may do so through granzymes. How the granzymes might induce this damage is not known but it is not by simple external contact with the plasma membrane (SHI et al. 1992a; SHIVER et al. 1992).

Reduction in CTL killing has been attempted by the transfection of antisense granzyme A in mouse cytolytic T cells lines (TALENTO et al. 1992). It was found that a line stably expressing antisense with reduced granzyme A was less able to damage target cell membrane and is consistent with a role for granzyme A in the induction of membrane damage. However, the CTL line expressing antisense granzyme A was also unable to kill sheep red blood cells (SRBC), and erythrocyte membrane has been shown to be primarily sensitive to the action of perforin suggesting that there may be some other defect in the cell line.

Evidence for a granzyme being important for membrane damage was recently addressed using mice bearing a homozygous null mutation of granzyme B. HEUSEL et al. (1994) found that CTLs from granzyme B –/– mice generated by mixed lymphocyte culture (MLC) stimulation were less efficient at inducing ^{51}Cr release in several target cells, but still retained significant activity suggesting that other molecules such as perforin, other granzymes or Fas were also contributing.

The question of how granzymes might be responsible for membrane damage is unanswered at this time. Granzymes induced nuclear damage, and apoptosis (see below) might lead secondarily to plasma membrane disruption. Granzymes may act intracellularly to initiate membrane damage by, for example, affecting the integrity of the cellular cytoskeleton. Another possibility is that the granzymes may amplify the membrane damage induced by perforin without directly affecting the membrane itself by, for example, increasing intracellular calcium levels or inactivating membrane repair mechanisms. Alternatively, it has been suggested that granzymes are necessary for perforin processing (EWOLDT et al. 1992; HUDIG et al. 1991) and would in this case not be acting on the membrane at all.

2.2 Nuclear Damage and Apoptosis

The first major departure from the notion that target cells are killed by only damaging the cell membrane appeared with the work of John Russell who showed that CTLs induced a nuclear lesion in target cells (RUSSELL and DOBOS 1980; RUSSELL et al. 1980, 1982; RUSSELL 1983). These results were later confirmed by Duke, Cohen and coworkers (COHEN et al. 1985; DUKE et al. 1983, 1985), who observed very early nuclear changes with the impressive feature of fragmentation of DNA, which led to the hypothesis that CTLs induced an internal disintegration of target cells which began with the nucleus (RUSSELL 1983). Thus in seeking molecular mediators of CTL cytotoxicity, they established the principle that one must be able to account for both the membrane and nuclear damage observed in the target cells.

Following the availability of the purified protein, it became clear that perforin was unable to reproduce either the DNA fragmentation or the intense chromatin condensation seen in dying targets following CTL attack (SHI et al. 1992a). Thus, it was hypothesized by a number of groups that the induction of nuclear damage by CTLs must be due to another molecule or molecules (see GOLSTEIN et al. 1991). Evidence accumulating in the last three years suggests that CTL-mediated apoptosis may be induced by one of three candidate molecules. Two types of granule proteins, the CTL serine proteases or granzymes (HAYES et al. 1989; SHI et al. 1992a, b, 1994; SHIVER et al. 1992) and a poly(A)-binding protein TIA-1 (TIAN et al. 1991; KAWAKAMI et al. 1992), as well as the activation of Fas (ROUVIER et al. 1993; VIGNAUX and GOLSTEIN 1994; P. Golstein, personal communication) have been observed to induce apoptosis. Both TIA-1 and Fas will be reviewed in other chapters of this volume and we will discuss here only the evidence for a role for granule serine proteases in apoptosis.

2.2.1 Granzymes and Apoptosis

The existence of a CTL-specific serine protease was first reported by PASTERNACK et al. (1985). In the following year several groups reported the cloning of unique cDNAs encoding a family of CTL/NK-specific serine proteases (LOBE et al. 1986; GERSHENFELD and WEISSMAN 1986; BRUNET et al. 1986; JENNE et al. 1989). The nomenclature of the family is varied but one usually refers to these proteases by the commonly accepted name of granzymes (JENNE et al. 1989). The proteases are localized in the cytoplasmic granules of CTLs (OJCIUS et al. 1991) and can be released by stimulation of granule exocytosis (PASTERNACK et al. 1985). Recent evidence indicates that granzyme B can be synthesized and released directly without first being directed to the granules (GRIFFITHS 1994). The induction of synthesis of the proteases occurs during stimulation of lymphocytes to differentiate into cytotoxic effector cells (LOBE et al. 1986; GERSHENFELD and WEISSMAN 1986). They are expressed in vivo at sites of graft rejection (MUELLER et al. 1988; GRIFFITHS et al. 1991; LIPMAN et al. 1991; HAMEED et al. 1991; CLÉMENT et al. 1994; WIJNGAARD et al. 1993). Several members of the granzyme family have been purified to homogeneity and characterized as active proteases using synthetic substrates (ODAKE et al. 1991; POE et al. 1988, 1991; SHI et al. 1992b; SMYTH et al. 1992).

The possible role of proteases in killing was first suggested by the work of CHANG and EISEN (1980) and REDELMAN and HUDIG (1980), who found that protease inhibitors suppressed CTL- and NK-mediated cytolysis. How the proteases might contribute to cell death was not immediately obvious, particularly with the evidence that perforin could induce membrane damage. The picture that has recently emerged is that at least a portion of the granzyme family can initiate nuclear damage which is indistinguishable from apoptosis; however, they are inactive on their own and have an absolute requirement for perforin (HAYES et al. 1989; SHI et al. 1992a, b, 1994; SHIVER et al. 1992).

Several types of experiments have now confirmed the importance of granzymes in apoptosis. Purified NK granule proteases can induce extensive chromatin condensation and oligonucleosomal DNA fragmentation (SHI et al. 1992a) and DNA solubilization (HAYES et al. 1989; SHI et al. 1992a,b) in the presence of perforin. Mast cells expressing granzyme A and perforin in granules can induce DNA fragmentation in tumors after stimulating their degranulation (SHIVER et al. 1992), and antisense suppression of granzyme A expression inhibits DNA solubilization (TALENTO et al. 1992). Most recently, mice bearing a homozygous null mutation of granzyme B have been constructed which are profoundly defective in rapid CTL-induced apoptosis (HEUSEL et al. 1994).

Granzyme A immunoaffinity purified from murine CTL granules can induce DNA damage measured by the release of soluble ^{125}I-labeled DNA in target cells whose membranes had been permeabilized with detergent or perforin (HAYES et al. 1989). We were able to purify three proteases that could induce all of the features of apoptosis in the presence of perforin, which we called fragmentins (Table 1) (SHI et al. 1992a,b). The most potent was fragmentin-2, which on microsequencing proved to be almost identical to the deduced sequence of the RNKP-1 cDNA, the rat orthologue of granzyme B. We have recently found very similar specific activity and kinetics with purified human granzyme B (Shi, Trapani and Greenberg, unpublished). Fragmentin-1 and -3, were slower acting, had lower specific activity, and were homologous to granzyme A and 3, respectively. The conclusions drawn from the experiments using mast cells doubly transfected with granzyme A and cytolysin (perforin) are similar to those taken from the

Table 1. Characteristics of apoptosis-inducing granzymes

Granzyme	Molecular mass	Kinetics of apoptosis	Class of protease	Specificity of amino acid hydrolysis	Specificity of induction of apoptosis
A	47 kDa	Slow	Tryptase	Arginine/Lysine	Arginine
B	31 kDa	Fast	Asp-ase	Aspartic acid	Aspartic acid
3	27 kDa	Slow	Tryptase	Arginine/Lysine	Arginine

[1]From SHI et al. 1992b.

in vitro data using purified granzymes and perforin. In transfected mast cells, perforin or granzyme A expression alone has no effect on the target cell nucleus, while the expression of both can induce DNA damage (SHIVER et al. 1992). However, mast cells did not reach the level of killing seen in CTLs, which suggests that granzyme A may not be sufficient. Similar conclusions can be drawn from the expression of antisense granzyme A in mouse cytolytic T cells lines which reduced granzyme A by a maximum of 50% in its ability to degrade target cell DNA (TOLENTO et al. 1992). Compelling evidence for the importance of granzyme B in CTL-induced apoptosis comes from experiments using granzyme B deficient CTLs from the homozygous null mutant mice. These cells lost their ability to induce DNA fragmentation and apoptosis in short-term killing assays (HEUSEL et al. 1994). Interestingly, when assayed over longer term incubations, some restoration of apoptosis was observed. This residual activity would be consistent with the activity of slower acting granzymes A and 3 (SHI et al. 1992b). The alternative possibility of Fas activity is less likely as Fas stimulation has been reported to act quite quickly to kill cells (ROUVIER et al. 1993).

Although the work on purified granzyme B and the granzyme B null mutant argue strongly for an important role for this particular granzyme in CTL-induced apoptosis, it remains to be determined if all CTLs or only those generated by allogeneic stimulation preferentially utilize this granzyme. The expression of different granzymes in CTLs is not uniform (GARCIA-SANZ et al. 1990; PRENDERGAST et al. 1992; VELOTTI et al. 1992) so that the protease or proteases used to induce apoptosis may depend on the predominant types induced in a given CTL line or population.

To date, our attempts to find other proteases in the RNK cell cytoplasmic granules that have granzyme-like activity have proved unsuccessful. This is not due to the absence of other proteases in the granule, as we can easily identify several by their proteolytic activity on various substrates. As well, we know other granzymes are expressed in these cells, although at low level, on the basis of reverse transcriptase polymerase chain reaction (RT-PCR) using conserved primers of the granzyme family (S. Willems and A. H. Greenberg, unpublished data). Thus, not all granule proteases and possibly not all members of the granzyme family are able to induce apoptosis.

2.2.2 How do Granzymes Attack the Target Cell?

Incubation of intact target cells with the rat orthologues of granzyme A, B, 3 (SHI et al. 1992a,b), murine granzyme A (HAYES et al. 1989) or human granzyme B (Shi, Trapani and Greenberg, unpublished data) does not induce apoptosis. Degranulation of mast cells expressing granzyme A to cross-linked targets also has no effect on the targets (SHIVER et al. 1992). Thus, it seems that granzymes, whether in simple solution or vectorally delivered at the intercellular site of contact, are not sufficient to initiate apoptosis. This observation virtually eliminates the possibility that the proteases activate a membrane receptor such as Fas. However, it does not mean that they have no effect at the membrane. One report indicates that, when incubated with rat granzyme B, some hepatoma or carcinoma cells reduce their growth and begin rounding (SAYERS et al. 1992). Granzymes have been shown to hydrolyze matrix proteins and may conceivably affect growth by reducing cell adhesion (SIMON et al. 1987, FROELICH et al. 1993).

The requirement for perforin to induce granzyme-mediated apoptosis is clear, but as perforin itself cannot induce apoptosis, its role in this process is less evident. One possibility is that perforin aids the entry of the protease into the cell. Perforin is able to promote granzyme B or granzyme A-induced apoptosis (HAYES et al. 1989; SHI et al. 1992a,b) not only at doses that induce membrane damage but also at sublytic levels (SHI et al. 1992a). Thus, the CTL proteases are not simply passing through a perforin-shattered membrane, which suggests that they use perforin in an active way, perhaps as a pore through which they can enter the cell. If this were the case then one should be able to reproduce apoptosis by microinjecting granzymes into the cell. We have attempted this experiment and have produced only transient changes in the cell and nucleus, but have not seen a nucleus with fully condensed chromatin as one observes after the application of perforin and granzymes to the plasma membrane (Kemp, Shi and Greenberg, unpublished). Perforin appears to participate in some other active way without directly inducing apoptosis itself. This might be related to the requirement for calcium during granzyme B-induced apoptosis (SHI et al. 1992b). Cytoplasmic granules cause large increases in intracellular Ca^{2+} measured by the cytoplasmically trapped fluorescence indicators fura-2 and indo-1 in target cells (ALLBRITTON et al. 1988). An additional and not mutually exclusive possibility comes from the observation that microtubule inhibitors can block granzyme-induced apoptosis without affecting perforin activity (SHI et al. 1992a). One explanation for this observation is that perforin stimulates pinocytosis which then takes the proteases into the cell and perhaps to a specific site. Studies are currently underway in our laboratory to identify granzyme targeting in the cell.

Assuming that the granzymes enter the cell, it is completely unknown how they can initiate apoptosis. Only nucleolin has been identified as a potential intracellular target for granzyme A as it both binds and is cleaved by the protease, and it has been suggested that it may transport granzyme A to the nucleus (PASTERNACK et al. 1991). However, it is not known if nucleolin degradation or granzyme binding occurs during CTL- or granzyme-induced apoptosis. We have hypothesized that granzymes inappropriately activate cell cycle regulatory molecules which then participate in the apoptotic process (SHI et al. 1994). This will be discussed below (see Sect. 3), but it is worth noting here that $p34^{cdc2}$ and associated cyclin B must be transported to the nucleus before initiating mitotic catastrophe, a phenomenon with many similarities to apoptosis (KREK and NIGG 1991; HEALD et al. 1993).

2.2.3 Granzyme Specificity in Apoptosis

Using protease nomenclature and identified by the ability to cleave tripeptide thiobenzyl ester substrates, the human, murine and rat granzyme A are tryptases and can cleave after arginine or lysine (ODAKE et al. 1991; SHI et al. 1992b). Human, murine and rat granzyme B have an unusual Asp-ase specificity, hydrolyzing after aspartic acid (ODAKE et al. 1991; POE et al. 1991; SHI et al. 1992b). Human and rat granzyme 3 are also tryptases (HAMEED et al. 1988; SHI et al. 1992b). These are the only three granzymes that have been identified as effectors of apoptosis; however, granzymes with other specificities have been identified in granules as well as purified (MASSON and TSCHOPP 1987; BLEACKLEY et al. 1988; PRENDERGAST et al. 1992; SMYTH et al. 1992), and studies with protease inhibitors suggest that some may participate in the lytic process (HUDIG et al. 1991; EWOLDT et al. 1992). A comparative study of inhibition of hydrolysis of the tripeptide thiobenzyl esters and of DNA damage by the rat granzymes was done using the tripeptide chloromethyl ketone inhibitors which irreversibly bind to the protease catalytic site (POWERS 1977). We found a close correlation between the assays in which the Phe-Pro-Arg-CH_2Cl blocked the two tryptases while having no effect on the Asp-ase and the Ala-Ala-Asp-CH_2Cl blocked the Asp-ase but not the tryptases (SHI et al. 1992b). Interestingly, when we analyzed target cell susceptibility, each of the three granzymes exhibited a distinct pattern indicating that the intracellular effects were more complex than recognition of these particular peptide sequences, as the two tryptases differed from each other as well as the Asp-ase. The additional observation that, when assayed together, the Asp-ase granzyme B and tryptase granzyme 3 synergistically amplify their apoptotic activity (SHI et al. 1992a) argues that they may act on complementary pathways.

It has been recently observed that another protease, the interleukin-1 converting enzyme (ICE), has the ability to induce apoptosis when transiently transfected into the Rat-1 fibroblast (MIURA et al. 1993). ICE shares with granzyme B the unusual proteolytic specificity for cleaving after aspartic acid (HOWARD et al. 1991; THORNBERRY et al. 1992). Although ICE was first described as the protease responsible for processing the precursor form of interleukin-1, YUAN and colleagues (1993) identified it as a possible mediator of apoptosis by its homology to the ced-3 gene of C. elegans and by showing that a specific inhibitor of ICE, called crmA, which is derived from the cowpox virus (RAY et al. 1992), can inhibit ICE-induced apoptosis (MIURA et al. 1993) and block neuronal cell death following withdrawal of nerve growth factor (GAGLIARDINI et al. 1994). The ced-3 gene was described originally on the basis of mutational work as a gene encoding a possible effector protein on the pathway regulating programmed cell death in C. elegans development (ELLIS et al. 1991). The activation of the p45 precursor of ICE requires cleavage at several aspartic acid residues into p22 or p20 and p10 proteins, and active ICE forms as a homodimer each of which contains a pro and a p10 subunit (WILSON et al. 1994; WALBER et al. 1994). Pro-ICE may activate by autoprocessing although it is still unclear how it is activated in vivo (THORNBERRY et al. 1992). It is interesting to speculate that the Asp-ase granzyme may induce apoptosis by cleaving the precursor form of ICE thus either directly activating or initiating a self-amplifying autoactivation of ICE (Fig. 1). However, the activation of ICE by an Asp-ase cannot be a general mechanism of initiating apoptosis, as it would not account for the activity of the tryptase granzymes. Alternatively, the Asp-ase and

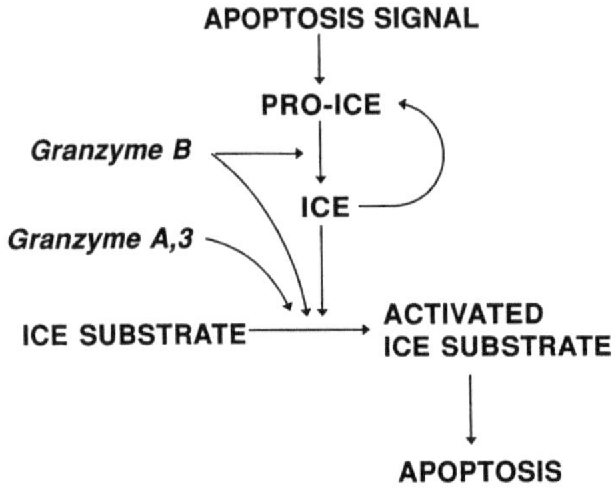

Fig. 1. Hypothetical activation of interleukin-1 converting enzyme (ICE) pathway by granzymes. The conversion of the p45 pro-ICE to the active $(p20/p10)_2$ homodimer requires processing by an Asp-ase. This could be accomplished by autoprocessing as ICE is an Asp-ase, or by the introduction of the Asp-ase granzyme B into the cell by CTLs. Granzyme A and 3 are tryptases and would more likely act downstream of ICE on an ICE substrate. Granzyme B could also act downsteam of ICE cleaving the substrate with the same specificity and consequence as ICE

tryptases may hit the target against which the activated ICE is directed on the apoptosis pathway. This target is presumably proteolytically processed by ICE to an active form and this processing may be mimicked by the granzymes. However, before taking this discussion further, these ideas will need to be critically tested.

3 Apoptosis and Mitotic Catastrophe

3.1 Posttranscriptional Regulation of Apoptosis

Although PCD is considered to require new protein and RNA synthesis (Wyllie et al. 1984, for reviews see Golstein et al. 1991 and Cohen et al. 1992), this observation has been contrasted with protein synthesis-independent CTL apoptosis (Duke et al. 1983; Landon et al. 1990). Consistent with these observations in cell-mediated apoptosis, granzyme/fragmentin-induced apoptosis (Shi et al. 1992a) proceeds with the same kinetics and intensity in target cells preincubated with inhibitors such as cycloheximide. Of interest, tumor necrosis factor (TNF-α) activation of its receptor and the structurally related Fas ligand and receptor also induce apoptosis without new protein synthesis (Ruff and Gifford 1981; P. Golstein, personal communication). These observation suggested to us that some forms of apoptosis may be regulated by a post-translational mechanism involving, for instance, the control of enzyme activity through phosphorylation. In this regard, it is of interest that mitotic catastrophe, a type of severe mitotic dysregulation, bears many similarities to apoptosis and results from the inappropriate regulation of cyclin-dependent protein kinases (CDKs) that are critical to the orderly progression of the cell cycle.

3.2 Cell Division and Mitotic Catastrophe

3.2.1 Control of the G_2/M Transition

In recent years, there has been a dramatic increase in the understanding of the molecular mechanisms that control cell division (reviewed by MALLER 1991; FORSBURG and NURSE 1991; NORBURY and NURSE 1992; FEILOTTER et al. 1992). In particular, biochemical and genetic studies in a variety of different organisms converged on the conclusion that a highly conserved protein serine/threonine kinase, $p34^{cdc2}$, has an indispensible role in regulating the G_2/M transition (reviewed by FORSBURG and NURSE 1991; FEILOTTER et al. 1992). LEE and NURSE (1987) subsequently isolated a cDNA encoding a human homolog of the *S. pombe* cdc2 and demonstrated functional conservation between the human and yeast proteins. It is now well established that the protein kinase activity of $p34^{cdc2}$ is dramatically elevated as cells traverse the G_2/M transition and that this protein kinase activity is a universal regulator of mitosis in eukaryotic cells (DRAETTA and BEACH 1988; LANGAN et al. 1989).

To ensure that daughter cells receive all necessary components for their survival, it is important that cell division be precisely controlled. In this regard, the activity of $p34^{cdc2}$ is regulated through its interactions with a regulatory protein (i.e., cyclin) and by stimulatory or inhibitory phosphorylation events. Cyclins were initially identified in fertilized sea urchin eggs as proteins that are synthesized continuously throughout the cell cycle and are abruptly degraded during mitosis (EVANS et al. 1983). Original studies demonstrated the existence of two distinct classes of cyclins, A and B. Additional classes of cyclins (including D, E and perhaps C) have been subsequently characterized (reviewed by PINES 1991, 1993). In proliferating cells, $p34^{cdc2}$ is present throughout the cell cycle but cannot be activated until it becomes associated with cyclin A or with cyclin B, the latter cyclin accounting for the majority of active $p34^{cdc2}$ in mitotic cells (DRAETTA et al. 1989; GAUTIER et al. 1990; PINES and HUNTER 1990a, 1991; NORBURY and NURSE 1992; PINES 1993). Although $p34^{cdc2}$ is defined as a cyclic dependent kinase (CDK), because cyclin binding is necessary for its activity, cyclin binding by itself is insufficient to activate $p34^{cdc2}$. Therefore, even though cyclin B synthesis and complex formation with $p34^{cdc2}$ begins in S phase and continues to accumulate through G_2, the complex remains inactive. Upon binding to cyclin B, $p34^{cdc2}$ is phosphorylated at highly conserved residues (i.e. Tyr-15, Thr-161 and in higher eukaryotes Thr-14 as well) and remains inactive until Thr-14 and Tyr-15 are dephosphorylated at the G_2/M interface (GOULD and NURSE 1989; KREK and NIGG 1991; GOULD et al. 1991; NORBURY et al. 1991) (Fig. 2). Interestingly, Thr-14 and Tyr-15 of $p34^{cdc2}$ are located within a motif that contributes to ATP binding and is conserved in all members of the protein kinase family (GOULD and NURSE 1989; HANKS and QUINN 1991). Phosphorylation of these residues does not significantly impair ATP binding, but does inhibit phosphate transfer (ATHERTON-FESSLER et al. 1993). Since these residues are phosphorylated immediately upon cyclin binding, the activity of $p34^{cdc2}$ remains suppressed until they are dephosphorylated

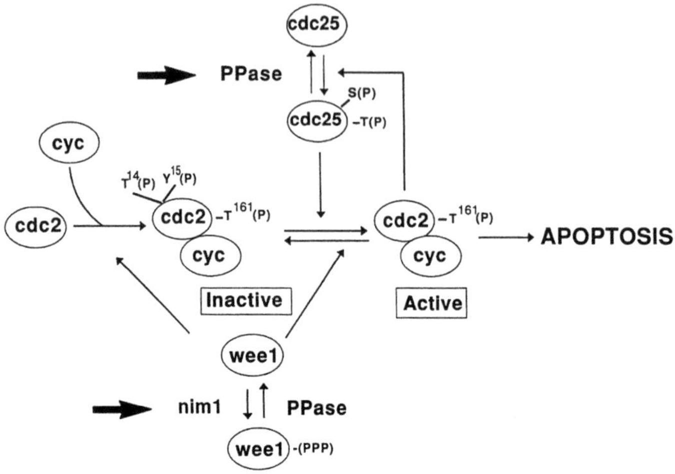

Fig. 2. Regulation of $p34^{cdc2}$ phosphorylation during apoptosis. The $p34^{cdc2}$ kinase is regulated by cyclin binding and phosphorylation. Upon binding to cyclin A or B, $p34^{cdc2}$ is phosphorylated on two residues (threonine-14 and tyrosine-15) that inhibit its activity and at an additional residue (threonine-61) that is essential for $p34^{cdc2}$ activity. The inhibitory phosphorylation of tyrosine-15 is controlled by the wee 1 protein kinase that is itself regulated by reversible phosphorylation mediated by the nim1 protein kinase and an as yet unidentified protein phosphatase (*PPase*). Activation of $p34^{cdc2}$ occurs when threonine-14 and tyrosine-15 are dephosphorylated by cdc25. The cdc25 phosphatase is phosphorylated and activated by $p34^{cdc2}$ forming an autoamplification loop for the abrupt activation of $p34^{cdc2}$. Potential targets of granzymes that result in continuously activated $p34^{cdc2}$ through inactivation of wee 1 or activation of cdc25 are indicated with *arrows*

abruptly at the G_2/M transition. By comparison with these residues, phosphorylation of Thr-161 is required for the activation of $p34^{cdc2}$ (Gould et al. 1991; Ducommun et al. 1991; Solomon et al. 1992).

Genetic studies with S. pombe had indicated that $p34^{cdc2}$ is positively regulated by cdc25 and negatively regulated by wee 1, which is itself negatively regulated by nim1 (Fig. 2) (reviewed by Forsburg and Nurse 1991; Feilotter et al. 1992). A functional homolog of wee1, known as mik1, was also identified in *S. pombe* (Lundgren et al. 1991). Human homologs of wee1 (Igarashi et al. 1991) and cdc25 (Sadhu et al. 1990) have been identified and appear to influence $p34^{cdc2}$ in the same manner as in yeast. Biochemical studies have shown that wee1 phosphorylates Tyr-15 (Parker and Piwnica-Worms 1992; McGowan and Russell 1993) while cdc25 is a phosphatase that catalyzes the dephosphorylation of this residue and is likely to dephosphorylate Thr-14 as well (Strausfeld et al. 1991; Gautier et al. 1991; Dunphy and Kumagai 1991; Lee et al. 1992; Millar et al. 1991). Nim1 from *S. pombe* is also a protein kinase that phosphorylates and inactivates the wee1 protein kinase (Parker et al. 1993; Wu and Russell 1993; Coleman et al. 1993). Although homologs of nim1 have not yet been identified in vertebrates, there are indications in frogs that the activity of wee1 is regulated by mechanisms that do involve phosphorylation (Tang et al. 1993). As yet, the definitive identification of an enzyme that phosphorylates Thr-14 has not been reported. An enzyme that phosphorylates Thr-161 and is referred to as the $p34^{cdc2}$ activating kinase (CAK) has recently been purified from *Xenopus* or starfish oocytes and appears to be the same as $p40^{MO15}$, an enzyme with some sequence similarity to $p34^{cdc2}$ that had been previously identified in frogs (FESQUET et al. 1993; POON et al. 1993; SOLOMON et al. 1993). Although its precise regulation remains poorly defined, CAK is critical for the activation of $p34^{cdc2}$ since mutation of Thr-167 (Thr-167 of *S. pombe* corresponds to Thr-161 of mammalian $p34^{cdc2}$) to a nonphosphorylatable residue results in loss of activity (GOULD et al. 1991; DUCOMMUN et al. 1991).

Although cyclin binding and phosphorylation of Thr-161 are essential for $p34^{cdc2}$ activation, the event that triggers the abrupt activation of the enzyme is the dephosphorylation of Tyr-15 (and Thr-14) by cdc25 (Fig. 2). In interphase cells, cdc25 activity is relatively low thus allowing wee1 to maintain $p34^{cdc2}$/cyclinB in an inactive state. The phosphatase activity of cdc25 is dramatically increased at the G_2/M transition resulting in activation of $p34^{cdc2}$ (IZUMI et al. 1992; KUMAGAI and DUNPHY 1992). Biochemical evidence suggests that $p34^{cdc2}$ contributes to its own activation through an amplification process in which it phosphorylates and activates cdc25 (HOFFMAN et al. 1993; IZUMI and MALLER 1993). Furthermore, the inactivation of wee1 by nim1 may parallel the activation of cdc25, again shifting the balance in favor of dephosphorylation and activation of $p34^{cdc2}$. $p34^{cdc2}$ is indispensible for the initiation of cell division, but must be inactivated to permit the completion of mitosis. Inactivation is achieved through the abrupt degradation of cyclin B (MURRAY et al. 1989; GLOTZER et al. 1991) and dephosphorylation of Thr-161 (LORCA et al. 1992).

3.2.2 Protein Kinases Related to $p34^{cdc2}$

In recent years, a number of protein kinases that share structural or enzymatic similarities to $p34^{cdc2}$ have been identified (MEYERSON et al. 1992; MATSUSHIME et al. 1992; LEW et al. 1992; BUNNELL et al. 1990; reviewed by PINES 1993). To this point, $p34^{cdc2}$ (which is also known as CDK1) is one of six distinct mammalian CDKs that are believed to have functions in regulating different transition points within the cell cycle. For example, CDK2 which exhibits 65% amino acid identity with $p34^{cdc2}$ complexes with cyclins A, D and E and appears to have an important role in the regulation of the G_1/S transition (FANG and NEWPORT 1991; VAN DEN HEUVEL and HARLOW 1993). CDK4, CDK5, and CDK6 all complex with cyclins of the D family suggesting that they are involved in the regulation of events during G_1 (XIONG et al. 1992; SHERR 1993; VAN DEN HEUVEL and HARLOW 1993). Other enzymes (including CDK3) exhibit sequence similarity to $p34^{cdc2}$ but as yet have not been shown to have a cyclin partner (PINES 1993). There are also indications that some of the CDKs as well as other protein kinases including members of the mitogen activated protein (MAP) kinase family (reviewed by COBB et al. 1991; THOMAS 1992; PELECH et al. 1993) have similar substrate specificities to those of $p34^{cdc2}$ (PETER et al. 1992). In general, the extent to which any of these enzymes could have redundant functions with $p34^{cdc2}$ remains to be established.

3.2.3 Mitotic Catastrophe

The onset of mitosis is associated with a number of alterations in cellular architecture including breakdown of the nuclear lamina, termination of transcription and chromosome condensation, disassembly of the nucleolus, reorganization of microtubules to form the mitotic spindle as well as changes in other cytoskeletal assemblies to permit cell rounding and cytokinesis. There are numerous indications that these cellular events are brought about by changes in protein phosphorylation that are mediated directly, and perhaps in some cases indirectly, by $p34^{cdc2}$. It has been observed that a burst of protein phosphorylation accompanies the G_2/M transition in a number of experimental systems. Although the list of physiological substrates that have been identified to date is no doubt incomplete, many proteins that are phosphorylated in mitotic cells have been characterized as likely cellular substrates of $p34^{cdc2}$ (reviewed by PINES and HUNTER 1990b; NIGG 1993). Probable physiological substrates include nuclear lamins that disassemble upon phosphorylation (PETER et al. 1990a; WARD and KIRSCHNER 1990; HEALD and MCKEON 1990), chromosome associated proteins, i.e., histone H1 (LANGAN et al. 1989) and several transcription factors (NIGG 1993), that could contribute to the termination of transcription and condensation of chromosomes; nucleolar proteins including nucleolin and NO38 (PETER et al. 1990b); and several cytoskeletal proteins associated with intermediate filaments or microfilaments, including vimentin (CHOU et al. 1990), caldesmon (YAMASHIRO et al. 1990; MAK et al. 1991), and myosin light chain (SATTERWHITE et al. 1992) that presumably participate in cytoskeletal alterations which occur during mitosis. The observation that protein kinases, i.e., pp $60^{c\text{-}src}$ (MORGAN et al. 1989; SHENOY et al. 1989), $p150^{c\text{-}abl}$ (KIPREOS and WANG 1990), casein kinase II (LITCHFIELD et al. 1991, 1992), and MKK1 (ROSSOMANDO et al. 1994), are potential physiological targets of $p34^{cdc2}$ suggests that protein kinases which are controlled by $p34^{cdc2}$ may also contribute to the burst of phosphorylation that occurs as cells enter mitosis.

To ensure that the daughter cells will be viable, it is critical that mitosis be initiated only after other cellular events such as chromosome replication and sufficient cell growth have been completed (reviewed by FORSBURG and NURSE 1991; FEILOTTER et al. 1992). In this regard, a number of studies have demonstrated that perturbations in the regulatory mechanisms that govern the activation of $p34^{cdc2}$ can have rather severe consequences on a cell. For example, genetic studies in *S. pombe* demonstrated that inactivation of wee 1 results in cells entering mitosis prematurely with the result that the daughter cells are abnormally small, exhibiting the wee phenotype (reviewed by FORSBURG and NURSE 1991). By comparison, inactivation of cdc25 or overproduction of wee results in mitotic delay with the result that daughter cells are larger than normal. In yeast overproducing cdc25 and lacking wee 1 or in yeast lacking both wee 1 and mik 1, the result is an abortive attempt at division known as mitotic catastrophe (FORSBURG and NURSE 1991; LUNDGREN et al. 1991). The events of mitosis including breakdown of the nuclear envelope and chromosome condensation are initiated but the cell has not undergone sufficient growth nor completed chromosome replication so it is unable to successfully divide and survive. Premature mitosis and in some cases mitotic catastrophe have also been observed in mammalian cells when the regulation of $p34^{cdc2}$ is perturbed. Overexpression of $p34^{cdc2}$ harboring mutations at Thr-14 and/or Tyr-15 induced premature mitosis in HeLa cells (KREK and NIGG 1991). In these experiments, the most significant effects were observed in the double mutants. Subsequent experiments in BHK cells by HEALD et al. (1993) demonstrated that a high proportion of mitotic catastrophes resulted from cotransfection of combinations of $p34^{cdc2}$ with cyclin B, $p34^{cdc2}$ with cyclin A or cdc25 with cyclin B while a very low level of mitotic catastrophes was observed when only one of the components was transfected into the cells. Interestingly, transfection of a cDNA encoding wee 1 into these cells dramatically reduced the frequency of observed mitotic catastrophes. It is noteworthy that the transfected wee 1 was localized within the nucleus and effectively suppressed the protein kinase activity of $p34^{cdc2}$ complexes in the nucleus but not in the cytoplasm.

Given the plethora of protein kinases that have similar properties to $p34^{cdc2}$ with respect to association with cyclins, amino acid sequence or substrate specificity, it is probable that the unregulated expression of some of these enzymes could also lead to cell cycle perturbations. In this vein, a $p34^{cdc2}$-related enzyme with no known cyclin partner and which is known as $p58^{GTA}$ or PITSLRE produced mitotic abnormalities when it was overexpressed in CHO cells (BUNNELL et al. 1990).

3.3 Cyclin-Dependent Kinases in Apoptosis

With the observation that dysregulation of the cell cycle through the inappropriate expression of CDKs can lead to cell death resembling apoptosis, the question of whether these kinases are induced during apoptosis can be raised. More specifically, since the expression of $p34^{cdc2}$ activity outside of G_2/M can induce mitotic catastrophe then perhaps its activation is linked to apoptosis. This issue was recently examined in our laboratory using the model of apoptosis induced by fragmentin-2/granzyme B (SHI et al. 1994). This model has the advantages that it is protein synthesis-independent and extremely rapid, reaching completion in 45–60 min (SHI et al. 1992a,b). To successfully identify an activation of $p34^{cdc2}$ in cells undergoing apoptosis, it is necessary that a large proportion of the cells enter apoptosis simultaneously. Prolonged activation periods will result in only a small proportion of the cells undergoing apoptosis at any given time. Consequently, unless the kinase activation is sustained, the total amount of active kinase in the population would be low and difficult to detect.

To measure $p34^{cdc2}$ activity, we isolated the enzyme by immunoprecipitation with an antibody directed against its unique COOH-terminal. Kinase activity in immune complexes was measured using synthetic peptides modeled after known $p34^{cdc2}$ phosphorylation sites within the β-subunit of casein kinase II (LITCHFIELD et al. 1991) and nucleolin (PETER et al. 1990b). Concomitant with an

Table 2. Activation of Cyclin-Dependent Kinases in Apoptosis and Mitotic Catastrophe

CDK	Cyclin	Method	Cell	Reference
cdc2^{A14,F15}	Unknown	Transfection	HeLa	KREK and NIGG 1991
cdc2	Cyclin A/B1	Transfection	BHK	HEALD et al. 1993
cdc2	Unknown	Etoposide	CHO	LOCK and ROSS 1990
cdc2	Unknown	Fragmentin-2/ granzyme B	YAC-1 Lymphoma	SHI et al. 1994
cdc2/cdk2	Cyclin A	Staurosporine	HeLa	MEIKRANTZ et al. 1994
unknown	Cyclin D1	NGF withdrawal	Sympathetic neurone	FREEMAN et al. 1994

CDK, cyclin-dependent Kinase; NGF, nerve growth factor.

increase in p34^{cdc2} activity observed in cells induced to undergo apoptosis was a decrease in its tyrosine phosphorylation. There were no indications that p34^{cdc2} is a proteolytic substrate for fragmentin-2/granzyme B since we did not observe any dramatic shift in its electrophoretic mobility during apoptosis. Significant increases in p34^{cdc2} activity were observed within 15 min, and in fact at 1 h the activity was very close to that observed in cells arrested in metaphase by 8 h of nocodazole treatment.

Activation of p34^{cdc2} during apoptosis has now been seen by other groups (Table 2). Robert Schlegel's laboratory examined the induction of apoptosis by drugs (including caffeine, staurosporine, 6-dimethylaminopurine or okadaic acid) that induce premature mitosis in HeLa cells that have been arrested at G_1/S by treatment with hydroxyurea (MEIKRANTZ et al. 1994). Apoptosis was observed within 2 h and was accompanied by increases in cyclin A expression as well as cyclin A-dependent histone H1 kinase activity with evidence of activation of p34^{cdc2} and CDK2. There were no indications that p34^{cdc2} was associated with cyclin B2. They also observed that transfection of Bcl-2 into HeLa cells, which had previously been shown to block staurosporine-induced apoptosis (JACOBSON et al. 1993), reduced the nuclear localization of p34^{cdc2} and CDK2 and perhaps prevented interaction of these kinases with critical nuclear substrates. Translocation of p34^{cdc2}/cyclin B to the nucleus normally occurs at G_2/M (PINES and HUNTER 1991; MEIKRANTZ et al. 1994) and is required for overexpressed p34^{cdc2}/cyclin B to induce mitotic catastrophe (HEALD et al. 1993). Increases in p34^{cdc2} kinase activity have also been observed in CHO cells following etoposide-induced cell death (LOCK and ROSS 1990) and fibroblasts induced to undergo apoptosis by staurosporine (Jacobson, Raff, Rubin, Brooks, personal communication). Induction of CDK activity following dexamethasone-induced apoptosis (Ucker 1994; M. Shröter, M.C. Peitsch, P. Gallant, F.A. Nigg, J. Tschopp, submitted) and T cell receptor (TCR) triggered apoptosis has also been observed (UCKER 1994) although the identity of the kinases has not yet been determined.

The activation of p34^{cdc2} outside of its normal restriction to G_2/M suggests, but does not prove, its requirement for the induction of apoptosis. Ideally, the critical test of the importance of p34^{cdc2} in apoptosis would be to examine the induction of apoptosis in cells that do not express, or have functionally suppressed, $p34^{cdc2}$ activity. Since $p34^{cdc2}$ is indispensible for normal cell division, we utilized two approaches to produce short-term reductions in $p34^{cdc2}$ activity (SHI et al. 1994). In the first experiment, the synthetic peptide containing the $p34^{cdc2}$ phosphorylation site on the β-subunit of casein kinase II (LITCHFIELD et al. 1991) was used as a competitive inhibitor of $p34^{cdc2}$ activity. In the second experiment, we used the FT210 cell line, which has a temperature sensitive $p34^{cdc2}$ that loses at least 80% of its kinase activity within 16 h of incubation at the restrictive temperature (TH'NG et al. 1990). In both experiments, we observed a significant inhibition of apoptosis, observations that are consistent with the requirement of $p34^{cdc2}$ in inducing nuclear damage associated with apoptosis. Although these experiments show that $p34^{cdc2}$ is required for apoptosis, we do not know the mechanism by which the enzyme was dephosphorylated at Tyr-15 and activated, or how $p34^{cdc2}$ produced a substained effect on the cell. The latter point would appear to be important because there are indications that an increase in $p34^{cdc2}$ activity by itself is not sufficient to induce mitotic perturbation. For example, microinjection of active $p34^{cdc2}$/cyclin B into mammalian fibroblasts induces many of the alterations in cytoskeletal architecture and cell morphology that are observed in mitotic cells, but fails to bring about disassembly of the nuclear envelope (LAMB et al. 1990). Furthermore, the studies of HEALD et al. (1993) indicate that wee 1 resides in the nucleus where it maintains $p34^{cdc2}$ in an inactive state. Therefore, it seems likely that if apoptosis/mitotic catastrophe is to occur, a state must exist that not only activates $p34^{cdc2}$ but also prevents its inactivation. Such a state might be induced by either inactivation of wee 1, thus preventing $p34^{cdc2}$. inactivation or by the persistent activation of cdc25, thus maintaining $p34^{cdc2}$ in its active dephosphorylated state (HEALD et al. 1993). Figure 2 suggests possible sites of action for the sustained activation of $p34^{cdc2}$.

Additional observations are consistent with a role for $p34^{cdc2}$ or perhaps other CDKs in apoptosis. NISHIOKA and WELSH (1994) recently observed that 3T3 fibroblasts become resistant to the induction of nuclear damage by CTLs when arrested at G_0 by serum starvation. This result is consistent with a requirement for $p34^{cdc2}$ in the induction of apoptosis because, although the activity of $p34^{cdc2}$. was not measured directly in this study, it is not normally expressed in G_0. FREEMAN et al. (1994) have recently examined PCD which occurs in postmitotic sympathetic neurons following withdrawal of nerve growth factor. Since these cells do not express $p34^{cdc2}$, CDK2, and cyclin A, it seems likely that $p34^{cdc2}$ may not be important in apoptosis of some noncycling cells. Following nerve growth factor withdrawal, increases in cyclin D1 expression were observed but no changes in the expression of kinase activities were detected. These cells do express CDK4 and CDK5, known binding partners for cyclin D1 (MATSUSHIME et al. 1992; XIONG et al. 1992), but it was not clear if they were present in sufficient quantity to be active or if they could act to trigger apoptosis. Nevertheless, these observations do raise the interesting possibility that a CDK pathway that is distinct from pathways involving $p34^{cdc2}$ can trigger apoptosis.

4 Conclusion

Over the last 3 years, major advances have been made in the understanding of CTL-mediated apoptosis. Evidence for an important and perhaps critical role for the granzymes and particularly granzyme B in the process seems exceptionally strong. Attention must now turn to the mechanism by which these CTL proteases can induce apoptosis in the target cell. We have suggested that the deregulation of CDKs, most notably $p34^{cdc2}$, is critical to this process. Elucidation of the mechanism by which the granzymes initiate $p34^{cdc2}$ activation could reveal biochemical events that are fundamental to many forms of apoptosis.

Acknowledgements. We thank Pierre Golstein for helpful discussions and comments on this manuscript. We are grateful to Bob Schlegel, Martin Raff, David Ucker, Gillian Griffiths, and Jürg Tschopp for sharing unpublished data with us. This work was supported by the Medical Research Council and National Cancer Institute of Canada. AHG is an NCIC Terry Fox Scientist and DL is an NCIC Scholar.

Note added in proof: Fotedar et al. (Mol. Cell. Biol, in press,1994) have recently shown that $p34^{cdc2}$/cyclin B is required for activation-induced apoptosis following T cell receptor cross linking.

References

Allbritton NL, Verret CR, Wolley RC, Eisen HN (1988) Calcium ion concentrations and DNA fragmentation in target cell destruction by murine cloned cytotoxic T lymphocytes. J Exp Med 167: 514–527

Atherton-Fessler S, Parker LL, Geahlen RL, Piwnica-Worms H (1993) Mechanisms of $p34^{cdc2}$ regulation Mol Cell Biol 13: 1675–1685

Bleackley RC, Lobe CG, Dugan B, Ehrman N Fregeau C, Meier M, Letellier M, Havele C, Shaw J, Paetkau V (1988) The isolation and characterization of a family of serine protease genes expressed in activated cytotoxic T lymphocytes. Immunol Rev 103: 5–19

Brown DG, Sun S-M, Cohen GM (1993) Dexamethasone-induced apoptosis involves cleavage of DNA to large fragments prior to internucleosomal fragmentation. J Biol Chem 268: 3037–3039

Brunet J-F, Dosseto M, Denizot F, Mattei M-G, Clark WR, Haqqi TM, Ferrier P, Nabholz M, Schmitt Verhulst A-M, Luciani M-F, Golstein P (1986) The inducible cytotoxic T lymphocyte-associated gene transcript CTLA-1 sequence and gene localization to mouse chromosome 14. Nature 322: 268–271

Bunnell BA, Heath LS, Adams DE, Lahti JM, Kidd VJ (1990) Increased expression of a 58kDa protein kinase leads to changes in the CHO cell cycle. Proc Natl Acad Sci USA 87: 7467–7471

Chang TW, Eisen HN (1980) Effects of N^3-tosyl-L-lysyl-chloromethylketones on the activity of cytotoxic T lymphocytes. J Immunol 1245: 1028–1033

Chou YH, Bischoff JR, Beach D, Goldman, RD (1990) Intermediate filament reorganization during mitosis is mediated by $p34^{cdc2}$ phosphorylation of vimentin. Cell 62: 1063–1071

Clément M-E, Legros-Maïda S, Israël-Biet D, Carnot F, Soulié A, Reynaud P, Guillet J, Gandjbakch I, Sasportes M (1994) Perforin and granzyme B expression is associated with severe acute refection. Transpantation 57: 322–326

Cobb M, Boulton TG, Robbins DJ (1991) Extracellular signal-regulated kinases: ERKs in progress. Mol. Biol Cell 2: 965–978

Cohen GM, Sun X-M, Snowden RT, Dinsdale D, Skilleter DN (1992) Key morphological features of apoptosis may occur in the absence of internucleosomal DNA fragmentation. Biochem J 286: 331

Cohen JJ. Duke RC, Chervenak R, Sellins KS, Olson LK (1985) DNA fragmentation in targets of CTL: an example of programmed cell death in the immune system. Adv Exp Med Biol 184: 493–508

Cohen JJ, Duke RC, Fadok VA, Sellins KS (1992) Apoptosis and programmed cell death in immunity. Annu Rev Immunol 10: 267–293

Coleman TR, Tang Z, Dunphy WG (1993) Negative regulation of the wee 1 protein kinase by direct action of the nim1/cdr1 mitotic inducer. Cell 72: 919–929

Draetta G, Beach D (1988) Activation of cdc2 protein kinase during mitosis in human cells: cell cycle-dependent phosphorylation and subunit rearrangement Cell 54: 17–26
Draetta G, Luca F, Westendorf J, Brizuela L, Ruderman J, Beach D (1989) cdc2 protein kinase is complexed with both cyclin A and B: evidence for proteolytic inactivation of MPF. Cell 56: 829–838
Ducommun B, Brambill P, Felise MA, Franza BR, Karsenti E, Draetta G (1991) cdc2 phosphorylation is required for its interaction with cyclin. EMBO J 10: 3311–3319
Duke, RC, Chervenak R, Cohen JJ (1983) Endogenous edonuclease-induced DNA fragmentation: an early event in cell-mediated cytolysis. Proc Natl Acad Sci USA 80: 6361–6365
Duke, RC, Cohen JJ, Chervenak R (1986) Differences in target cell DNA fragmentation induced by mouse cytotoxic T lymphocytes and natural killer cells. J Immunol 137: 1442–1447
Dunphy WG, Kumagai A (1991) The cdc25 protein contains an intrinsic tyrosine phosphatase activty. Cell 67: 189–196
Ellis RE, Yuan J, Horvitz HR (1991) Mechanisms and functions of cell death. Annu Rev Cell Biol 7: 663–698
Evans T, Rosenthan ET, Youngblom J, Distal D, Hunt T (1983) Cyclin: a protein specified by maternal mRNA in sea urchin eggs that is destroyed at each cleavage division. Cell 33: 389–396
Ewoldt GR, Winkler U, Powers JC, Hudig D (1992) Sulfony1 fluoride serine protease inhibitors inactivate RNK-16 lymphocyte granule proteases and reduce lysis by granule extracts and perforin. Mol Immunol 29: 713–721
Fang F, Newport JW (1991) Evidence that the G1-S and G2-M transitions are controlled by different cdc2 proteins in higher eukaryotes. Cell 66: 731–742
Feilotter H, Lingner C, Rowley R, Young PG (1992) Regulation of the G2-mitosis transition. Biochem. Cell Biol 70: 954–971
Fesquent D, Labbe JC, Derancourt J, Capony JP, Galas S, Girard F, Lorca T, Shuttleworth J, Doree M, Cavadore JC (1993) The MO15 gene encodes the catalytic subunit of a protein kinase that activates cdc2 and other cyclin-dependent kinases (CDKs) through phosphorylation of Thr-161 and its homologues. EMBO J 12: 3111–3121
Forsburg SL, Nurse P (1991) Cell cycle regulation in the yeasts saccharomyces cerevisiae and schizosaccharomyces pombe. Annu. Rev. Cell Biol 7: 227–256
Freeman RS, Estus S, Johnson EM (1994) Analysis of cell cycle-related gene expression in postmitotic neurons: selective induction of cyclin D during programmed cell death. Neuron 12: 343–355
Froelich CJ, Zhang X, Turbov J, Hudig D, Winkler U, Hanna WL (1993) Human granzyme B degrades aggrecan proteoglycan in matrix synthesized by chondrocytes. J Immunol 151: 7161–7171
Gagliardini V, Fernandez P-A, Lee RKK, Drexler HCA, Rotello RJ, Fishman MC, Yuan J (1994) Prevention of vertebrate neuronal death by the crmA gene. Science 263: 826–828
Garcia-Sanz JA, McDonald HR, Jenne DE, Tchopp J, Nabholz M (1990) Cell specificity of granzyme gene expression. J Immunol 145: 3111–3118
Gautier J, Minshull J, Lohka M, Glotzer M, Hunt T, Maller JL (1990) Cyclin is a component of MPF from Xenopus. Cell 60: 487–494
Gautier J, Solomon MJ, Booher RN, Bazan JF, Kirschner MW (1991) cdc25 is a specific tyrosine phosphatase that directly activates $p34^{cdc2}$. Cell 67: 197–211
Gérshenfeld HK, Weissman IL (1986) Cloning of a cDNA for a T cell-specific serine protease from a cytotoxic T lymphocyte. Science 232: 854–858
Glotzer M, Murray AW, Kirschner MW (1991) Cyclin is degraded by the ubiquitin pathway. Nature 349: 132–138
Golstein P, Ojcius DM, Young JD-E (1991) Cell death mechanisms and the immune system. Immunol Rev 121: 29–65
Gould KL, Nurse P (1989) Tyrosine phosphorylation of the fission yeast $cdc2^{+}$ protein kinase regulates entry into mitosis. Nature 342: 39–45
Gould KL, Moreno S, Owen DJ, Sazer S, Nurse P (1991) Phosphorylation at Thr167 is required for Schizosaccharomyces pombe $p34^{cdc2}$ function. EMBO J 10: 3297–3309
Griffiths GM (1994) TCR triggers release of lytic proteins via two pathways during killing by cytotoxic T lymphocytes (submitted)
Griffiths GM, Namikawa R, Mueller C, Liu D-D, Young JD-E, Billingham M, Weissman I (1991) Granzyme A and perforin as markers for rejection in cardiac transplantation. Eur J Immunol 21: 687–692
Hameed A, Lowrey DM, Lichtenheld M, Podack ER (1988) Characterization of three serine esterases isolated from human IL-2 activated killer cells. J Immunol 141: 3142–3147
Hameed A, Truong LD, Price V, Krahenbuhl O, Tschopp (1991) Immunohistochemical localization of granzyme B antigen in cytotoxic cells in human tissues. Am J Pathol 138: 1069–1075

Hanks SK, Quinn Am (1991) Protein kinase catalytic domain sequence database: identification of conserved features of primary structure and classification of family members. methods Enzymol 200: 38–62

Hayes MP, Bexrrebi G, Henkart PA (1989) Induction of target cell DNA release by the cytotoxic T lymphocyte protease granzyme A. J Exp Med 170: 933–946

Heald R, McKeon F (1990) Mutations of phosphorylation sites in lamin A that prevent nuclear lamina disassembly in mitosis. Cell 61: 579–589

Heald R, McLoughlin M, McKeon F (1993) Human wee 1 maintains mitotic timing by protecting the nucleus from cytoplasmically activated cdc2 kinase. Cell 74: 463–474

Henkart P (1985) Mechanisms of lymphocytes mediated cytotoxicty. Annu Rev Immunol 3: 31–58

Henkart PA, Millard PJ, Reynold CW, Henkart MP (1984) Cytoytic activity of purified cytoplasmic granules from cytoxotic rat large granular lymphocyte tumors. J Exp Med. 160: 75–93

Heusel JW, Wesselschmidt RL, Shresta S, Russell JH, Ley TJ (1994) Cytotoxic lymphocytes require granzyme B for the rapid induction of DNA fragmentation and apoptosis in allogeneic target cells. Cell 76: 977–987

Hoffman I, Clarke PR, Marcote MJ, Karsenti E, Draetta G (1993) Phosphorylation and activation of human cdc 25-C by cdc2-cyclin B and its involvement in the self-amplification of MPF at mitosis. EMBO J 12: 53–63

Howard AD, Kostura MJ, Thornberry N, Ding GJF, Limjuco G, Weidner J, Salley JP, Hogquist KA, Chaplin DD, Mumford RA, Schmidt JA, Tocci MJ (1991) IL-1 converting enzyme requires asparic acid residues for processing of the IL-1β precursor at two distinct sites and does not cleave 31-kDa IL-1a. J Immunol 147: 2964–2969

Hudig D, Allison VJ, Pickett TN, Winkler V, Kam C, Powers J (1991) The function of lymphocyte proteases. Inhibition and restoration of granule mediated lysis with isocoumarin serine protease inhibitors. J Immunol 147: 1360–1368

Igarishi M, Nagata A, Jinno S, Suto K, Okayama H (1991) Wee 1-like gene in human cells. Nature 353: 80–83

Itoh N, Yonehara S, Ishii A, Yonehara M, Mizushima S-I, Samshima M, Hase A, Seto Y, Nagata S (1991) The polypeptide encoded by the cDNA for human cell surface antigen Fas can mediate apoptosis. Cell 66: 233–243

Izumi T, Maller JL (1993) Elimination of cdc2 phosphorylation sites in the cdc25 phosphatase blocks initiation of mitosis. Mol Biol Cell 4: 1337–1350

Izumi T, Walker DH, Maller JL (1992) Periodic changes in phosphorylation of the Xenopus cdc25 phosphatase regulates its activity. Mol Cell Biol 11: 3860–3867

Jacobson MD, Burne JF, King MP, Miyashita T, Reed JC, Raff MC (1993) Bcl-2 blocks apoptosis in cells lacking mitochondrial DNA. Nature 361: 365–369

Jenne DE, Masson D, Zimmmer M, Haefliger J-A, Li W-H, Tschöpp J (1989) Isolation and complete structure of the lymphocyte serine protease granzyme G, a novel member of the grazyme multigene family in murine cytolytic T lymphocytes. Evolutionary origin of lymphocyte proteases. Biochem 28: 7953–7961

Kägi D, Ledermann B, Bürki K, Seiler P, Odermatt B, Olsen KJ, Padack ER, Zinkernagel RM, Hengartner H (1994) Cytotoxicity mediated by T cells and natural killer cells is greatly impaired in perforin-deficient mice. Nature 369: 31–37

Kawakami A, Tian Q, Duan X, Streuli M, Schlossman SF, Anderson P (1992) Identification and characterization of a TIA 1-related nucleolysin. Proc Natl. Acad Sci USA 89: 8681–8685

Kipreos ET, Wang JYJ (1990) Differential phosphorylation of c-abl proto-oncogene product during interphase and mitosis by cdc2 kinase. Science 248: 217–220

Kraut RP, Bose D, Cragoe Jr EJ, Greenberg AH (1990) The influence of clacium, sodium and the Na^+/Ca^{2+} antiport on susceptibility to cytolysin/perfoin-mediated cytolysis. J Immunol 144: 3498–3505

Kraut RP, Bose R, Cragoe Jr EJ, Greenberg AH (1992) The Na^+/Ca^{2+} exchanger regulates cytolysin/ perforin-induced increases in intracellular Ca^{2+} and susceptibility to cytolysis. J Immunol 148: 2489–2496

Krek W, Nigg EA (1991) Mutation of $p34^{cdc2}$ phosphorylation sites induce premature mitoric events in HeLa cells: evidence for a double block to $p34^{cdc2}$ kinase activation in vertebrates. EMBO J 10: 3331–3341

Kumagai A, Dunphy WG (1992) Regulation of the cdc25 protein during the cell cycle in Xenopus extracts. Cell 70: 139–151

Lamb NJC, Fernandez A, Watrin A, Labbé JC, Cavadore JC (1990) Microinjection of $p34^{cdc2}$ kinase induces marked changes in cell shape, cytoskeletal organization and chromatin structure in mammalian fibroblasts. Cell 60: 151–165

Landon C, Nowicki M, Sugawara S, Dennert G (1990) Differential effects of protein synthesis inhibition on CTL and targets in cell-mediated cytotoxicity. Cell Immunol 128: 412–426
Langan TA, Gautier J, Lohka M, Hollingsworth R, Moreno S, Nurse P, Maller J, Scalfani RA (1989) Mammalian Growth-associated H1 histone kinase: a homolog of cdc2+/CDC28 protein kinases controlling mitotic entry in yeast and frog cells. Mol Cell Biol 9: 3860–3868
Lee MG, Nurse P (1987) Complementation used to clone a human homologue of the fission yeast cell cycle control gene cdc2. Nature 327: 31–35
Lee MS, Ogg S, Xu M, Parker LL, Donoghue DJ, Maller JL, Piwnica-Worms H (1992) $cdc25^{+}$ encodes a protein phosphatase that dephosphorylates $p34^{cdc2}$. Mol Biol Cell 3: 73–84
Lew J, Beaudette K, Litwin CME, Wang JH (1992) Purification and characterization of a novel proline-directed protein kinase from bovine brain. J Biol Chem 267: 13383–13390
Lipman ML, Stevens AC, Bleackley RC, Helderman H, McCune TR, Harmon WH, Shapiro ME, Rosen S, Strom TB (1991) The strong correlation of cytotoxic T-lymphocyte-specific serine protease gene transcripts with renal allograft rejection. Transplantation 53: 73–79
Litchfield DW, Lozeman FJ, Cicirelli MF, Harrylock M, Ericsson LH, Piening CJ, Krebs EG (1991) Phosphorylation of the β subunit of casein kinase II in A431 cells. J Biol Chem 266: 20380–20389.
Litchfield DW, Luscher B, Lozeman FJ, Eisenman RN, Krebs EG (1992) Phosphorylation of casein kinase II by $p34^{cdc2}$ in vitro and at mitosis. J Biol Chem 267: 13943–13951
Lobe CG, Finlay BB, Parachych W, Paetkau VH, Bleackley RC (1986) Novel serine proteases encoded by two cytotoxic T lymphocyte-specific genes. Science 232: 858–861
Lock RB, Ross WE (1990) Possible role for $p34^{cdc2}$ kinase in etoposide induced cell death of Chinese hamster ovary cells. Cancer Res. 50: 3767–3771
Lorca T, Labbé JC, Devault A, Fesquet D, Capony JP, Cavadore JC, Le Bouffant F, Dorée M (1992) Dephosphorylation of cdc2 on threonine 161 is required for cdc2 kinase inactivation and normal anaphase. EMBO J 11: 2381–2390
Lundgren K, Walworth N, Booher R, Dembski M, Kirschner M, Beach D (1991) mik 1 and wee 1 cooperate in the inhibitory phosphorylation of cdc2. Cell 64: 1111–1122
Mak AS, Watson MH, Litwin CE, Wang JH (1991) Phosphorylation of caldesmon by cdc2 kinase. J Biol Chem 266: 6678–6681
Maller JL (1991) Mitotic control. Curr Op in Cell Biol 3: 269–275
Masson D, Tschopp J (1987) A family of serine esterases in lytic granules of cytolytic T lymphocytes. Cell 49: 679–685
Matsushime H, Ewen ME, Strom DK, Kato JY, Hanks SK, Roussel MF, Sherr CJ (1992) Identification and properties of an atypical catalytic submit (p34PSK-J3/cdk4) for mammalian D type G1 cyclins. Cell 71: 323–334
McGowan CH, Russell P (1993) Human wee 1 kinase inhibits cell division by phosphorylating $p34^{cdc2}$ exclusively on tyr 15. EMBO J 12: 75–85
Meier M, Kwong PC, Fregeau CJ, Atkinson EA, Burrington M, Erhman N, Sorensen O, Lin CC, Wilkins J, Bleackley RC (1990) Cloning of a gene that encodes a new member of the human cytotoxic cell protease family. Biochemistry 29: 4042–4049
Meikrantz W, Gisselbrecht S, Tam SW, Schlegel R (1994) Activation of cyclin A-dependent protein kinase during apoptosis. Proc Natl Acad Sci USA 91 (in press)
Meyerson M, Enders GH, Wu CL, Su LK, Gorka C, Nelson C, Harlow E, Tsai LH (1992) A family of human cdc-2 related protein kinases. EMBO J 11: 2902–2917
Millar JBA, McGowan CH, Lenaers G, Jones R, Russell P (1991) $p80^{cdc25}$ mitotic inducer is the tyrosine phosphatase that activates $p34^{cdc2}$ in fission yeast. EMBO J 10: 4301–4309
Miura M, Zhu H, Rotello R, Hartwieg EA, Yuan J (1993) Induction of apoptosis in fibroblasts by IL-1β-converting enzyme, a mammalian homolog of the C.elegans cell death gene ced-3. Cell 75: 653–660
Morgan DO, Kaplan JM, Bishop JM, Varmus HE (1989) Mitosis-specific phosphorylation of $p60^{c\text{-}src}$ by $p34^{cdc2}$ associated protein kinase. Cell 57: 775–786
Mueller C, Gershenfeld HK, Lobe CG, Okada DY, Bleackley RC, Weissman IL (1988) A high proportion of T lymphocytes that infiltrate H-2 incompatible heart allografts in vivo express genes encoding cytotoxic cell-specific serine proteases, but do not express the MEL-14-defined lymph node homing receptor. J Exp Med 167: 1254–1136
Murray AW, Solompon MJ, Kirschner MW (1989) The role of cyclin synthesis and degradation in the control of maturation promoting factor activity. Nature 339: 280–286
Nigg EA (1993) Targets of cyclin-dependent protein kinases. Curr Opin Cell Biol 5: 187–193
Nishioka WK, Welsh RM (1994) Susceptibility to cytotoxic T lymphocyte-induced apoptosis is a function of the proliferative status of the target. J Exp Med 179: 769–774

Norbury C, Blow J, Nurse P (1991) Regulatory phosphorylation of the p34^{cdc2} protein kinase in vertebrates. EMBO J 10: 3321–3329

Norbury C, Nurse P (1992) Animal cell cycles and their control Annu Rev Biochem 61: 441–470

Oberhammer F, Wilson JW, Dive C, Morris ID, Hickman JA, Wakeling AE, Walker PR, Sikorska M (1993) Apoptotic death in epithelial cells: cleavage of DNA to 300 and/or 50 kb fragments prior to or in the absence of internucleosomal fragmentation. EMBO J 12: 3679–3684

Ocha-Orbea H, Scarpellino L, Herig S, Dupuis M, Tshcöpp J (1990) Inhibition of lymphocyte mediated cytotoxicity by perforin antisense oligonucleotides. EMBO J 9: 3815–3819

Odake S, Kam G-M, Narasimhan L, Pal M, Blake JT Krähenbül O, Tschopp J, Powers JC (1991) Human and murine cytotoxic lymphocyte serine proteases: substrate mapping with peptide thioester substrates and inhibition of enzyme activity and cytolysis by isocoumarins. Biochemistry 30: 2217–2227

Ojcius DM, Zheng LM, Sphicas EC, Zychlinsky A, Young JD-E (1991) Subcellular localization of perforin and serine esterase in lymphokine-activated killer cells and cytotoxic T cells by immunogold labeling. J Immunol 146: 4427–4432

Orrenius S, McConkey DJ, Bellomo G, Nicotera P (1989) Role of Ca^{2+} in cell killing. Trends Pharmocol Sci 10: 281

Parker LL, Piwnica-Worms H (1992) Inactivation of the p34^{cdc2}-cyclin B complex by the human Wee 1 tyrosine kinase. Science 257: 1955–1957

Parker LL, Walter SA, Young PA, Piwnica-Worms, H (1993) Phosphorylation and inactivation of the mitotic inhibitor wee 1 by the nim1/cdr1 kinase. Nature 363: 736–738

Pasternack MS, Eisen HN (1985) A novel serine esterase expressed by cytolytic T lymphocytes. Nature 314: 743–745

Pasternack MS, Bleier KJ, McInerney TN (1991) Granzyme A binding to target cell proteins. Granzyme A binds to and cleaves nucleolin in vitro. J Biol Chem 266: 14703–14708

Pelech SL, Charest DL, Mordret GP, Siow YL, Palaty C, Campbell D, Charlton L, Samiei M, Sanghera JS (1993) Networking with mitogen-activated protein kinases. Mol Cell Biochem 127/128: 157–169

Peter M, Nakagawa J, Dorée M, Labbé JC, Nigg EA (1990a) In vitro disassembly of the nuclear lamina and M phase-specific phosphorylation of lamins by cdc2 kinase. Cell 61: 591–602

Peter M, Nakagawa J, Dorée M, Labbé JC, Nigg EA (1990b) Identification of major nucleolar proteins as candidate mitotic substrates of cdc2 kinase. Cell 60: 791–801

Peter M, Sanghera JS, Pelech SL, Nigg EA (1992) Mitogen-activated protein kinases phosphorylate nuclear lamins and display sequence specificity overlapping that of mitotic protein kinase p34^{cdc2}. Eur J Biochem 205: 287–294

Pines J (1991) Cyclins: wheels within wheels. Cell Growth differ. 2: 305–310

Pines J (1993) Cyclins and cyclin-dependent kinases: take your partners. Tends Biochem Sci 18: 195–197

Pines J, Hunter T (1990a) Human cyclin A is adenovirus E1-A associated protein p60 and behaves differently from cyclin B. Nature 346: 760–763

Pines J, Hunter T (1990b) p34^{cdc2}: the S and M kinase? New Biol 2: 389–401

Pines J, Hunter T (1991) Human cyclins A and B 1 are differentially located in the cell and undergo cell cycle-dependent nuclear transport. J Cell Biol 115: 1–17

Podack, ER, Young, JD-E, Cohn ZA (1985) Isolation and biochemical characterization of perforin 1 from cytolytic T cell granules. Proc Natl Acad Sci USA 82: 8629–8633

Podack ER, Hengartner H, Lichtenheld MG (1991) A central role of perforin in cytolysis? Annu Rev Immunol 9: 129–157

Poe M, Bennett CD, Biddison WE, Blake JT, Norton GP, Rodkey JA, Sigal NH, Turner RV, Wu JK, Zweerink HJ (1988) Human cytotoxic lymphocyte tryptase. Its purification from granules and the characterization of inhibitor and substrate specificity. J Biol Chem 263: 13215

Poe M, Blake JT, Boulton DA, Gammon M, Sigal NH, Wu JK, Zweerink HJ (1991) Human cytotoxic lymphocyte granzyme B. Its purification from granules and the characterization of substrate and inhibitor specificity. J Biol Chem 266: 98–103

Poenie M, Tsien RY, Schmitt-Verhulst AM (1987) Sequential activation and lethal hit measured by $(Ca^{2+})_i$ in individual cytolytic T cells and targets. EMBO J 6: 2223–2232

Poon RYC, Yamashita K, Addamczewski JP, Hunt T, Shuttleworth J (1993) The cdc2-related protein p40^{MO15} is the catalytic subunit of a protein kinase that can activate p33^{cdk2} and p34^{cdc2}. EMBO J 12: 3123–3132

Powers JC (1977) Haloketone inhibitors of proteolytic enzymes. In: Weinstein B (ed) Chemistry and biochemistry of amino acids. Dekker, New York, pp. 65–178 (Peptides and proteins, vol 4)

Prendergast JA, Helgason CD, Bleackley RC (1992) Quantitative polymerase chain reaction analysis of cytotoxic cell proteinase gene transcripts in T cells. Pattern of expression^{con} is dependent on the nature of the stimulus. J Biol Chem 267: 5090–5095

Raff M (1992) Social controls on cell death and survival. Nature 356: 397–400

Ray CA, Black RA, KronheimSR, Greenstreet TA, Sleath PR, Salvesen GS, Pickup DJ (1992) Viral inhibition of inflammation: Cowpox virus encodes an inhibitor of the interleukin-1β converting enzyme. Cell 69: 597–604

Redelman D, Hudig (1980) The mechanisms of cell-mediated cytotoxicity I. Killing by murine cytotoxic T lymphocytes requires cell surface thiols an activated protease. J Immunol 124: 870–878

Rossomando AJ, Dent P, Sturgill TW, Marshak DR (1994) Mitogen-activate protein kinase kinase 1 (MKK1) is negatively by threonine phosphorylation. Mol Cell Biol 14: 1594–1602

Rouvier E, Luciani M-F, Golstein P (1993) Fas involvement in Ca^{2+}-indepenent T cell-mediated cytotoxicity. J Exp Med 177: 195–200

Ruff MR, Gifford GE (1981) Rabbit tumor necrosis factor: mechanism of action. Infect Immun 31: 380–385

Russell JH (1983) Internal disintegration model of cytotoxic lymphocyte-induced target damage. Immunol Rev 72: 97–118

Russell JH, Dobos CB (1980) Mechanisms of immune lysis. II. CTL-induced nuclear disintegration of the target begins within minutes of cell contact. J Immunol 125: 1256–1261

Russell JH, Basakowski VR, Dobos CB (1980) Mechanisms of immune lysis I. Physiological distinction between target cell death mediated by cytotoxic T limphocytes and antibody plus complement. J Immunol 124: 1100–1105

Russell JH, Masakowski V, Rucinsky T, Phillips G (1982) Mechanims of immune lysis III. Characterization of the nature and kinetics of the cytotoxic T lymphocyte-induced nuclear lesion in the target. J Immunol 128: 2087–2094

Sadhu K, Reed SI, Richardson H, Russell P (1990) Human homolog of fission yeast cdc25 mitotic inducer is predominantly expressed in G2. Proc. Natl Acad Sci USA 87: 5139–5143

Satterwhite LL, Lohka MJ, Wilson KL, Scherson TY, Cisek LJ, Corden JL, Pollard TD (1992) Phosphorylation of myosin II regulatory light chain by cyclin-p34^{cdc2}: a mecharism for the timing of cytokinesis. J Cell Biol 118: 595–605

Sayers TJ, Wiltrout TA, Sowder R, Munger WL, Smyth MJ, Henderson LE (1992) Purification of a factor from the granules of a rat natural killer cell line (RNK) that reduces tumor cell growth and changes tumor morphology. Molecular identity with a granule serine protease (RNKP-1) J Immunol 148: 292–300

Schwartz LM, Osborne BA (1993) Programmed cell death, apoptosis and killer genes. Immunol Today 14: 5582–5590

Schwartz LM, Smith SE, Jones MEE, Osborne BA (1993) Do all programmed cell deaths occur via apoptosis? Proc. Natl Acad Sci USA 90: 980–984

Shenoy S, Choi J, Bagrdia S, Copeland TD, Maller JL, Shalloway D (1989) Purified maturation promoting factor phosphorylates pp60$^{c\text{-}src}$ at the sites phosphorylated during fibroblast mitosis. Cell 57: 763–774

Sherr CJ (1993) Mammalian G1 cyclins. Cell. 73: 1059–1065

Shi L, Kraut RP, Aebersold R, Greenberg AH (1992a) A natural killer cell granule protein that induces DNA fragmentation and apoptosis. J Exp Med 175: 553–566

Shi L, Kam C-M, Powers JC, Aebersold R, Greenberg AH (1992b) Purification of three cytotoxic lymphocyte granule serine proteases that induce apoptosis through distinct substrate and target cell interactions. J Exp Med 176: 1521–1529

Shi L, Nishioka WK, Th'ng J, Bradbury EM, Litchfield DW, Greenberg AH (1994) Premature p34^{cdc2} activation required for apoptosis. Science 263: 1143–1145

Shiver JW and Henkart PA (1991) A noncytotoxic mast cell tumor line exhibits potent IgE-dependent cytotoxicity after transfection with the cytolysin/perforin gene. Cell 64: 1175–1181

Shiver JW, Su L, Henkart PA (1992) Cytotoxicity with target DNA breakdown by rat basophilic leukemia cells expressing both cytolysin and granzyme A. cell 71: 315–322

Simon MM, Simon H, Fruth U (1987) Cloned cytolytic T-effector cells and their malignant counterparts produce an extracellular matrix degrading trypsin-like serine protease. Immunology 60: 219–230

Solomon MJ, Lee T, Kirschner MW (1992) Role of phosphorylation in p34^{cdc2} activation: identification of an activating kinase. Mol Biol Cell 3: 13–27

Solomon MJ, Harper JW, Shuttleworth J (1993) CAK, the p34^{cdc2} activating kinase, contains a protein identical or closely related to p40^{MO15}. EMBO J 12: 3133–3142

Smyth MJ, Wiltrout T, Trapani JA, Ottaway KS, Sowdr R, Henderson LE, Kan C-M, Powers JC, Young HA, Sayers TJ (1992) Purification and cloning of a novel serine protease, RNK-met-1, from the granules of a rat natural killer cell leukemia. J Biol Chem 267: 24418–24425

Strausfeld U, labbé JC, Fesquet D, Cavadore JC, Picard A, Sadhu K, Russell P, Dorée M (1991) Dephosphorylation and activation of a $p34^{cdc2}$/cyclin B complex in vitro by human cdc25 protein. Nature 351: 242–245

Suda T, Takahashi T, Golstein P, Nagata δ (1993) Molecular cloning and expression of the Fas ligand, a novel member of the tumor necrosis factor family. Cell 75: 1169–1178

Sun DY, Jiang S, Zheng L-M, Ojcius DM, Young JD-E (1994) Separate metabolic pathways leading to DNA fragmentation and apoptotic chromatin condensation. J Exp Med 179: 559–568

Talento A, Nguyen M, Lae S, Wu JK, Poe M, Blake JT, Patel M, Wu T-J, Manyak CL, Siberklang M, Mark G, Springer M, Sigal NH, Weissman IL, Bleakley RC, Podack ER, Tkykocinski ML, Koo GC (1992) Transfection of mouse cytotoxic T lymphocyte with an antisense granzyme A vector reduces lytic activity. J Immunol 12: 4009–4015

Tang Z, Coleman TR, Dumphy WG (1993) Two distinct mechanisms for negative regulation of the wee 1 protein kinase. EMBO J 12: 3427–3436

Th'ng JPH, Wright PS, Hamaguchi J, Lee MG, Norbury CJ, Nurse P, Bradbury, EM (1990) The FT210 cell line is a mouse G2 phase mutant with a temperature-sensitive CDC2 gene product. Cell 63: 313–324

Thomas G (1992) MAP kinase by any other name smells just as sweet. Cell 68: 3–6

Thornberry NA, Bull HG, Calaycay JR, Chapman KT, Howard AD, Kostura NJ, Miller DK, Molineaus SN, Weidner JR, Aunins J, Elliston KO, Ayala JN, Casano FJ, Chin J, Ding GJ-f, Egger LA, Gaffney EP, Limjuco G, Palyha OC, Raju SN, Rolando An, Salley JP, Yamin T-T, Lee TD, Shively JE, MacCroll M, Mumford RA Schmidt JA Tocci NJ (1992) A novel heterodimeric cysteine protease is required for interleukin-1β processing in monocytes. Nature 356: 768–774

Tian Q, Streuli M, Saito H, Schlossman F, Anderson P (1991) A polyadenylate binding protein localized to the granules of cytolytic lymphocytes induces DNA fragmentation in target cells. Cell 67: 629–639

Tomei LD, Shapiro JP, Cope FO (1993) Apoptosis in C3H/10T1/2 mouse embryonic cells: evidence for internuclosomal DNA modification in the absence of double-strand cleavage. Proc Natl Acad Sci USA 90: 833

Ucker DS (1994) Physiological cell death is an abortive mitosis. Proc Soc Exp Büiol Bed (in press)

Ucker DS, Obermiller PS, Eckhart W, Apgar JR, Berger NA, Meyers J (1992) Genome digestion is a dispensable consequence of physiological cell death emdiated by cytotoxic T lymphocytes. Mol Cell Biol 12: 3060–3069

Van den Heuvel S, Harlow E (1993) Distinct roles for cyclin-dependent kinases in cell cycle control. Science 262: 2050–2054

Velotti F, Palmieri G, D'Ambrosio D, Piccoli M, Frati L, Santtoni A (1992) DIfferential expression of granzyme A and grnazyme B proteases and their secretion by fresh rat natural killer cells (NK) and lymphokine-activated killer cells with NK phenotype (LAK-NK). Eur J Immunol 22: 1049–1053.

Vignaux F, Golstein P (1994) Fas-based lymphocyted-mediated cytotoxicity against syngeneic activated lymphocytes: a regulatory pathway? Eur J Immunol 24: 923–927

Walker NP, Talanian RV, Brady KD, Dang LC, Bump NJ, Ferenz CR, Franklin S, Ghayur T, Hackett MC, Hammill LD, Herzog L, Hugunin M, Houy W, Mankovich JA, McGuiness L, Orlewicz E, Paskind M, Pratt CA, Reiss P, Summani A, Terranova M, Welch JP, Xiong L, Möller A, Tracey DE, Kamen R, Wong WW (1994) Crystal structure of the cysteine protease interleukin-1 beta converting enzyme: a $(p20/p10)_2$ homodimer. Cell 78: 343–352

Ward GE, Kirschner MW (1990) Identification of cell cycle-regulated phosphorylation sites on lamin-C. Cell 61: 561–577

Wijngaard PLJ, Tuijnman WB, Meyling FHJG, van der Muelen A, Huytink M, Jambroes G, Schuurman J-J (1993) endomyocardial biopsies after heart transplantation. The presence of markers indicative of activation. Transplantation 55: 103–110

Williams GT, Smith CA (1993) Molecular regulation of apoptosis: genetic controls on cell death. Cell 74: 777–779

Wilson KP, Black JAF, Thomson SA, Kim EE, Griffiths JP, Nava MA, Murcko MA, Chambers SP, Aldape RA, Raybuck SA, Livingston DJ (1994) Structure and mechanism of interleukin-1β converting enzyme. Nature 370: 270–275

Wu L, Russell P (1993) Niml kinase promotes mitosis by inactivating wee 1 tyrosine kinase. Nature 363: 738–741

Wyllie AH (1987) Cell death. Int Rev Cytol [Suppl] 17: 775–785

Wyllie AH, Kerr FR, Currie AR (1980) Cell death: the significance of apoptosis. Int Rev Cytol 68: 251–356
Wyllie AH, Morris RG, Smith AL, Dunlop D (1984) Chromatin cleavage in apoptosis: association with condensed chromatin morphology and dependence on macromolecular synthesis. J Pathol 142: 67–77
Xiong Y, Zhang H, Beach D (1992) D type cyclins associate with multiple protein kinase and the DNA replication and repair factor PCNA. Cell 71: 505–514
Yamashiro S, Yamakita, Y, Ishikawa R, Matsumura F (1990) Mitosis specific phosphorylation causes 83K non-muscle caldesmon to dissociate from microfilaments. Nature 344: 675–678
Young JD-E, Nathan CF, Podack ER, Palladino MA (1986) Functional channel formation associated with cytotoxic T cell grnaules. Proc Natl Acad Sci USA 83: 150
Yuan J, Shaham S, Ledoux S, Ellis HN, Horvitz HR (1993) The C.elegans cell death gene ced-3 encodes a protein similarr to mammalian interleukin-1β-converting enzyme. Cell 75: 641–652
Zakeri ZF, Quaglino D, Latham T, Lockshin RA (1993) Delayed internucleosomal DNA fragmentation in programmed cell death. FASEB J 7: 470–478
Zychlinsky A, Zheng LM, Liu C-C, Young JD-E (1991) Cytolytic lymphocytes induce both apoptosis and necrosis in target cells. J Immunol 146: 393–400

Perforin, Killer Cells and Gene Transfer Immunotherapy for Cancer

E.R. PODACK

1 Introduction

The science of immunology has entered the stage at which a reasonably detailed understanding of molecular mechanisms allows the application of immunological concepts to the treatment of certain diseases. Cytotoxic mechanisms of lymphocytes have been clarified to the point where cogent decisions can be made as to how to exploit them for the treatment of viral diseases and cancer. Granule exocytosis and perforin have been identified as major cytolytic mechanisms in vivo and in vitro in $CD8^+$ T cells and natural killer (NK) cells by the creation of perforin knock-out mice and analysis of their cytotoxic potential. In addition Fas and Fas-ligand-mediated cytolytic mechanisms, which are granule-independent, are mediated by cytotoxic T cells (CTLs). The expression of Fas on the target, however, is required. Independent of Fas and perforin, a third, protein synthesis-independent, Ca^{2+}-dependent cytotoxic mechanism exists in CTLs. The molecular nature of this pathway is not known. Finally, a protein synthesis-dependent, Ca^{2+}-dependent pathway appears to be present in lympholuine-activated killer (LAK) cells.

Since perforin, packaged in granules, is required for a major part of cytotoxic activity, it is necessary to consider its induction synthesis and maintenance in cytotoxic T-cells and NK-cells if they are to be used for tumor therapy. This review will attempt to give a summary of efforts directed at gene transfer immunotherapy of cancer taking advantage of the generation of cytotoxic lymphocytes with specific emphasis toward perforin mediated mechanisms.

Department of Microbiology and Immunology, University of Miami, School of Medicine, 1600 NW 10th Ave (R-138), Miami, FL 33136, USA

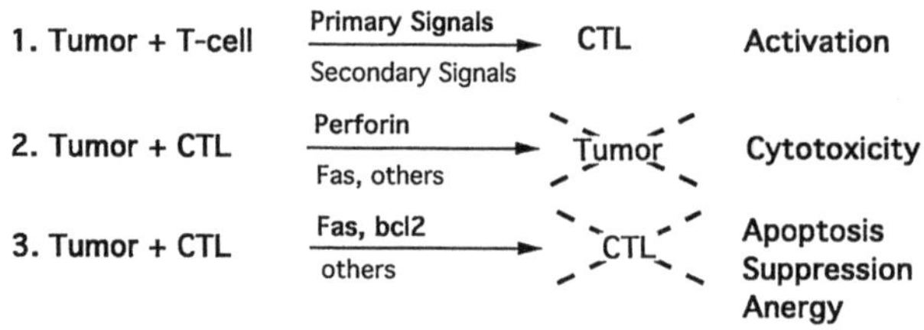

Fig. 1. Signals and molecules involved in the generation, effetor function and regulation of cytotoxic cells

2 Principles

Immunotherapy of cancer requires consideration of at least three operationally separate stages (Fig. 1). First, tumor specific T cells or NK cells must be activated and clonally expanded. Second, they must acquire cytotoxic activity of sufficient intensity to destroy tumor cells. Third, the tumor-specific cytotoxic response must be of sufficient duration to eliminate all tumor cells even at remote sites. The latter requirement dictates manipulation that prevent suppression of cytotoxic cells or functional or physical elimination, i.e., anergy or apoptosis, respectively.

3 Activation and Clonal Expansion of Tumor-Specific Effector Cells

Tumor antigens presentable as peptides by class I or class II MHC have been identified in some tumors (van der Bruggen et al. 1991). It is evident, however, from the fact that tumors arise, that host T cells are not sufficiently alerted by the tumor to induce its elimination. This may stem from several selective mechanisms during tumorigenesis. Even though antigenic peptides may be present, only tumors will grow that have been able to suppress antigen presentation or that prevent costimulatory signals required for T cell activation. Alternatively, mutations in oncogene or tumor suppressor genes have been selected to be tumorigenic but not presentable by any of the patient's MHC alleles. In the latter case no T cell immunogenicity can be expected.

Recent studies by numerous groups (e.g., Colombo and Forni 1994) have confirmed original reports (Tepper et al. 1989; Fearon et al. 1990; Gansbacher et al. 1990) demonstrating that many tumors become immunogenic when they produce T cell costimulatory molecules introduced to the tumor by gene transfer. Lewis lung carcinoma (LLC), for instance, is a nonimmunogenic, spontaneous tumor that, upon transplantation of 10^4 or more cells, results in 100% tumor formation. Interleukin-2 (IL-2) transfection of LLC causes an almost complete loss of tumorigenicity in immunocompetent mice (Ohe et al. 1993). However, in mice lacking functional NK cells or T cells IL-2 transfected LLC continue to form tumors, indicating that an intact T cell and NK repertoire is required for the rejection of tumors producing IL-2 as costimulatory molecules. It is of interest that a variety of cytokines can fulfill the costimulatory role, but that there is also a degree of tumor variability (COLOMBO and FORNI 1994).

Tumor-specific cytotoxic activity, although inferred from studies above, is not always easy to demonstrate directly in vitro. In most cases, protective immunity against challenge with the wild-type tumor can be generated by immunization with the transfected tumor, indicating the generation of a memory response. In contrast, curative immunity, resulting in the rejection of already established tumors by immunization with transfected tumors, has only infrequently been observed (LINSLEY et al. 1991; TEPPER and MULE 1994). Apparently, the cytotoxic activity generated by transfected tumor cells can be readily suppressed by an already established tumor. This phenomenon may be similar to the observation (NORTH and DIGIACOMO 1986) that small tumors initially generate an immune response that appears to become suppressed when the tumor reaches a certain size.

4 Generation of Tumor-Specific Cytotoxic Activity

Cytotoxic mechanisms of CTLs and NK cells have been intensively studied and debated over the last decade. It is universally accepted that multiple killing pathways exist. Figure 2 shows in schematic form the cytolytic pathways uncovered to date (for review PODACK 1991). Secretory (HENKART et al. 1984; PODACK and KONIGSBERG 1984; MASSON and TSCHOPP 1985) and nonsecretory pathways are engaged in target cell lysis. The nonsecretory pathways utilize the Fas-ligand (SUDA et al. 1993; ROUVIER et al. 1993) and membrane tumor necrosis factor (TNF) (RATNER and CLARK 1993) or lymphotoxin β (BROWNING et al. 1993) on the killer cell. Appropriate receptors, TNF receptor I and Fas, on the target cell obviously are required to render these killing mechanisms effective. The Fas-ligand-mediated killing mechanism is the only pathway that is operative in the absence of Ca^{2+} (ROUVIER et al. 1993; TRENN et al. 1989; OSTERGAARD et al. 1987). Its contribution in CTL killing ranges from 5% to 30% of the total cytotoxic activity, depending on the target.

Constitutive secretory killing mechanisms are expected to be sensitive to protein synthesis inhibitors (JU et al. 1990). The exception to this rule is the release of membrane TNF or lymphotoxin (LT) from the killer cell probably by proteolysis. In contrast, inducible secretion of preformed granules is not affected by protein synthesis blockers. Unlike constitutive secretion, however, inducible secretion requires Ca^{2+} signals and is inhibited by cyclosporin A (LANCKI et al. 1989; JU 1991). The uniquely intense killing activity of perforin may be attributed to its synthesis and storage in granules in sufficient quantity prior to the lytic event. Perhaps more importantly, perforin does not require specific proteinaceous receptors on the target membrane for its action. Ca^{2+} and phospholipids are the only requirements for perfroin binding, insertion, polymerization and trans-

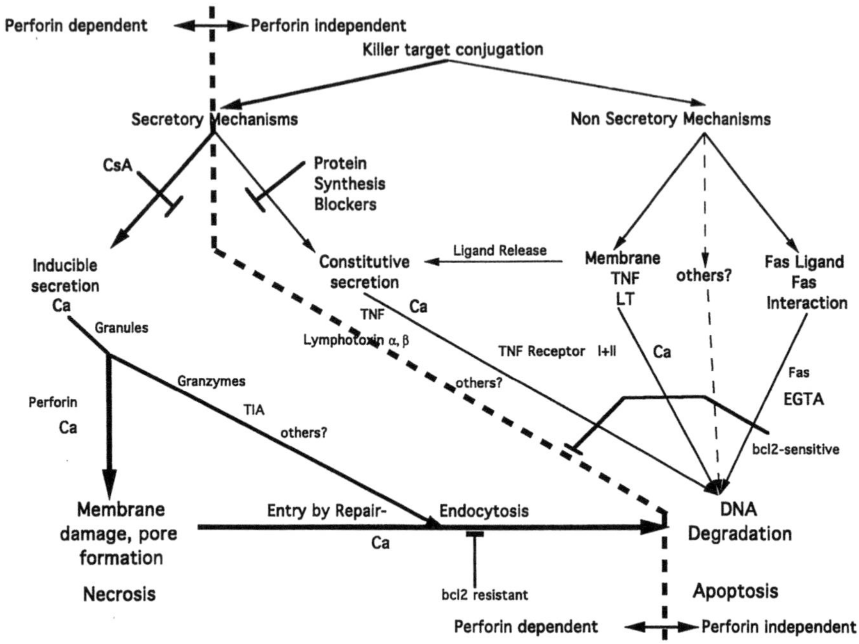

Fig. 2. Cytotoxic pathways mediated by cytotoxic T lymphocytes (CTLs). The *bold dashed line* separates perforin-dependent from perforin-independent pathways. It is interesting to note that all theoretical possibilities of mediating cytotoxicity by secretory and nonsecretory pathways have been implemented and are used by CTLs. *CsA*, cyclosporin A; *Ca*, calcium; *TNF*, tumor necrosis Factor; *LT*, lymphotoxin

membrane channel formation (Podack et al. 1985, 1988; Tschopp et al. 1989). The target cell cannot evade perforin attack by the down-regulation of membrane receptors as it might do for all the other cytolytic pathways.

Finally, perforin comprises the only pathway that permeabilizes the target membrane by the creation of transmembrane pores of varying size. This property is crucial for the uptake of other molecules by repair endocytosis (Dennert and Podack 1983; Podack 1991) In particular the uptake of granzymes, which constitute about 80% of the total granule protein, is important for subsequent ensuing DNA degradation (Shi et al. 1992; Shiver et al. 1992; Heusel et al. 1994). Although the details of the mechanism of this apoptotic death are not understood, it is clear that it makes an important contribution to the total perforin dependent pathway (Duke and Cohen 1988; Russell 1983).

Data from perforin deficient mice indicate a crucial importance of perforin in the biology of the cytolytic responses to intracellular pathogens (Kagi et al. 1994) Table 1. Viral infections that are cleared by normal immunocompetent mice through the activation of CD8$^+$ CTLs cause lethal infections in perforin-deficient mice. Similarly, tumors that are normally rejected grow and kill perforin-deficient mice. As a corollary of this lack of in vivo cytotoxic function, cytotoxicity in vitro is

Table 1. Biological effects of perforin deficiency in mice

Impaired	
In vivo	Viral clearance absent
	Tumor rejection reduced
	Local inflammatory response impaired
In vitro	NK cytolytic activity absent
	CTL MHC restricted cytolytic activity to fibroblasts and epithelial cells absent
	CTL MHC restricted cytolytic activity to hemopoietic cells strongly reduced
	LAK activity absent
Unimpaired	
In vivo	Thymic selection and differentiation
	Peripheral T cell distribution
	T cell activation and proliferation
	Pregnancy and fertility
In vitro	CTL activity in redirected lysis (CD3ε)
	LAK activity in redirected lysis reduced
	Non-MHC-restricted Fas-mediated cytolytic activity

CTL, cytotoxic T lymphocyte; NK, natural killer cell.

also diminished. Virtually no MHC-restricted CTL activity and no NK activity is detectable in perforin-deficient lymphocytes, whereas redirected lysis using anti-ε antibodies to the T cell receptor seems unimpaired (Spielman et al. 1994, unpublished). In addition, the Fas-mediated, MHC nonrestricted killing mechanisms, operating in the presence of EGTA, are normal.

The studies in perforin-deficient mice and with perforin-deficient killer cells leave little doubt about the essential role of perforin for $CD8^+$ CTL-mediated cytolytic function in vivo. The importance of perforin for $CD4^+$ T cell-mediated effector functions in vivo and in vitro remains to be investigated.

It is of interest that cytolytic functions other than perforin can be elicited in perforin-deficient CTL and LAK cells via the system of redirected lysis using the anti-ε-antibody to the T cell receptor. This finding confirms previous studies indicating several pathways of cytolysis (Podack 1991; Ostergaard and Clark 1987; Ju 1991). Perforin-independent pathways in CTLs are: the Fas Fas-ligand pathway, which is MHC unrestricted, CD3-independent and Ca^{2+}-independent; a Ca^{2+}-dependent, protein synthesis-independent, cyclosporin A-resistant pathway. The molecular mediators of these pathways have not been defined. In spite of the existence of perforin-independent pathways, the above cited in vivo experiments in perforin-deficient mice demonstrate the crucial essence of perforin, and presumably granule exocytosis, in physiological and pathophysiological situations.

For the purposes of immunotherapy of cancer using cytolytic killing mechanisms it would therefore seem important to impose a regimen during T cell activation by gene transfer to tumor cells that results in the strong induction of perforin. Table 2 lists inducers of perforin in T cells (Lu et al. 1992; Salcedo et al. 1993; Smyth et al. 1990a, b; Lichtenheld et al. 1988). Gene transfer of the cDNAs of any of these inducers to tumor cells for the purpose of immunization would

Table 2. Inducers and suppressors of perforin in T cells

Inducers
Ca^{2+} + accessory signal
TCR + CD28
IL-2
IL-2 + IL-6
IL-12
Suppressors
TGF-β

TCR, T cell receptor; IL, interleukin; TGF, transforming growth factor.

appear to be potentially useful for tumor prevention and tumor therapy. The benefit of IL-2 and B7 (CD80) transfer in many tumor models may at least in part be explained through this mechanism.

5 Cytotoxic T Lymphocyte Suppression or Apoptosis

An alternating clonal expansion and contraction of specific immune T-cells is essential for a functioning immune system. Thus, during an antigenic challenge, especially with a rapidly replicating pathogen, it is essential that specific and effective effector cells expand rapidly and control the pathogen before it overwhelms the host. The example of Lymphocytic Choriomeningitis (LCM) virus elimination in normal mice as compared to perforin-deficient mice (KAGI et al. 1994) indicates that expansion of specific cells without the effector molecule perforin is not sufficient to clear the virus. Lack of cytotoxicity in specific cells despite appropriate proliferation thus may constitute one form of tolerance or active suppression. Lack of T cell expansion of premature elimination may constitute another important mechanism for T cell tolerance induced by tumors.

T cell expansion and contraction is largely controlled by two genes whose products seem to oppose each other (SENTMAN et al. 1991; WATANABE-FUKUNAGA et al. 1992). Whereas Fas is a mediator of apoptosis of T cells (TAKHASHI et al. 1994; TRAUTH et al. 1989), bcl2 acts to protect T cells from premature cell death in the absence of IL-2 (NUNEZ et al. 1990; DENG and PODACK 1993). In addition bcl2 can protect target cells from Fas-mediated killing via the Fas-ligand on CTLs (Lee et al., unpublished). In agreement with these observations Fas deficiency and bcl2 deficiency cause major disturbances in the balance of proliferation and apoptosis of peripheral T cells. Whereas T cell development is apparently normal in these deficiencies, drastic changes in the peripheral T cell repertoire develop during adolescence and adulthood of deficient mice (Fig. 3). Fas deficiency causes uncontrolled T cell expansion resulting in a lymphoproliferative disorder associated with autoimmunity. In contrast bcl2 deficiency is associated with fulminant apoptosis and disappearance of T cells from the periphery.

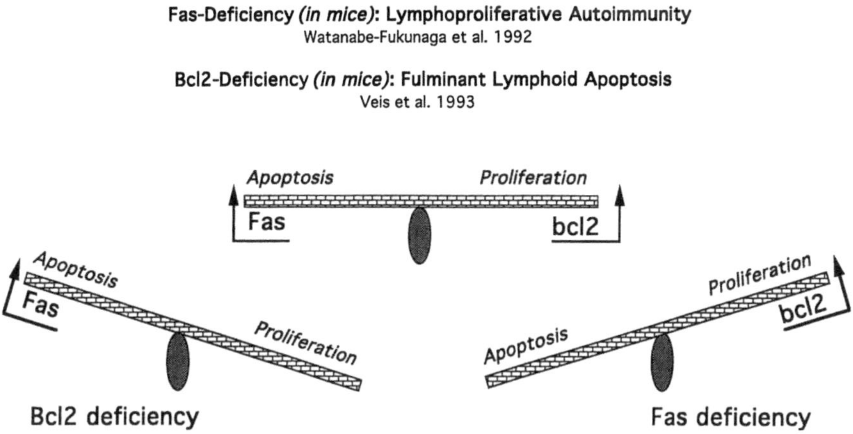

Fig. 3. The roles of Fas and bcl2 in the expansion and contraction of T cells during and following an immune response

Whether tumors are capable of influencing the process of T cell apoptosis by inappropriate T cell activation and subsequent apoptosis remains to be seen. T cell expansion is accompanied by the expression or up-regulation of late T cell activation antigens, e.g., CD30 (Bowen et al. 1993) and Fas (Fig. 4). It appears that these membrane receptors, belonging to the new family of nerve growth factor receptor homologues, deliver important signals for negative growth control and function. It is noteworthy that activated T cells express both Fas and the Fas-ligand (Takahashi et al. 1994; Watanabe-Fukunaga et al. 1992). Hence they are equipped to eliminate each other in an MHC-unrestricted fashion (Fig. 5). This form of fratricide could lead to a severe depression of responsiveness. In an effective immune response down-regulatory signals leading to T cell loss must be suppressed until the immune challenge has been eliminated. This is clearly not the case for growing tumors.

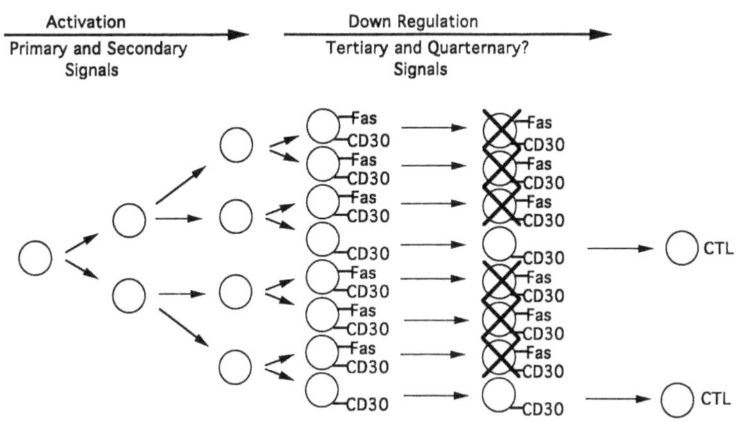

Fig. 4. The postulated role of the late T cell activation antigens Fas and CD30 in elimination of activated T cells clonal expansion activates signals for down-regulation

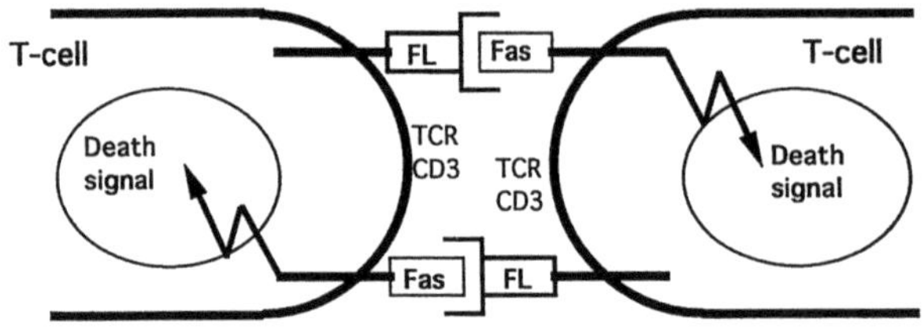

Fig. 5. Expression on activated T cells of both Fas and Fas-ligand may serve to down regulate T cell expansion by fratricide

Effective gene transfer immunotherapy therefore may not only require genetic engineering of tumor cells to enable the induction of a T cell response, but also gene transfer to responding T cells in order to maintain their reactivity and cytotoxicity until the tumor is eliminated. Several strategies towards this goal are underway and await evaluation.

Acknowledgements. The work by the author was supported by NIH grants CA39201, NCI CA59351 and NCI CA57904.

References

Bowen MA, Olsen KJ, Cheng L, Avila D, Podack ER (1993) Functional effects of CD30 on a large granular lymphoma cell line, YT. J Immunol 151: 5896–5906

Browning JL, Ngam-ed A, Lawton P, DeMarinis J, Tizard R, Chow EP, Hession C, O'Brine-Greco B, Foley SF, Ware CR (1993) Lymphotoxin β, a novel member of the TNF family that forms a heteromeric complex with lymphotoxin on the cell surface. Cell 72: 847–856

Colombo M, Forni G (1994) Cytokine gene transfer in tumor inhibition and tumor therapy: where are we now? Immunol Today 15: 48–51

Deng G, Podack ER (1993) Suppression of apoptosis in a cytotoxic T-cell line by interleukin 2-mediated gene transcription and deregulated expression of the protooncogene bcl-2. Immunology 90: 2189–2193

Dennert G, Podack ER (1983) Cytolysis by H-2-specific T killer cells. Assembly of tubular complexes on target membranes. J Exp Med 157: 1483–1495

Duke RC, Cohen JJ (1988) The role of nuclear damage in lysis of target cells by cytotoxic T lymphocytes. In: Podack ER (ed) Cytolytic lymphocytes and complement: effectors of the immune system, vol 2. CRC Press, Boca Raton, 235 pp

Fearon ER, Pardoll DM, Itaya T, Golumbek P, Levitsky HI, Simons JW, Karasuyama H, Vogelstein B, Frost P (1990) Interleukin-2 production by tumor cells bypasses T- helper function in the generation of an antitumor response. Cell 60: 397

Gansbacher B, Zier B, Daniels B, Cronin K, Banjerji R, Gilboa E (1990) Retroviral vector-mediated gamma-interferon gene transfer into tumor cells generates potent and long lasting antitumor immunity. Cancer Res 50: 7820–7825

Henkart PA, Millard PJ, Reynolds CW, Henkart MP (1984) Cytolytic activity of purified cytoplasmic granules from cytotoxic rat large granular lymphocyte tumors. J Exp Med 160: 75–93

Heusel JW, Wesselschmidt RL, Shresta S, Russell JH, Ley TJ (1994) Cytotoxic lymphocytes required granzyme B for the rapid induction of DNA fragmentation and apoptosis in allogeneic target cells. Cell 76: 977–987

Ju S (1991) Distinct pathways of CD4 and CD8 cells induce rapid target DNA fragmentation. J Immunol 146: 812–818

Ju S-T, Ruddle NH, Strack P, Dorf ME, DeKruyff RJ (1990) Expression of two distinct cytolytic mechanisms among murine CD4 subsets. J Immunol 144: 23–31

Kagi D, Ledermann B, Burki K, Seller P, Odermatt B, Olsen KJ, Podack ER, Zinkernagel RM, Hengartner H (1994) Cytotoxicity mediated by T cells and natural killer cells is greatly impaired in perforin-deficient mice. Nature 369: 31–37

Lancki DW, Kaper BP, Fitch FW (1989) The requirements for triggering of lysis by cytolytic T lymphocyte clones. II. Cyclosporin A inhibits TCR-mediated exocytosis but only selectively inhibits TCR-mediated lytic activity by cloned CTL. J Immunol 42: 416–424

Lichtenheld MG, Olsen KJ, Lu P, Lowrey DM, Hameed A, Hengartner H, Podack ER (1988) Structure and function of human perforin. Nature 335: 448–451
Linsley PS, Brady W, Grosmaire L, Aruffo A, Damle NK, Ledbetter JA (1991) Binding of the B cell activation antigen B7 to CD28 constimulates T cell proliferation and interleukin-2 mRNA accumulation. J Exp Med 173: 721–730
Lu P, Garcia-Sanz JA, Lichtenheld MG, Podack ER (1992) Perforin expression in human peripheral blood mononuclear cells. Definition of an IL-2-independent pathway of perforin induction in $CD8^+$ T cells. J Immunol 148: 3354
Masson D, Tschopp J (1985) Isolation of a lytic, pore-forming protein (perforin) from cytolytic T-lymphocytes. J Biol Chem 260: 9069–9072
North RJ, DiGiacomo A (1986) Generation and suppression of the immune response to immunogenic tumors. In: Steinman RM, North RJ (eds) Mechanisms of host resistance to infectious agents, tumors, and allografts. A conference in recognition of the Trudeau Institute Centennial. Rockefeller University Press, New York, pp 387–396
Nunez G, London L, Hockenbery D, Alexander M, McKearn JP, Korsmeyer SJ (1990) Deregulated bCl2-gene expression selectively prolongs survival of growth factor-deprived hemopoietic cell lines. J Immunol 144: 3602–3810
Ohe Y, Podack ER, Olsen KJ, Miyahara Y, Ohira T, Miura K, Nisho K, Saijo N (1993) Combination effect of vaccination with IL-2 and IL-4 cDNA transfected cells on the induction of a therapeutic immune response against Lewis lung carcinoma cells. Int J Cancer 53: 432–437
Ostergaard H, Clark WR (1987) The role of Ca^{2+} in activation of mature cytotoxic T-lymphocytes for lysis. J Immunol 139: 3573–3579
Ostergaard HL, Kane KP, Mescher MF, Clark WR (1987) Cytotoxic T lymphocyte mediated lysis without release of serine esterase. Nature 330: 71–72
Podack ER (1991) T-Cell effector functions: mechanisms for delivery of cytotoxicity and help. Annu Rev Cell Biol 7: 479–504
Podack ER, Konigsberg PJ (1984) Cytolytic T-cell granules: isolation, structural, biochemical and functional characterization. J Exp Med 160: 695–710
Podack ER, Lowrey DM, Lichtenheld M, Hameed A (1988) Function of granule perforin and esterases in T cell-mediated reactions. In: Batisto JR, Plate J, Shearer G(eds) Cytotoxic T cells. Biology and relevance to disease. 532: 292–302
Podack ER, Young JDE, Cohn ZA (1985) Isolation and biochemical and functional characterization of perforin I from cytolytic T-cell granules. Proc Natl Acad Sci USA 82: 8629–8633
Ratner A, Clark WR (1993) Role of TNF-α in $CD8^+$ cytotoxic T iymphocyte-mediated lysis. J Immunol 150: 4303–4314
Rouvier E, Luciani M-F, Golstein P (1993) Fas involvement in Ca^{2+} -independent T cell-mediated cytotoxicity. J Exp Med 177: 195–200
Russell JH (1983) Internal disintegration model of cytotoxic lymphocyte-induced target damage. Immunol Rev 72: 97–117
Salcedo T W, Azzoni L, Wolf SF, Perussia B (1993) Modulation of perforin and granzyme messenger RNA expression in human natural killer cells. J Immunol 151: 2511–2520
Sentman CI, Shutter JR, Hockenbery D, Kanagawa O, Korsmeyer SJ (1991) bcl-2 inhibits multiple forms of apoptosis but not negative selection in thymocytes. Cell 67: 879–888
Shi L, Kam C-M, Powers JC, Aebersold R, Greenberg AH (1992) Purification of three cytotoxic lymphocyte granule serine proteases that induce apoptosis through distinct substrate and target cell interactions. J Exp Med 176: 1521–1529
Shiver JW, Su L, Henkart PA (1992) Cytotoxicity with target DNA breakdown by rat basophilic leukemia cells expressing both cytolysin and granzyme A. Cell 71: 315–322
Smyth MJ, Ortaldo JR, Bere W, Yagita H, Okumura K, Young HA (1990a) IL-2 and IL-6 synergyze to augment the pore-forming protein gene expression and cytotoxic potential of human peripheral blood T cells. J Immunol 145: 1159
Smyth MJ, Ortaldo JR, Shinkai Y-I, Yagita H, Nakata M, Okumura K, Young HA (1990b) Interleukin-2 induction of pore forming protein gene expression in human peripheral blood $CD8^+$ T cells. J Exp Med 171: 1269
Suda T, Takahashi T, Goldstein P, Nagata S (1993) Molecular cloning and expression of the Fas ligand, a novel member of the tumor necrosis factor family. Cell 75: 1169–1178
Takahashi T, Tanaka M, Brannan CI, Jenkins NA, Copeland NG, Suda T, Shigekazu N (1994) Generalized Lymphoproliferative disease in mice, caused by a point mutation in the Fas ligand. Cell 76: 969–976
Tepper RI, Mule JJ (1994) Experimental and clinical studies of cytokine genemodified tumor cells. Hum Gene Ther 5: 153–164
Tepper RI, Pattengale PK, Leder P (1989) Murine interleukin-4 displays potent antitumor activity in vivo. Cell 57: 503

Trauth BC, Klas C, Peters AM, Matzku S, Moller P, Falk W, Debatin K-M, Krammer PH (1989) Monoclonal antibody-mediated tumor regression by induction of apoptosis. Science 245: 301–305
Trenn G, Takayama H, Sitkovsky MV (1989) Exocytosis of cytolytic granules may not be required for a target cell lysis by cytotoxic T-lymphocytes. Nature 330: 72–74
Tschopp J, Schafer S, Masson D, Peitsch MC, Heusser C (1989) Phosphorylcholine acts as a Ca^{2+}-dependent receptor molecule for lymphocyte perforin. Nature 337: 272–274
van der Bruggen P, Traversari C, Chomez P, Lurquin C, De Plaen E, Van den Eynde B, Knuth A, Boon T (1991) A gene encoding an antigen recognized by cytolytic T-lymphocytes on a human melanoma. Science 254: 1643
Watanabe-Fukunaga R, Brannan CI, Copeland NG, Jenkins NA, Nagata S (1992) Lympho-proliferation disorder in mice explained by defects in Fas antigen that mediates apoptosis. Nature 356: 314–317

TIA-1: Structural and Functional Studies on a New Class of Cytolytic Effector Molecule

P. ANDERSON

1 Introduction

Natural killer (NK) cells and cytotoxic T cells (CTLs) can recognize and eliminate a wide variety of virus-infected or transformed target cells (RITZ et al. 1988; TRINCHIERI 1989). Although the surface receptors expressed by these effector cells are distinct, they appear to use similar cytolytic mechanisms in the elimination of their respective target cells (HENKART 1985; PODACK et al. 1991; TSCHOPP and NABHOLZ 1990; YOUNG and LIU 1988). Both NK cells and CTLs are characterized by their inclusion of cytoplasmic granules that are released in response to target cell recognition. Cytolytic lymphocyte granules are morphologically complex organelles that can either fuse with the plasma membrane, like secretory granules, or with phagosomes, like primary lysosomes (BURKHARDT et al. 1990; PETERS et al. 1991). Although cytotoxic granules are somewhat heterogeneous in structure, they typically contain one or more dense cores which are surrounded by multiple small internal vesicles (BURKHARDT et al. 1990; PETERS et al. 1991). The dense cores are the major reservoirs of secretory proteins such as perforin, serine proteases, and chondroitin sulfate proteoglycans (BURKHARDT et al. 1990). Smaller amounts of these proteins are found in a multivesicular domain of the cytotoxic granule which contains lysosomal proteins such as lamp-1, lamp-2, lgp-120, acid

Division of Tumor Immunology, Dana Farber Cancer Institute, 44 Binney Street, Boston, MA 02115, USA phosphatase and cathepsin D (Burkhardt et al. 1990). The multivesicular domain is made up of small internal vesicles that express T cell receptors and adhesion receptors on their outer surfaces (Peters et al. 1989). Because granule exocytosis releases these vesicles into the cleft between effector cells and target cells, surface expression of T cell receptors (TCRs) has been postulated to allow a directed fusion with the target cell plasma membrane, effecting the delivery of putative cytolytic effector molecules including perforin and the serine proteases (Peters et al. 1990).

Perforin is a 68 kDa granule-associated protein that has been shown to be directly cytolytic (Hameed et al. 1989; Lichtenheld et al. 1988; Shiver and Henkart 1991). In the presence of Ca^{2+}, perforin inserts into target cell membranes, where it polymerizes to form nonspecific ion channels through which markers of intracellular compartments can readily pass (Tschopp et al. 1989; Young et al. 1986; Yue et al. 1987). The formation of these ion channels appears to be sufficient to induce the lysis of certain cell types. The recent demonstration that transfection of perforin cDNA into rat basophilic leukemia (RBL) cells confers the ability to lyse erythrocytes via a regulated secretory mechanism, supports a direct role for perforin in lymphocyte-mediated cytolysis (Shiver and Henkart 1991). The inability of perforin-transfected RBL cells to efficiently lyse nucleated cells, however, suggests that additional granule components are required for optimal lymphocyte-mediated killing. That perforin is not the only cytolytic effector molecule is supported by the ability of NK cells and CTLs to kill some target cells in the absence of extracellular Ca^{2+}, which is required for perforin activity (Tirosh and Berke 1985). Furthermore, cytolytic lymphocytes which express little or no perforin (e.g., $CD4^{+}$ CTL clones) have been shown to be potent cytolytic effector cells (Takayama et al. 1991). These results imply that perforin-independent cytolytic effector mechanisms contribute to at least some forms of target cell killing.

In addition to perforin-mediated lysis, CTLs have been shown to induce in target cells an endogenous pathway of programmed cell death known as apoptosis (Russell 1983). Characterized morphologically by nuclear condensation and fragmentation (Kerr et al. 1972), a convenient marker of apoptotic death is the digestion of genomic DNA into integer multiples of a 200 bp nucleosome-sized monomer (Duke et al. 1983; Wyllie 1980). The resulting "ladder" of DNA fragments is considered to be characteristic of this programmed suicide pathway. The observation that purified perforin induces cell lysis, but not DNA fragmentation (Duke et al. 1989), suggests that another granule component is likely to be responsible for the induction of programmed cell death. The granzymes, a family of granule-associated serine proteases, are candidate perforin-independent cytolytic effector molecules (Masson and Tschopp 1987; Pasternack and Eisen 1985). Although purified granzymes are not directly cytotoxic, the ability of protease inhibitors to block lymphocyte-mediated cytolysis suggests that they play a role in target cell killing (Lavie et al. 1985; Rodgers et al. 1988). The observation that granzyme A, the most abundant granule-associated serine protease, can induce DNA fragmentation in detergent permeabilized EL4 cells argues that these molecules might contribute to the induction of apoptosis in CTL targets (HAYES et al. 1989). In addition, the combination of perforin and either granzyme A or granzyme B has been shown to trigger DNA fragmentation in target cells (SHI et al. 1992, 1994). The physiologic relevance of these findings is underscored by the observation that RBL transformants expressing both perforin and granzymes trigger DNA fragmentation in target cells via a regulated secretory pathway (NAKAJIMA and HENKART 1994; SHIVER et al. 1992). Because RBL transformants expressing perforin and granzymes trigger apoptosis less efficiently that CTLs, it is likely that other mechanisms also contribute to cell-mediated apoptosis. Several cytotoxic stimuli other than granzyme A can also induce apoptosis in target cells. Lymphotoxin and tumor necrosis factor are soluble mediators of apoptosis (MARTZ and HOWELL 1989), whereas interactions between the cell surface molecules Fas and Fas ligand can trigger cell-mediated apoptosis (ROUVIER et al. 1993; SUDA and NAGATA 1994). Although the relative importance of each of these pathways in CTL-triggered apoptosis is not known, it is likely that various external stimuli can converge on a common intracellular pathway that results in the morphologic and biochemical changes ascribed to apoptotic death.

2 Identification of TIA-1

We have identified another cytotoxic granule-associated protein that is a candidate toxin involved in triggering apoptotic death in CTL target cells. TIA-1 is a 15 kDa cytotoxic granule-associated protein whose cytoplasmic expression is restricted to CTLs and NK cells (ANDERSON et al. 1990). Upon cellular activation with mitogens, antibodies reactive with the TCR complex, or cytokines, the expression of TIA-1 increases dramatically (ANDERSON et al. 1990; CESANO et al. 1993). In activated T cells, immunoreactive TIA-1 isoforms are induced which migrate at 28 kDa (p28-TIA-1), 40 kDa (p40-TIA-1) and 53 kDa (p53-TIA-1). The migration of p28-TIA-1 and p53-TIA-1 on two-dimensional nonreducing/reducing diagonal gels suggests that these species are composed of a p15-TIA-1:p15-TIA-1 homodimer and a p15-TIA-1: p40-TIA-1 heterodimer, respectively (unpublished observations). Molecular cloning of a cDNA encoding p40-TIA-1 revealed it to be a new member of the RRM (RNA recognition motif) family of RNA-binding proteins (TIAN et al. 1991). The 90 amino acid RRM motif confers the ability to bind to single-stranded RNA or DNA. RRM family members are modular proteins composed of one or more RRM motifs joined to one or more charged protein interaction domains (BANDZIULIS et al. 1989; MATTAJ 1989). The structural and functional diversity of these proteins is truly extraordinary (Fig. 1). For example, the polyadenylate binding protein (PABP) has been shown to regulate the size of poly(A) tails on mRNAs by binding to, and activating, an endogenous ribonuclease (SACHS and DEARDORFF 1992). The Sex lethal protein helps to determine the sex of Drosophila by regulating the alternative splicing of the transformer RNA (another

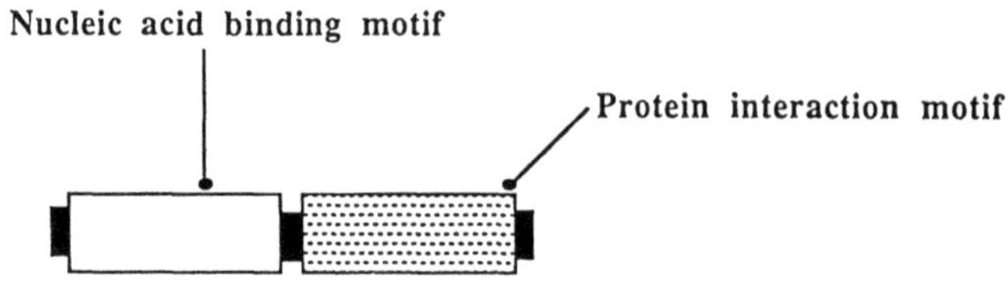

RRM member	#RRM motifs	Localization	Proposed function
PABP	4	Cytoplasm Nucleus	Regulates poly(A) size
Sex lethal	2	Nucleus	Sex-specific RNA splicing
La	1	Nucleus Cytoplasm	RNA pol III termination Translational regulator
PTB	1	Nucleus Cytoplasm	RNA-splicing Translational regulator
MTS1	1	Mitochondria	Mitochondrial targeting
Hel-N1	3	Cytoplasm	mRNA stability
Nucleolin	4	Cytoplasm Nucleus	RNA transport

Fig. 1. RNA recognition motif (RRM) proteins. *PABP*, poly(A) binding protein; *La*, autoantigen originally identified by a patient sera designated La; *PTB*, polypyrimidine tract binding protein; *MTS1*, mitochondrial targeting suppressor 1; *Hel-N1*, human elav-like neuronal protein 1

RRM-type RNA binding protein) (Bell et al. 1988; Inoue et al. 1990). One function of the nuclear autoantigen La is to facilitate termination of RNA polymerase III transcripts (Gottlieb and Steitz 1989; Mathews and Francoeur 1984), but following poliovirus infection, La is recruited to the cytoplasm where it facilitates translation of viral proteins (Meerovitch et al. 1993). The polypyrimidine tract binding protein (PTB) also has more than one function, having been identified as a splicing factor (Mulligan et al. 1992) and as a regulator of protein translation (Hellen et al. 1993). MTS1, a mitochondrial protein, has been shown to be involved in the delivery of cytoplasmic proteins to mitochondia (Ellis and Reid 1993). Hel-N1 is one of several RRM-type proteins that bind to "AUUUUA" destabilizing elements found in the 3' untranslated regions of short lived mRNAs (Levine et al. 1993). Finally, nucleolin is an RRM-type RNA binding protein that regulates RNA polymerase I transcription (Borer et al. 1989). It is also a candidate substrate for granzyme A, the granule-associated effector of CTL killing (Pasternack et al. 1991). TIA-1 is the only known RNA-binding proteins that is delivered to the secretory pathway where it can act as an intracellular effector molecule.

TIA-1 belongs to a subgroup of RRM-type RNA-binding proteins that contain three RRM motifs and glutamine/asparagine rich protein interaction domains (Fig. 2). This group includes two gene products postulated to be involved in neuronal maturation (elav in Drosophila (Robinow et al. 1988) and HuD in humans (Szabo et al. 1991)) and a yeast poly(A)-binding protein (PUB1) (Anderson et al.

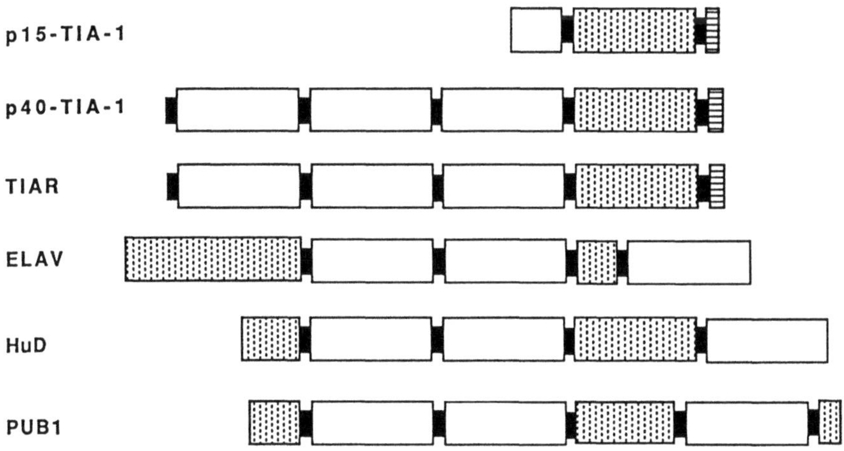

Fig. 2. TIA-1 related RNA binding proteins. *TIA-1*, T cell intracellular antigen 1; TIAR, TIA-1-related; *ELAV*, embryonic lethal, abnormal vision; *HuD*, autoantigen originally identified by patient sera; *PUB1*, poly(U) binding protein 1; *open boxes*, RNA binding motif; *dotted boxes*, glutamine/asparagine rich; *striped boxes*, lysosome targeting motif

1993; MATUNIS et al. 1993). The 15 kDa TIA-1 isoform, the major species present in cytotoxic granules, is identical to the COOH-terminal protein interaction domain of p40-TIA-1. Whereas recombinant p40-TIA-1 has been shown to bind to poly(A) and poly(U)-homopolymers, p15-TIA-1, which does not possess a complete RNA-binding domain, does not bind to nucleic acids. Although TIA-1 does not have a signal peptide or a transmembrane domain, immunoelectron

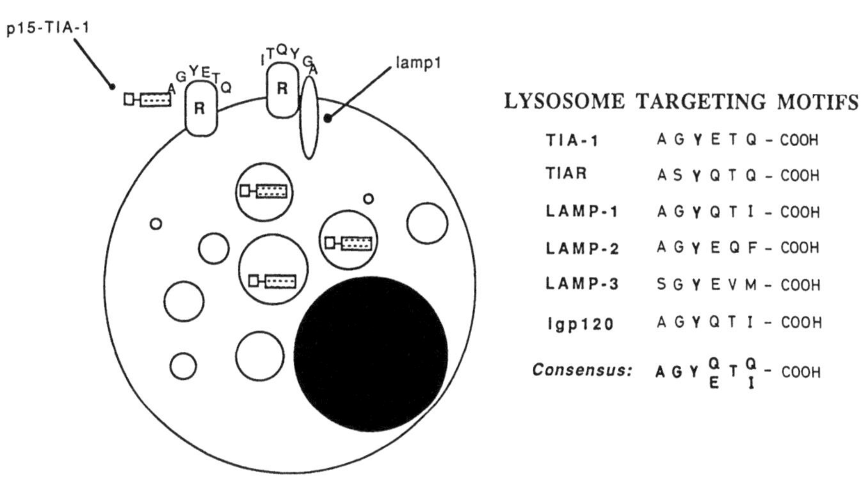

Fig. 3. Targeting of TIA-1 and lamp-1 to cytotoxic granules. *Upper panel* depicts an intermediate granule containing a condensed core (*black circle*) and multiple small vesicles (*open circles*). A putative receptor for the lysosomal targeting motif is labeled *R*. Following receptor binding, p15-TIA-1 is postulated to enter the granule by internal budding from the outer limiting membrane

microscopy has shown it to decorate the cytoplasmic face of the outer limiting membranes of both type I and type II granules (ANDERSON et al. 1990). In type II granules, TIA-1 is also found within the small vesicles that arise by budding inward from the outer limiting membrane (unpublished observations). Both p15-TIA-1 and p40-TIA-1 have a lysosome targeting motif (LTM) in their COOH-terminals that has been shown to deliver lamp-1, lamp-2, and lgp-120 to the granule membrane (FUKUDA 1991). We have proposed that TIA-1 is delivered to the secretory pathway by the mechanism schematized in Fig. 3. After binding to a putative LTM receptor oriented towards the cytoplasmic face of the outer granule membrane, TIA-1 is taken up into the granule by internal budding, placing it in the interstices of small intragranular vesicles. TIA-1-containing vesicles are then secreted following target cell recognition. Thus, TIA-1, with its restricted tissue distribution and subcellular localization, resembles several known or suspected cytolytic effector molecules.

Functional characterization of TIA-1 suggests that it may contribute to target cell death via apoptosis. Purified, recombinant p15-TIA-1 or p40-TIA-1 induces DNA fragmentation in digitonin permeabilized thymocytes and, to a lesser extent, in peripheral blood lymphocytes (TIAN et al. 1991). In the absence of permeabilization, the nucleolytic activity of TIA-1 is markedly reduced, suggesting that it must enter the target cell in order to effect its nucleolytic activity. Localization of TIA-1 to the membranes of cytotoxic granules is consistent with its possible delivery to target cells following granule exocytosis. It will be necessary to confirm the toxic effect of TIA-1 in more physiologic circumstances. To date, we have been unable to stably express p15-TIA-1 in mammalian cells. Whether this results from the inherent toxicity of p15-TIA-1, or the tight regulation of its expression remains to be determined. Given the lack of direct evidence implicating TIA-1 in CTL killing, it will be important to also consider other functions for this granule-associated molecule.

3 Clinical Correlates of TIA-1 Expression

Cytotoxic lymphocytes expressing TIA-1 have been identified in inflammatory infiltrates at sites of immune-mediated tissue distruction. Thus, the infiltating lymphocytes responsible for epidermal cell death in human graft vs host disease are largely $CD8^+$, $TIA\text{-}1^+$ CTLs (SALE et al. 1992). In these cells, TIA-1 protein was shown to be associated with cytoplasmic granules which were oriented towards adjacent target cells. TIA-1 protein was also observed between epithelial cells, implying its specific release following target cell recognition. TIA-1 was similarly observed in lymphocytes infiltrating the intestine in patients with Celiac disease (RUSSEL et al. 1993) and in the CTLs infiltrating B cell follicular lymphomas (LEGER-RAVET et al. 1994). In the latter study, TIA-1 expression correlated well with expression of perforin and granzymes, suggesting that multiple toxins may cooperate in CTL-mediated killing of target cells. Although $TIA\text{-}1^+$ CTLs are abundant in hyperplastic lymph nodes taken from HIV-infected individuals, these cells are not found in patients with clinical AIDS (TENNER-RACZ et al. 1993). This result suggests that an aggresive host CTL response is somehow subverted in the course of HIV infection. Taken together, these results implicate TIA-1⁺ CTLs as important effector cells in both autoimmune disease and viral infection.

4 Structural Organization of TIA-1

TIA-1 was initially identified using a monoclonal antibody reactive with a 15 kDa protein expressed in the granules of cytotoxic lymphocytes (ANDERSON et al. 1990). Immunoselection of a lambda cDNA library derived from activated T cells identified two related cDNAs (λT4T8.9-5 and λ2G9.4) that encoded 15 kDa and 40 kDa proteins corresponding to the granule-associated, and activation-induced species (TIAN et al. 1991). The sequence of cDNAs encoding p15-TIA-1 (λT4T8.9-5) was identical to the 3' region of cDNAs encoding p40-TIA-1 (λ2G9.4), suggesting that they might be partial cDNAs derived from the mRNA encoding p40-TIA-1. In this case, p15-TIA-1 could be produced from p40-TIA-1 by proteolysis, a possibility that was supported by the ability of permeabilized lymphocytes to cleave the 40 kDa isoform to produce a 15 kDa COOH-terminal fragment (TIAN et al. 1991). It was also possible, however, that cDNAs encoding p15-TIA-1 reflected a discrete mRNA species initiated from a second promoter within the TIA-1 gene. Our attempts to identify such a species using S1 mapping analysis have been unsuccessful (KAWAKAMI et al. 1994). This result supports the possibility that p15-TIA-1 is derived from p40-TIA-1 by proteolytic processing.

The TIA-1 gene contains 13 exons separated by 12 introns (Fig. 4) (KAWAKAMI et al. 1994). Sequences corresponding to exon 5 are found in some cDNAs, but not in others, suggesting that it is an alternatively spliced exon. Reverse transcription-Polymerase chain reaction (RT-PCR) analysis indicates that both transcripts are produced in a variety of hematopoeitic and nonhematopoeitic cells (KAWAKAMI et al. 1992). It appears that transcripts including exon 5 may be more prevalent than transcripts without exon 5 (KAWAKAMI et al. 1992). Because the addition of 11 amino acids at the junction between the first and second RNA-binding domains may influence the binding specificity of TIA-1, it will be important to determine the binding specificity of each TIA-1 isoform. Comparison of the intron-exon organization of TIA-1 with that of another RNA-binding protein whose genomic structure is known, nucleolin, reveals several important differences (Fig. 5). The four RNA-binding domains of nucleolin are similarly arranged (SRIVASTAVA et al. 1990). Each RNA-binding domain is made up of two exons of similar size. In each case, the RNP1 sequences are interrupted by an intervening sequence. This uniform organization is not a feature of the TIA-1 gene in which the first RNA binding domain is assembled from 4 exons, the second from 3 or 4 exons, and the third from 3 exons. None of the RNP1 domains are interrupted by an intervening sequence. The strikingly different organization of these structurally and function-

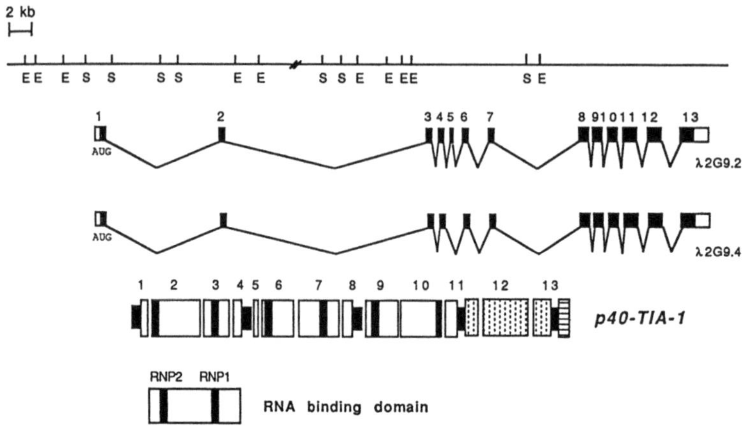

Fig. 4. Intron-exon organization of TIA-1. The relative positions of *Eco*RI (*E*) and *Sac*I (*S*) restriction sites in the genomic DNA are shown on the *upper line*. The relative positions of the 13 exons spliced together to form the indicated cDNAs are as shown. Untranslated regions are shown as *open bars*, translated regions shown as *closed bars*. A schematic representation of the location of all 13 TIA-1 exons relative to the RNA-binding domains, COOH-terminal domain and lysosomal targeting motif of TIA-1 (see Fig. 2) is shown at the *bottom* of the figure

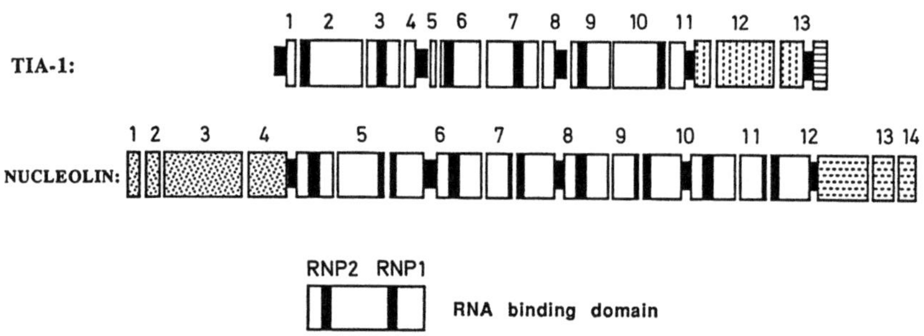

Fig. 5. Comparison of the exon organization of TIA-1 and nucleolin

ally conserved domains is surprising. Although both genes are on human chromosome 2, (nucleolin is on the long arm, and TIA-1 on the short arm), their different structures make it unlikely that they arose from a recent gene duplication. The TIA-1 gene is located at 2p13, near the chromosome 2 breakpoint encoding the immunoglobulin light chain locus in the t(2;8) variant form of Burkitt's lymphoma. It seems unlikely, however, that TIA-1 contributes to cellular transformation in this condition.

5 Phylogeny of TIA-1

The deduced amino acid sequence of the murine homologue of p40-TIA-1 has been determined (Boothby and Anderson, unpublished). There is a remarkable conservation of both amino acid and nucleotide sequence. Both human and murine TIA-1 possess three NH_2-terminal RNA binding domains, a COOH-terminal protein interaction domain, and a LTM. Although functional studies have not yet been done using murine TIA-1, its conserved structure suggests that it will function similarly to human TIA-1. A TIA-1 related gene has recently been identified in Drosophila (Brand and Bourbon 1993). Like human TIA-1, DmTIAR possesses three NH_2-terminal RNA binding domains, a COOH-terminal, glutamine-rich protein interaction domain, and a putative LTM. Two distinct DmTIAR mRNAs were identified which differ in their 3' untranslated regions. The shorter mRNA is expressed in the ovary, and during early embryogenesis (before 4 h). The larger form is first expressed in larvae, and persists in adult flies. Although the function of DmTIAR has not been determined, the striking structural conservation extending from humans to fly implies an important functional role for this molecule.

6 TIAR: A Candidate Substrate for TIA-1

The demonstration that secretory granules from CTLs contain the isolated protein interaction domain of an RNA binding protein suggests a possible mechanism of action for p15-TIA-1. By competing with p40-TIA-1 for binding to a protein substrate, p15-TIA-1 might disrupt the formation of an essential ribonucleoprotein complex. Although the expression of TIA-1 appears to be restricted to CTLs a closely related protein (TIAR) is widely expressed in hematopoetic and nonhematopoetic cells (Kawakami et al. 1992). Like p40-TIA-1, TIAR possesses three NH_2-terminal RNA binding domains and a COOH-terminal protein interaction domain. The RNA binding domains of TIA-1 and TIAR are approximately 90% identical at the amino acid level, whereas the protein interaction domains are 50% identical. Unlike TIA-1, the COOH-terminal protein interaction domain of TIAR does not appear to be processed proteolytically. The structural similarity between TIA-1 and TIAR suggests the possibility that p15-TIA-1 might disrupt the function of TIAR by competing for binding to essential protein substrates. As such, the toxic effects of TIA-1 might be mediated by disrupting the function of TIAR. We have used a genetic screen to identify a cDNA encoding a 70 kDa protein capable of binding to the protein interaction domain of both TIA-1 and TIAR (Tian and Anderson, unpublished). Its deduced amino acid sequence is not significantly related to any protein in the EMBL database. Functional characterization of this candidate substrate for TIA-1 and TIAR might shed light on the molecular mechanism by which TIA-1 contributes to CTL killing.

7 Conclusion

Our results indicate that CTL granules contain a 15 kDa protein derived from the protein interaction domain of an RNA-binding protein. Because the expression of p15-TIA-1 is limited to CTLs, antibodies reactive with TIA-1 have proven to be useful reagents for the recognition of CTLs in patients with inflammatory cellular infiltrates. The subcellular localization of TIA-1 within the small vesicles contained in type II granules suggests a mechanism for its delivery to target cells. Granule exocytosis has been shown to release vesicles from type II granules into the cleft between the effector cell and the target cell. Because these vesicles express T cell receptors and adhesion molecules on their surfaces, they have the potential to bind to target cells in an antigen specific manner. Delivery of TIA-1 to the cytoplasm of the target cell would result if TIA-1 containing vesicles fused with the plasma membrane of the target cell. Alternatively, endocytosis of TIA-1 containing vesicles, followed by vesicle-endosome fusion would similarly result in the delivery of TIA-1 to target cells. Because the nucleolytic activity of TIA-1 requires its delivery to the interior of the cell, vesicle-mediated delivery of TIA-1 would be essential. Although the molecular mechanism of TIA-1 action is not yet known, the identification of candidate substrates that bind to the protein interaction domains of both TIA-1 and TIAR provide a direction for future research. Figure 6 shows a working model of how TIA-1 might be delivered to the secretory pathway, and how it might interact with target cell substrates to contribute to cell-mediated cytotoxicity.

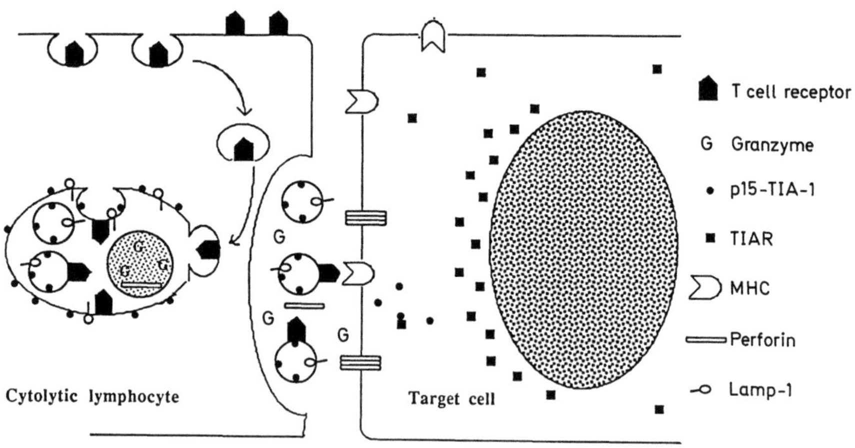

Fig. 6. Delivery of TIA-1 to target cells (From PODACK et al. 1991)

References

Anderson JT, Paddy MR, Swanson MS (1993) PUB1 is a major nuclear and cytoplasmic polyadenylated RNA-binding protein in Saccharomyces cerevidsiae. Mol Cell Biol 13: 6102–6113

Anderson P, Nagler-Anderson C, O'Brien C, Levine H, Watkins S, Slayter HS, Blue M-L, Schlossman SF (1990) A monoclonal antibody reactive with a 15-kDa cytoplasmic granule-associated protein defines a subpopulation of CD8$^+$ T lymphocytes. J Immunol 144: 574–582

Bandziulis RJ, Swanson MS, Dreyfuss G (1989) RNA-binding proteins as developmental regulators. Genes Dev 3: 431–437

Bell LR, Maine EM, Schedl P, Cline TW (1988) Sex-lethal, a Drosophila sex determinaton switch gene, exhibits sex-specific RNA splicing and sequence similarity to RNA binding proteins. Cell 55: 1037–1046

Borer R, Lehner C, Eppenberger H, Nigg E (1989) Major nucleolar proteins shuttle between nucleus and cytoplasm. Cell 56: 379–390

Brand S, Bourbon H (1993) The developmentally-regulated Drosophila gene rox8 encodes an RRM-type RNA binding protein structurally related to human TIA-1-type nucleolysins. Nucleic Acids Res 21: 3699–3704

Burkhardt JK, Hester S, Lapham CK, Argon Y (1990) The lytic granules of natural killer cells are dual-function organelles combining secretory and pre-lysosomal compartments. J Cell Biol 111: 1227–1240

Cesano A, Visonneau S, Clark SC, Santoli D (1993) Cellular and molecular mechanisms of activation of MHC nonrestricted cytotoxic cells by IL-12. J Immunol 151: 2943–2957

Duke RC, Chervenak R, Cohen JJ (1983) Endogenous endonuclease-induced DNA fragmentation: an early event in cell-mediated cytolysis. Proc Natl Acad Sci USA 80: 6361–6365

Duke RC, Persechini PM, Chang S, Liu CC, Cohen JJ, Young JDE (1989) Purified perforin induces target cell lysis but not DNA fragmentation. J Exp Med 170: 1451–1456

Ellis EM, Reid GA (1993) The Saccharomyces cervisiae MTS1 gene encodes a putative RNA-binding protein involved in mitochondial protein targeting. Gene 132: 175–183

Fukuda M (1991) Lysosomal membrane glycoproteins. J Biol Chem 266: 21327–21330

Gottlieb E, Steitz J (1989) Function of the mamalian La protein: evidence for its action in transcription termination by RNA polymerase III. EMBO J 8: 851–861

Hameed A, Osen KJ, Lee MK, Lichtenheld MG, Podack ER (1989) Cytolysis by Ca-permeable transmembrane channels. Pore formation causes extensive DAN-degradation and cell lysis. J Exp Med 169: 765–777

Hayes MP, Berrebi GA, Henkart PA (1989) Induction of target cell DNA release by the cytotoxic T lymphocyte granule protease granzyme A. J Exp Med 170: 933–946

Hellen C, Witherell G, Schmid M, Shin S, Pestova T, Gil A, Wimmer E (1993) A cytoplasmic 57 kDa protein that is required for translation of picornavirus RNA by internal ribosomal entry is identical to the nuclear pyrimidine tract binding protein. Proc Natl Acad Sci USA 90: 7642–7646

Henkart PA (1985) Mechanism of lymphocyte-mediated cytotoxicity. Annu Rev Immunol 3: 31–58

Inoue K, Hoshijima K, Sakamoto H, Shimura Y (1990) Binding of the drosophila sex-lethal gene product to the alternative splice site of transformer primary transcript. Nature 344: 461–463

Kawakami A, Tian Q, Duan X, Streuli M, Schlossman S, Anderson P (1992) Identification and functional characterization of a TIA-1 related nucleolysin. Proc Natl Acad Sci USA 89: 8681–8685

Kawakami A, Tian Q, Streuli M, Poe M, Edelhoff S, Disteche C, Anderson P (1994) Intron-exon organization and chromosomal localization of the human TIA-1 gene. J Immunol (in press)

Kerr J, Whllie A, Currie A (1972) Apoptosis: a basic biological phenomenon with wide-ranging implications in tissue kinetics. Br J Cancer 26: 239–257

Lavie G, Leib Z, Servadio C (1985) The mechanism of human NK cell mediated cytotoxicity. Mode of action of surface-associated proteases in the early stages of the lytic reaction. J Immunol 135: 1470–1476

Leger-Ravet M, Devergne O, Peuchmaur P, Solal-Celigny P, Brousse N, Gaulard P, Galanaud P, Emilie D (1994) In situ detection of activated cytotoxic cells in follicular lymphomas. Am J Pathol (in press)

Levine T, Gao F, King P, Andrews L, Keene J (1993) Hel-N1: an autoimmune RNA-binding protein with specificity for uridylate-rich 3' untranslated regions of growth factor mRNAs. Mol Cell Biol 13: 130–136

Lichtenheld MG, Olsen KJ, Lu P, Lowrey DM, Hameed A, Hengartner H, Podack ER (1988) Structure and function of human perforin. Nature 335: 448–451

Martz E, Howell DM (1989) CTL: virus control cells first and cytolytic cells second? Immunol Today 10: 79–86
Masson D, Tschopp J (1987) A family of serine esterases in lytic granules of cytolytic T lymphocytes. Cell 49: 679–685
Mathews MB, Francoeur AM (1984) La antigen recognizes and binds to the 3'-oligouridylate tail of a small RNA. Mol Cell Biol 4: 1134–1140
Mattaj IW (1989) A binding consensus: RNA-protein interactions in splicing, snRNPs, and sex. Cell 57: 1–3
Matunis MJ, Matunis EL, Dreyfuss G (1993) PUB1: a major yeast poly(A)$^+$ RNA-binding protein. Mol Cell Biol 13: 6114–6123
Meerovitch K, Svitkin Y, Lee H, Lejbkowicz F, Kenan D, Chan E, Agol V, Keene J, Sonenberg N (1993) La autoantigen enhances and corrects aberrent translation of polivirus RNA in reticulocyte lysate. J Virol 67: 3798–3807
Mulligan GJ, Guo W, Wormsley S, Helfman DM (1992) Polypyrimidine tract binding protein interacts with sequences involved in alternative splicing of β-Tropomyosin pre-mRNA. J Biol Chem 267: 25480–25487
Nakajima H, Henkart P (1994) Cytotoxic lymphocyte granzymes tirgger a target cell internal disintigration pathway leading to cytolysis and DNA breakdown. J Immunol 152: 1057–1063
Pasternack, MS, Eisen, HN (1985) A novel serine esterase expressed by cytotoxic T lymphocytes. Nature 314: 743–745
Pasternack MS, Bleier KJ, McInerney TN (1991) Granzyme A binding to target cell proteins: granzyme A binds to and cleaves nucleolin in vitro. J Biol Chem 266: 14703–14708
Peters PJ, Geuze HJ, Van der Donk HA, Slot JW, Griffith JM, Stam NJ, Clevers HC, Borst J (1989) Molecules relevant for T cell-target cell interaction are present in cytolytic granules of human T lymphocytes. Eur J Immunol 19: 1469–1475
Peters PJ, Geuze HJ, Van der Donk HA, Borst J (1990) A new model for lethal hit delivery by cytotoxic T lymphocytes. Immunol Today 11: 28–31
Peters PJ, Borst J, Oorschot V, Fukuda M, Krahenbuhl O, Tschopp J, Slot JW, Geuze HJ (1991) Cytotoxic T lymphocyte granules are secretory lysosomes, containing both perforin and granzymes. J Exp Med 173: 1099–1109
Podack ER, Hengartner H, Lichtenheld MG (1991) A central role of perforin in cytolysis? Annu Rev Immunol 9: 129–157
Ritz J, Schmidt RE, Michon J, Hercend T, Schlossman SF (1988) Characterization of functional surface structures on human natural killer cells. Adv Immunol 42: 181–211
Robinow S, Campos AR, Yao K-M, White K (1988) The elav gene product of drosophila, required in neurons, has three RNP consensus motifs. Science 242: 1570–1572
Rodgers KE, Grayson MH, Ware CF (1988) Inhibition of cytotoxic T lymphocyte and natural killer cell-mediated lysis by O, S, S-trimethyl phoosphorodithioate is at an early post-recognition step. J Immunol 140: 564–570
Rouvier E, Luciani M, Golstein P (1993) Fas involvement in Ca^{2+}– independent T cell-mediated cytotoxicity. J Exp Med 177: 195–201
Russel G, Nagler-Anderson C, Anderson P, Bahn A (1993) Cytotoxic potential of intraepithelial lymphocytes. Presence of TIA-1, the cytolytic granule-associated protein, in human IELs in normal and diseases intestine. Am J Pathol 143: 350–354
Russell, JH (1983) Internal desintigration model of cytotoxic lymphocyte- induced target damage. Immunol Rev 72: 97–118
Sachs AB, Deardorff JA (1992) Translation initiation requires the PAB-dependent poly(A) ribonuclease in yeast. Cell 70: 961–973
Sale G, Anderson P, Browne M, Myerson D (1992) Evidence of cytotoxic T-cell destruction of epidermal cells in human graft-vs-host disease. Arch Pathol Lab Med 116: 622–625
Shi L, Kraut RP, Aebersold R, Greenberg AH (1992) A natural killer cell granule protein that induces DNA fragmentation and apoptosis. J Exp Med 175: 553–566
Shi L, Nishioka WK, Th'ng J, Bradbury EM, Litchfield DW, Greenberg AH (1994) Premature $p34^{cdc2}$ activation required for apoptosis. Science 263: 1143–1145
Shiver J, Su L, Henkart P (1992) Cytotoxicity with target DNA breakdown by rat basophilic leukemia cells expressing both cytolysin and granzyme A. Cell 71: 315–322
Shiver JW, Henkart PA (1991) A noncytotoxic mast cell tumor line exhibits potent IgE-dependent cytotoxicity after transfection with the cytolysin/perforin gene. Cell 64: 1175–1181
Srivastava M, McBride O, Fleming PJ, Pollard H, Burns A (1990) Genomic organization and chromosomal localization of the human nucleolin gene. J Biol Chem 25: 14922–14931

Suda T, Nagata S (1994) Purification and characterization of the Fasligand that induces apoptosis. J Exp Med 179: 873–879

Szabo A, Dalmau J, Manley G, Rosenfeld M, Wong E, Henson J, Posner JB, Furneaux HM (1991) HuD, a paraneoplastic encephalomyelitis antigen, contains RNA-binding domains and is homologous to Elav and sex-lethal. Cell 67: 325–333

Takayama H, Shinohara N, Kawasaki A, Someya Y, Hanaoka S, Kojima H, Yagita H, Okumura K, Shinkai Y (1991) Antigen-specific directional T target cell lysis by perforin-negative T lymphocyte clones. Int Immunol 3: 1149–1156

Tenner-Racz K, Racz P, Thome C, Meyer C, Anderson P, Schlossman S, Letvin N (1993) Cytotoxic effector cell granules recognized by the monoclonal antibody TIA-1 are present in $CD8^+$ lymphocytes in lymph nodes of human immunodeficiency virus-1-infected patients. Am J Pathol 142: 1750–1758

Tian Q, Streuli M, Saito H, Schlossman SF, Anderson P (1991) A polyadenylate binding protein localized to the granules of cytolytic lymphocytes induces DNA fragmentation in target cells. Cell 67: 629–639

Tirosh R, Berke G (1985) T lymphocyte mediated cytolysis as an excitatory process of the target. (1) Evidence that the target may be the site of calcium action. Cell Immunol 75: 113–123

Trinchieri G (1989) Biology of natural killer cells. Adv Immunol 47: 187–376

Tschopp J, Nabholz M (1990) Perforin mediated target cell lysis by cytolytic T lymphocytes. Annu Rev Immunol 8: 279

Tschopp J, Schafer S, Masson D, Peitch MC, Heusser C (1989) Phosphorylcholine acts as a Ca-dependent receptor molecule for lymphocyte perforin. Nature 337: 272–274

Wyllie AH (1980) Glucocorticoid-induced thymoycte apoptosis is associted with endogenous endonuclease activation. Nature 284: 555–556

Young JDE, Liu CC (1988) Multiple mechanisms of lymphocyte-mediated killing. Immunol Today 9: 140–144

Young JDE, Nathan CF, Podack ER, Palladino MA, Cohn Z (1986) Functional channel formation associated with cytotoxic T cell granules. Proc Natl Acad Sci USA 83: 150–154

Yue CC, Reynolds CW, Henkart PA (1987) Inhibition of cytolysin activity in large granular lymphocyte granules by lipids: evidence for a membrane insertion mechanism in lysis. Mol Immunol 24: 647–653

Calreticulin: A Granule-Protein by Default or Design?

R.C. Bleackley, E.A. Atkinson, K. Burns, and M. Michalak

1 Introduction

Calreticulin is a major calcium binding protein normally found in the lumen of the endoplasmic reticulum (ER). When T lymphocytes were activated we observed an increase in the levels of calreticulin mRNA and protein (Burns et al. 1992). Initially we thought this induction might relate to the changes in intracellular Ca^{2+} levels that have been associated with signal transduction, ultimately leading to specific gene transcription and cell activation. However, subsequent experiments revealed that calreticulin itself can also directly influence patterns of steroid hormone-dependent gene expression by binding to the receptor protein (Burns et al. 1994a). This left us with somewhat of a conundrum: in order to influence

Department of Biochemistry, University of Alberta, Edmonton, Alberta, T6G 2H7 Canada gene expression calreticulin would have to find its way from the ER to the nucleus, possibly via the cytoplasm. We were very surprised to discover that, when we immunolocalized calreticulin in activated cytotoxic T cells (CTLs) the major positive organelles were none other than our old friends the cytoplasmic granules. Concomitantly a report appeared on the NH_2-terminal sequence analysis of a 60 kDa granule-associated protein that copurifies with perforin. The peptide was identical to calreticulin (Dupuis et al. 1993).

What is calreticulin doing in an organelle that has been implicated in target cell destruction by CTLs? Is it merely a passenger that is carried along with bulk flow from the ER as granules are formed, or is it a specific component that not only belongs in the granules, but plays an integral part in their biogenesis and perhaps in the killing process itself? The latter possibility is obviously the more interesting one and we believe that evidence is mounting in its favor. In this chapter we will attempt to justify this view, but first, as calreticulin is likely an unfamiliar molecule to most readers of this book, we will review some of the properties and features of this protein that have influenced our considerations of its potential role in cytolysis. For more detailed reviews on the biochemistry of calreticulin the reader is referred to other recent publications (Michalak et al. 1992; Burns et al. 1994b).

2 Calreticulin—A Multifunctional Protein

2.1 A Conserved Calcium Binding Protein

Calreticulin is a 46 kDa protein (60 kDa in SDS-PAGE) with a high proportion of acidic amino acids (109 vs 52 basic), that was discovered in skeletal muscle sarcoplasmic reticulum over 20 years ago (Ostwald and MacLennan 1974). It binds Ca^{2+} with both high affinity/low capacity (K_d = ~1 μM, B_{max} = 1 mole of Ca^{2+}/mole of protein) and low affinity/high capacity (K_d = ~250 μM, B_{max} = 25 moles of Ca^{2+}/mole of protein) (Ostwald and MacLennan 1974; Ostwald et al. 1974; Baksh et al. 1992; Baksh and Michalak 1991; Michalak et al. 1991). Given this high binding capacity, it is believed that calreticulin functions as a Ca^{2+} storage protein in the lumen of the ER (Michalak et al. 1992), and therefore, likely plays an important role in the control of cytoplasmic Ca^{2+} concentrations.

The protein has been detected in a wide variety of different cell types (Michalak et al. 1991, 1992; Opas et al. 1991; Tharin et al. 1992), and cDNAs encoding calreticulin have been isolated from human, mouse, rabbit, rat, *Xenopus, Aplysia, Drosophila, C. elegans and S. mansoni* cDNA libraries (Fliegel et al. 1989 a,b; Kennedy et al. 1992; Liu et al. 1993; Mazzarella et al. 1992; McCauliffe et al. 1990a; Murthy et al. 1990; Nakamura et al. 1993; Rokeach et al. 1991; Smith and Koch 1989; Smith 1992; Treves et al. 1992; Unnasch et al. 1988; Waterston et al. 1992; Hawn et al. 1993). The amino acid sequence of calreticulin has been deduced from the nucleotide sequences of the isolated cDNAs, and all sequences have been found to be extremely similar. Thus, in addition to being ubiquitous, calreticulin is also highly conserved.

2.2 A Three Domain Structure

Structural analysis of the amino acid sequence of calreticulin indicates that the protein can be divided into at least three distinct domains (MICHALAK et al. 1992; BAKSH and MICHALAK 1991; SMITH and KOCH 1989). The N-domain of calreticulin is predicted to form anti-parallel β-strands with a helix-turn-helix motif at the extreme NH_2-terminal. In newly synthesized calreticulin, the NH_2-terminal is extended with a hydrophobic ER signal sequence (FLIEGEL et al. 1989a). The amino acid sequence of the N-domain shows no homology with any other protein sequence in current data bases, and it is also the most conserved domain among all calreticulins so far cloned. The N-domain of calreticulin binds the synthetic peptide KLGFFKR, the sequence of which is derived from the α-subunit of integrin (ROJIANI et al. 1991) and it interacts with the DNA binding domain of the glucocorticoid receptor (BURNS et al. 1994a; DEDHAR et al. 1994). The N-domain of calreticulin also appears to bind Zn^{2+} (Baksh and Michalak, unpublished observations). Although it does not contain a "Zn^{2+} finger" consensus sequence this domain is enriched in Cys and His residues implicated to be involved in Zn^{2+} binding to other proteins (BERG 1990). As yet, the in vivo significance of these observations remains to be determined.

The next domain of calreticulin has been designated the P-domain. This region of the protein is proline-rich and contains three sequence repeats of 17 amino acids (PxxIxDPDAxKPEDWDE). The amino acid sequence of the P-domain is highly conserved between species, and interestingly, an amino acid sequence very similar to this domain of calreticulin has been found in calnexin, an ER membrane protein proposed to function as a chaperone. Functionally, the P-domain binds ~1 mole of Ca^{2+} per mole of protein with a dissociation constant of ~1 μM indicating that this domain contains the high affinity Ca^{2+} binding site of calreticulin (BAKSH and MICHALAK 1991). The most well characterized high affinity Ca^{2+} binding proteins are members of the EF-hand family (KRETSINGER et al. 1988). Although the amino acid sequence of calreticulin contains no EF-hand consensus, the region of the P-domain containing the three repeats is predicted to form a helix-loop-helix motif, similar to that found in EF-hand Ca^{2+} binding sites (KRETSINGER et al. 1988). The P-domain of calreticulin also contains a nuclear localization signal (NLS) which may be relevant to the recent detection of calreticulin in the nucleus (OPAS et al. 1991). The N-and P-domains of calreticulin are the most conserved across different species. They might therefore be the areas of the most fundamental importance.

The C-domain contains 37 acidic residues and binds Ca^{2+} with high capacity/ low affinity. This region probably represents the Ca^{2+} storage site of the protein as it binds >25 moles of Ca^{2+}/mole of protein (BAKSH and MICHALAK 1991). One potential glycosylation site is found in the C-domain (residue 326), along with the ER retention signal sequence KDEL (PELHAM 1989). In addition to this, the C-domain of calreticulin interacts in vitro with a set of ER/SR proteins (BURNS and MICHALAK 1993), as well as with the blood clotting factors IX, X and prothrombin (BENEDICT et al. 1993; KUWABARA et al. 1993). The interactions of the C-domain with these blood clotting factors may be responsible for the recently observed antithrombotic activity of calreticulin.

The C-domain of calreticulin shares limited amino acid sequence similarity with calsequestrin, and also with other proteins resident in the ER (PDI, BiP, endoplasmin) (MICHALAK et al. 1992; FLIEGEL et al. 1989b). Interestingly, the greatest divergence of sequence identity between calreticulins occurs in the COOH–terminal region. In particular, the C-domain of calreticulin from *O. volvulus* is positively, rather than negatively, charged and also does not terminate with the KDEL ER retention signal (UNNASCH et al. 1988).

3 Intracellular Location

Most evidence has indicated that calreticulin is a resident ER protein (MICHALAK et al. 1991; OPAS et al. 1991; THARIN et al. 1992; KENNEDY et al. 1992; SMITH and KOCH 1989; WAISMAN et al. 1985). However, calreticulin may also be found outside of the ER compartment. For example, immunoreactive calreticulin has been detected in the nuclear envelope and in the nucleus of some cell types (OPAS et al. 1991). It has also been proposed that the protein may be localized to the cytoplasm (ROJIANI et al. 1991; CONRAD et al. 1991), but as yet, there is no direct evidence for this. Further, calreticulin has been isolated from human serum (SUEYOSHI et al. 1991), suggesting that it may be secreted into the blood stream.

Other recent evidence that calreticulin is subject to unusual intracellular trafficking was found during studies on spermatogenesis. NAKAMURA et al. (1993) observed calreticulin in spermatocytes, spermatids and Sertoli cells. The protein in these cells is nonglycosylated, binds Ca^{2+}, and has both an NH_2 terminal signal sequence and a COOH-terminal KDEL signal. Immunolocalization revealed that calreticulin was present in the sperm acrosome and, to a lesser extent, in the Golgi. NAKAMURA et al. (1993) proposed that calreticulin is incorporated into the acrosomal vesicles via the Golgi apparatus, during spermatogenesis. Therefore, in order for calreticulin to appear in the cytotoxic granules of CTL or in the acrosomal vesicles of sperm cells, the protein would have to bypass the KDEL receptor and avoid the KDEL signal retention mechanism.

4 Similarity to Other Proteins

4.1 Calnexin

Recently, an integral ER membrane protein named calnexin has been identified as being very similar to calreticulin (reviewed in BERGERON et al. 1994). Human, dog, plant and *S. mansoni* cDNAs encoding calnexin have been isolated and the amino acid sequence deduced from the nucleotide sequences (HAWN et al. 1993; WADA et al. 1991; HUANG et al. 1993; DE VIRGILIO et al. 1993). The sequences reveal that calnexin shares several regions of similarity with calreticulin, ranging from 42% to 78% identity (WADA et al. 1991). Interestingly, several of the sequence features that are highly conserved among calreticulin species are present in calnexin. The most striking of these conserved sequences is in the *P-domain* of calreticulin; two of three repeats KPEDWD found in the *P-domain* of calreticulin are entirely preserved in calnexin.

The sequence similarity between calnexin and calreticulin suggests that these distinct ER proteins may have common functions. Calnexin associates with several different multimeric protein complexes, including MHC class I, the T cell receptor (TCR), and membrane immunoglobulin (AHLUWALIA et al. 1992; GALVIN et al. 1992; HOCHTENBACH et al. 1992; OU et al. 1992). As a result of this, it has been proposed to act as a chaperone in the assembly of these complexes (AHLUWALIA et al. 1992; GALVIN et al. 1992; HOCHTENBACH et al. 1992; OU et al. 1992). OU et al. (1993) showed that calnexin associates transiently and selectively with newly synthesized glycoproteins and proposed that it might, therefore, play a role in the synthesis and secretion of glycoproteins. Calreticulin has been shown to interact with a number of ER/SR proteins, and recently NIGAM et al. (1994) demonstrated that it may indeed function as an ER membrane, Ca^{2+}- and ATP-dependent chaperone.

4.2 Autoantigens

Calreticulin has been suggested to be the Ro/SS-A antigen (MCCAULIFFE et al. 1990a,b; COLLINS et al. 1989; LIEU et al. 1988, 1989), however, this has been disputed by other studies (ROKEACH et al. 1991). Antibodies directed against Ro/SS-A antigen(s) are found in the majority of patients with Sjögren's syndrome and with systemic lupus erythematosus, and occur in mothers of infants with neonatal lupus and congenital heart block (TAN 1989). These antibodies do not cross react with recombinant calreticulin, and anti-calreticulin antibodies have failed to immunoprecipitate Ro/SS-A RNA (ROKEACH et al. 1991). Despite this evidence that calreticulin may not be the Ro/SS-A antigen, autoantibodies against calreticulin are found in onchocerciasis patients (LIEU et al. 1989; LUX et al. 1992; LU et al. 1993; MEILOF et al. 1993). Onchocerciasis, or river blindness, is caused by the filarial parasite *O. volvulus*, which expresses the RAL-1 antigen, a protein 64%

homologous to calreticulin (Unnasch et al. 1988). The Ro/ss-A antigen was also recently shown to be similar to the B50 melanoma antigen (Gersten et al. 1991).

5 Interaction of Calreticulin with KXFFKR-Like Peptides

In studies designed to identify proteins which interact with the α subunit of integrins, Rojiani et al. (1991) demonstrated that calreticulin binds to the KLG-FFKR peptide, a highly conserved sequence found in all α integrin subunits. Calreticulin and its cytoplasmic homologue molbiferrin were subsequently shown to coimmunoprecipitate with α integrins (Conrad et al. 1993). Interestingly very similar peptides are also found in a number of other proteins including the inositol-3-phosphate ($InsP_3$) receptor, nuclear receptors, cyclophilins and perforin. As discussed in subsequent sections both nuclear receptors and perforin have also been found to associate with calreticulin. The diversity of molecules which contain this peptide and their widespread distribution within the cell suggests the intriguing possibility that calreticulin may interact with these proteins and thus be involved at multiple steps in T cell activation and functioning.

5.1 Calreticulin and Nuclear Receptors

The nuclear receptors are a superfamily of ligand activated transcription factors which modulate a number of cellular processes (Truss and Beato 1993). Members of this family include steroid receptors such as the glucocorticoid receptor, and the progesterone and androgen receptors (Laudet et al. 1991). They are organized in a modular fashion such that the molecule can be divided into three structurally and functionally distinct regions: NH_2 terminal domain, DNA-binding domain and a ligand-binding domain. The DNA binding domain is the most highly conserved and is comprised of two Zn^{2+} fingers that are coordinated by highly conserved cysteine resides. Crystallography of the DNA binding domain of the glucocorticoid receptor with DNA revealed that their association is mediated through the peptide which lies between the two Zn^{2+} fingers (Luis et al. 1991). The highly conserved nature of this peptide (consensus sequence KXFFKR) suggests that other nuclear receptors similarly contact the DNA via this sequence.

Recently calreticulin has been demonstrated to associate with the DNA binding domains of both the glucocorticoid and androgen receptors (Burns et al. 1994a; Dedhar et al. 1994). This interaction is believed to be mediated through the KXFFKR peptide and has been demonstrated by gel mobility shift assays to prevent the receptors from interacting with DNA. Inhibition of receptor DNA interactions by calreticulin in this manner is postulated to account for the observed modulation of receptor activity observed in vivo. Specifically, the expression of genes that are up regulated by steroids was shown to be inhibited in cell lines producing elevated levels of calreticulin and correspondingly to be super induced in cell lines with reduced amounts of the protein. These results point to a role for calreticulin as a modulator of the activities of the nuclear receptor family. In order to mediate such effects, however, calreticulin must first find its way to the same compartment as the receptor. There are a number of ways in which this might occur via intracellular trafficking (BURNS et al. 1994b), but the presence of calreticulin in granules offers an unusual opportunity for the protein to be delivered into the cytoplasm of the target cell, and thus influence downstream events such as apoptosis.

6 Calreticulin as a Regulator of Calcium Levels in T Lymphocytes?

The synthesis of calreticulin is induced during T cell proliferation following mitogen stimulation (BURNS et al. 1992). In resting splenocytes and peripheral blood lymphocytes calreticulin mRNA and protein is virtually undetectable. However, mitogen stimulation leads to increased expression and synthesis of calreticulin. In concanavalin A (ConA)-stimulated mouse splenocytes, for example, the steady state levels of calreticulin mRNA were found to gradually increase over a period of days, plateau and then decline. There was also a corresponding increase in protein although not as dramatic as the increases observed for calreticulin mRNA.

Various ligands that bind T cell receptors induce rapid hydrolysis of inositol phospholipids (TAYLOR et al. 1984; BERRIDGE 1990). This ultimately leads to the $InsP_3$–induced release of Ca^{2+} from the ER-associated intracellular store. Recent work from several laboratories suggests that the Ca^{2+} content of an internal Ca^{2+} store may also effect the Ca^{2+} permeability of the plasma membrane and result in enhanced Ca^{2+} influx via plasma membrane Ca^{2+} channels (PUTNEY 1990; MASON et al. 1991; CLEMENTI et al. 1992; DEMAUREX et al. 1992). This coupling mechanism may be important in mediating the effects of mitogen and other agonists in the control of Ca^{2+} homeostasis in T cells. Ca^{2+} binding proteins, especially those associated with the intracellular Ca^{2+} stores, would therefore be expected to play a major role in controlling intracellular Ca^{2+} concentrations during this process, but very little is known about their involvement.

The increased synthesis of calreticulin may be directly related to the T cell's Ca^{2+} homeostasis during mitogenic stimulation. By virtue of its cellular localization and its ability to bind Ca^{2+} calreticulin may modulate Ca^{2+}-dependent events in the ER of stimulated T lymphocytes. Undoubtedly calreticulin could influence levels of Ca^{2+} in whatever location it finds itself, the cytoplasm, a granule or a recipient target cell.

7 Calreticulin in Granules

After it was established that calreticulin was a constituent of the cytoplasmic granule we reanalyzed our induction data (Burns et al. 1992). A small amount of protein was present in resting cells, presumably in the ER, but upon T cell activation, either with antigen or mitogen, the levels of mRNA and protein rapidly increased. In the mouse we saw two species of protein (~60 kDa and 63 kDa) but in granules only the smaller form was detected. This would explain the observation that only the 60 kDa molecule is released upon TCR stimulation (Dupuis et al. 1993). The kinetics of accumulation of mRNA and 60 kDa protein closely paralleled the development of cytolytic activity in the cultures. A pattern very reminiscent of other granule-associated molecules, notably the granzymes and perforin which are key players in the cytolytic mechanism.

We were concerned, however, that as granules were forming they might incorporate various proteins nonspecifically from the ER and Golgi. If this were the case there should be evidence for other ER resident proteins within granules. So far this does not seem to be the case as we have been unable to detect three such molecules (BiP, PDI, GRP96) in a variety of preparations of granules. Thus calreticulin appears to be the only KDEL-protein selectively relocated into granules as they develop.

Why is calreticulin not retained in the ER by virtue of its KDEL sequence? Clearly the KDEL receptors are present and functional as the other KDEL proteins are still lumenal. Perhaps the signal is masked in calreticulin through interaction with another protein, or maybe the KDEL sequence is removed to yield a protein that is then not returned to the ER but is free to pass right through the Golgi into granules. This latter possibility could be envisioned to occur either by differential splicing or posttranslational processing in order to generate calreticulin with a modified COOH-terminal. This seemed particularly appealing in the mouse as two species had been detected and a potential tetrabasic proteinase cleavage site (Yoshimasa et al. 1990) is present upstream from KDEL. On the basis of results obtained using a number of antisera, however, we have concluded that the majority of the granule-associated calreticulin is very similar, if not identical, to the ER form and that it still has a KDEL sequence (Bleackley and Michalak, unpublished).

The masking hypothesis is certainly appealing given calreticulin's propensity to interact with other proteins. Indeed, Tschopp originally identified calreticulin as a granule protein by virtue of its association with perforin (Dupuis et al. 1993). This interaction appears to be of fairly high affinity, as in collaboration with Dr. D. Hudig we have found calreticulin in "highly purified" perforin preparations. There might also be an association with the granzymes but this appears to be weaker. Although we have analyzed preparations of granzyme B that contain calreticulin, we have also seen enzymatically active granzyme B preparations that are devoid of any immunoreactive calreticulin.

The interaction with perforin is particularly satisfying as perforin contains a sequence KVFF (residues 438-442) that is very similar to the peptide previously defined as a binding site for calreticulin on α-integrin (ROJIANI et al. 1991) and subsequently shown to mediate binding to the steroid hormone receptors (BURNS et al. 1994a; DEDHAR et al. 1994). This would certainly fit in with the model of calreticulin as a chaperone. Preliminary evidence also indicates that the interaction is via the N-domain of calreticulin and is calcium-dependent (Michalak and Bleackley, unpublished). This latter point would explain very nicely the known Ca^{2+} sensitivity of perforin's action as a pore forming molecule. It remains to be seen whether perforin can displace calreticulin from the KDEL receptor during biogenesis of granules or whether the two nascent proteins become associated immediately after translation. The latter possibility would explain, in part, why newly synthesized calreticulin is produced upon T cell activation. However, it should be noted that when we have looked at CTLs stained with calreticulin antibody by confocal microscopy we observed material primarily in granules with none, or at least very little, in the ER.

In summary, calreticulin appears to be one of a group of proteins that are newly synthesized upon CTL activation, and these include molecules already characterized as key players in the killing mechanism. In contrast to other ER proteins, the newly synthesized calreticulin, and perhaps the resident protein, relocates from its normal residence within the lumen of the ER into cytoplasmic granules. Within these granules it is found in tight association with at least one of the key players in cytotoxicity, perforin, and furthermore this interaction appears to be dependent upon Ca^{2+}. We would submit that these facts constitute persuasive, albeit circumstantial, evidence that calreticulin is directly involved as a player in the overall killing machine rather than being present as an unsuspecting passenger in granules. The earlier descriptions of the properties of calreticulin leave us with no shortage of suggestions as to how and where the molecule may act.

8 Possible Role(s) of Calreticulin in Cytolysis

The earlier description of the biochemistry and biology of calreticulin clearly indicate that this molecule has the potential to be multifunctional. Of particular interest to the mechanism of granule-mediated lysis are calreticulin's ability to bind Ca^{2+} (and other divalent cations) and thus influence local concentrations of the metal, its chaperone-like properties, and the potential of the molecule to bind to hormone receptors and thus control steroid dependent responses. These suggest that calreticulin could operate at one or a number of possible key steps in killing.

8.1 Within Granules

Given that calreticulin may possess chaperone-like activity, it seems reasonable to suggest that it might accompany monomeric perforin along the secretory pathway from the ER to the granules. It must be remembered that perforin is potentially a very dangerous protein, not only to the target cells which it is designed to penetrate, but also to the CTLs in which it is made and stored. Such a protein must surely be handled with exquisite care. It must not be allowed to polymerize prematurely, or it could become permanently inactivated and useless. Worse yet, polymerization of perforin at an inopportune time could cause damage to the membranes within the CTLs. It is therefore quite attractive to suggest that a protein like calreticulin may interact with perforin soon after it is translocated into the ER and carry it harmlessly to the lytic granules. That such a hypothetical chaperone should possess significant Ca^{2+} binding activity may be more than a coincidence. It would be extremely beneficial for perforin to associate with such a protein in order to be shielded from any free Ca^{2+} ions along the secretory pathway. Perhaps calreticulin plays a dual role in accompanying perforin to the granules: one as a chaperone to ensure that perforin remains unfolded and inactive, and another as a shield to further protect perforin from coming in contact with free Ca^{2+}.

Another possible function of granule-localized calreticulin could be to ensure that adequate amounts of Ca^{2+} are available upon granule exocytosis. It is conceivable that calreticulin could release any bound Ca^{2+} upon exocytosis in order for perforin to be able to polymerize within the target cell membrane. However, the pH of the granules is quite low, so it does not seem likely that calreticulin would be able to continue to store Ca^{2+} within the granules. It is quite possible, though, that calreticulin could bring significant amounts of Ca^{2+} with it to the granules and release it as the granule pH is lowered. At this low pH the Ca^{2+} probably would also not react with perforin, but could remain in the granules to be released with perforin upon exocytosis and a return to neutral pH. This would ensure a source of free Ca^{2+} for the perforin to polymerize within the target cell membrane.

8.2 After Exocytosis

Upon degranulation the perforin:calreticulin complex may be disrupted upon exposure to Ca^{2+} thus allowing perforin polymerization to occur. Perhaps calreticulin could even assist in targeting the perforin to the target cell membrane through its ability to bind other proteins.

Extracellular calreticulin could potentially also once again bind any free monomeric perforin that does not polymerize and/or insert into the target cell membrane. This could conceivably represent one mechanism that exists to protect the CTLs and innocent bystander cells from attack by rogue perforin molecules. The virtual absence of bystander killing that is routinely observed in

CTL lytic assays suggests a very stringent control of perforin's activity. That is, it must be limited both in the time that it is active and the distance that it can travel in an activated form upon exocytosis. Inactivation by Ca^{2+} has been assumed to be the mechanism involved in keeping this killer molecule on a short leash. It would not be surprising if additional safeguards were also in place in order to ensure that fugitive perforin molecules cannot go on a killing spree. In addition the exocytosed calreticulin could influence the properties of other factors that are often present at the site of an immune/inflammatory response. Although very little is known about effects of calreticulin in this regard it has recently been shown to bind to some of the blood clotting factors and also influence the course of thrombosis after vascular injury (Benedict et al. 1993; Kuwabara et al. 1993).

8.3 Into Target Cells

Another intriguing possibility is that calreticulin might actually get introduced into the cytoplasm of the target cells via the transmembrane channels formed by polymerized perforin. Although it is quite possible that cytoplasmic calreticulin may interact with target cell proteins and enhance the killing process, it is more exciting to imagine that its nuclear localization signal would target it to the nucleus, the site of many interesting events in the process of death by apoptosis of the target cell. Its ability to interact with a wide variety of proteins including the steroid receptor class of transcriptional activators could conceivably be utilized to aid in the killing process. This family of receptors has been repeatedly implicated as having a role in apoptosis (Schwartzman and Cidlowski 1993; Liu et al. 1994), so it is conceivable that if calreticulin is suddenly introduced into a cell by a CTL, it could have an influence on the ensuing suicide program of the target cell via its ability to interact with these proteins. Calreticulin might have additional effects on target cell viability by interacting with other types of nuclear proteins as well. In addition, it is known that the presence of free Zn^{2+} strongly inhibits the nuclear effects of apoptosis, probably by interfering with the nuclease responsible for the characteristic genome digestion (Schwartzman and Cidlowski 1993). Since calreticulin can bind a significant amount of Zn^{2+}, it might play a role in enhancing the apoptotic death of the target cell by being targeted to the nucleus and sequestering any free Zn^{2+} that might inhibit the necessary nuclease.

Elevated free intracellular Ca^{2+} levels have also been implicated in target cell destruction. When suboptimal levels of purified granules and calcium ionophore were added to cells, they were found to synergize to cause target cell lysis (Kraut et al. 1990). It seems possible that calreticulin could enter the target cell through the perforin channels and then release its bound Ca^{2+}, thereby contributing towards the lethal levels of intracellular Ca^{2+}. The signals responsible for calreticulin to either bind or release Ca^{2+} are as yet unknown. Perhaps there is a cytoplasmic factor that causes calreticulin to release its bound Ca^{2+}.

9 Concluding Remarks

Our aim in writing this review was to summarize the relevant biochemical properties of calreticulin, to convince the reader that calreticulin belongs in the cytolytic granules, and to speculate on the potential roles that this enigmatic protein may play in granule-mediated killing. Clearly calreticulin has a number of features that make it an interesting molecule and it does seem to belong in the granules. Its multifunctional potential indicates that it could be involved at, not just one, but a number of critical points in the activation and effector functions of cytolytic cells. With the recent developments in this area we are in a position to directly test a number of our suggestions. Time will tell how close to reality these speculations are, and in the not too distant future we hope to be writing the next installment of this story "Granule-Associated Calreticulin: A Key Player in the Cytolytic Effector Mechanism," notably with no question mark.

Acknowledgements. Experiments performed in our laboratories were supported by grants from the Medical Research Council and National Cancer Institute of Canada. K.B. is the recipient of a Province of Alberta Graduate Student Fellowship and E.A. holds a National Cancer Institute Studentship. M.M. is an MRC Scientist and a Scholar of the Alberta Heritage Foundation for Medical Research. RCB is a medical Scientist of the Alberta Heritage Foundation for Medical Research. Many of the unpublished results from our laboratories were performed by Jody Busaan, and Dr. Dorothy Hudig kindly provided us with some preparations of granules and perforin. The authors would also like to thank Mae Wylie for her assistance and patience during her typing of this review.

References

Ahluwalia N, Bergeron JJM, Wada I, Degen E, Williams DB (1992) The p88 molecular chaperone is identical to the endoplasmic reticulum membrane protein, calnexin. J Biol Chem 267: 10914–10918

Baksh S, Michalak M (1991) Expression of calreticulin in *Escherichia coli* and identification of its Ca^{2+} binding domains. J Biol Chem 66: 21458–21465

Baksh S, Burns K, Busaan J, Michalak M (1992) Expression and purification of recombinant and native calreticulin. Prot Exp Pur 3: 322–331

Benedict C, Kuwabara K, Todd G, Ryan J, Michalak M, Eaton D, Stern D (1993) Calreticulin is a novel antithrombotic agent: blockade of electrically-induced coronary thrombosis in a canine model. Clin Res 41: 275A

Berg J (1990) Zinc fingers and other metal binding domains. J Biol Chem 265: 6513–6516

Bergeron JGM, Brenner MB, Thomas DY, Williams DB (1994) Calnexin: a membrane-bound chaperone of the endoplasmic reticulum. TIBS 19: 124–128

Berridge MJ (1990) Calcium oscillations. J Biol Chem 265: 9583–9586

Burns K, Michalak M (1993) Interactions of calreticulin with proteins of the endoplasmic and sarcoplasmic reticulum membranes. FEBS Lett 318: 181–185

Burns K, Helgason CD, Bleackley RC, Michalak M (1992) Calreticulin in T-lymphocytes. Identification of calreticulin in T-lymphocytes and demonstration that activation of T-cells correlates with increased levels of calreticulin mRNA and protein. J Biol Chem 267: 19039–19042

Burns K, Duggan B, Atkinson EA, Famulski KS, Nemer M, Bleackley RC, Michalak M (1994a) Modulation of expression by calreticulin binding to the glucocorticoid receptor. Nature 367: 476–480

Burns K, Atkinson EA, Bleackley RC, Michalak M (1994b) Calreticulin: from Ca^{2+} binding to control of gene expression. Trends Cell Biol 4: 152–154

Clementi E, Scheer H, Zacchetti D, Fasolato C, Pozzan T, Meldolesi J (1992) Receptor-activated Ca^{2+} influx. J Biol Chem 267: 2164–2172

Collins JH, Xi Z, Alderson-Lang BH, Treves S, Volpe P (1989) Sequence homology of a canine brain calcium-binding protein with calregulin and the human Ro/SS-A antigen. Biochem Biophys Res Commun 164: 575–579

Conrad ME, Umbreit JN, Moore EG, Harper KP (1991) Molbiferrin, a homologue of Ro/SS-A autoantigen and calreticulin. Blood 78: 89a

Conrad ME, Umbreit JN, Peterson RDA, Moore EG, Harper KP (1993) Function of integrin in duodenal mucosal uptake of iron. Blood 81: 517–521

Demaurex N, Lew DP, Krause K-H (1992) Cyclopiazonic acid depletes intracellular Ca^{2+} stores and activates an influx pathway for divalent cations in HL-60 cells. J Biol Chem 267: 2318–2324

de Virgilio C, Bürckert N, Neuhaus JM, Boller T, Wiemken A(1993) Yeast sequencing reports. CNE1, a *Saccharomyces cerevisiae* homologue of the genes encoding mammalian calnexin and calreticulin. Yeast 9: 1850–1881

Dedhar S, Rennie PS, Shago M, Hagesteijn C-YL, Yang H, Filmus J, Hawley RG, Bruchovsky N, Cheng H, Matusik RJ, Giguère V (1994) Inhibition of nuclear hormone receptor activity by calreticulin. Nature 367: 480–483

Dupuis M, Schaerer E, Krause KH, Tschopp J (1993) The calcium-binding protein calreticulin is a major constituent of lytic granules in cytolytic T lymphocytes. J Exp Med 177: 1–7

Fliegel L, Burns K, MacLennan DH, Reithmeier RAF, Michalak M (1989a) Molecular cloning of the high affinity calcium-binding protein (calreticulin) of skeletal muscle sarcoplasmic reticulum. J Biol Chem 264: 21522–21528

Fliegel L, Burns K, Wlasichuk K, Michalak M (1989b) Peripheral membrane proteins of sarcoplasmic and endoplasmic reticulum. Comparison of carboxyl-terminal amino acid sequences. Biochem Cell Biol 67: 696–702

Galvin K, Krishna S, Ponchel F, Frohlich M, Cummings DE, Carlson R, Wands JR, Isselbacher KJ, Pillai S, Ozturk M (1992) The major histocompatibility complex class I antigen-binding protein p88 is the product of the calnexin gene. Proc Natl Acad Sci U S A 89: 8452–8456

Gersten DM, Bijwaard KE, Law LW, Hearing VJ (1991) Homology of the B50 murine melanoma antigen to the Ro/SS-A antigen of human systemic lupus erythematosus and to calcium-binding proteins. biochim Biophys Acta 1096: 20–25

Hawn TR, Tom TD, Strand M (1993) Molecular cloning and expression of SmIrV1, a *Schistosoma mansoni* antigen with similarity to calnexin, calreticulin, and OvRa11. J Biol Chem 268: 7692–7698

Hochtenbach F, David V, Watkins S, Brenner M (1992) Endoplasmic reticulum resident protein of 90 kilodaltons associates with the T- and B-cell antigen receptors and major histocompatibility complex antigens during their assembly. Proc Natl Acad Sci U S A 89: 4734–4738

Huang L, Franklin AE, Hoffman NE (1993) Primary strucutre and characterization of an *Arabidopsis thalina* calnexin-like protein. J Biol Chem 268: 6560–6566

Kennedy TE, Kuhl D, Barzilai A, Sweatt JD, Kandel ER (1992) Long-term sensitization training in Aplysia leads to an increase in calreticulin, a major presynaptic calcium-binding protein. Neuron 9: 1013–1024

Kraut RP, Bose D, Cragoe EJ Jr, Greenberg AH (1990) The influence of calcium, sodium, and the Na^+/Ca^{2+} antiport on susceptibility to cytolysin/perforin-mediated cytolysis. J Immunol 144: 3498–3505

Kretsinger RH, Moncrief ND, Goodman M, Czelusniak J (1988) Homology of calcium modulated proteins: their evolutionary and functional relationships. In: Morad M, Naylor WG, Kazda S, Schramm M (eds) The calcium channel, structure, function and implication. Springer, Berlin Heidelberg New York, pp 16–34

Kuwabara K, Benedict C, Todd G, Ryan J, Michalak M, Eaton D, Stern D (1993) Calreticulin is novel antithrombotic agent: Blockade of electrically-induced coronary thrombosis in a canine model. Thromb Haemost 69: 1362

Laudet V, Hänni C, Coll J, Catzeflis F, Stehelin D (1991) Evolution of the nuclear receptor gene superfamily. EMBO J 11: 1003–1013

Lieu T-S, Newkirk MM, Capra JD, Sontheimer RD (1988) Molecular characterization of human Ro antigen: amino terminal sequence of the protein moiety of human Ro antigen and immunological activity of a corresponding synthetic peptide. J Clin Invest 82: 96–101

Lieu TS, McCauliffe DP, Newkirk MM, Arnett FC, Lee LA, Deng JS, Capra JD, Sontheimer RD (1989) A major autoepitope is present on the amino terminus of the human Ro polypeptide. J Autoimmun 2: 367–374

Liu N, Fine RE, Johnson, RJ (1993) Comparison of cDNAs from bovine brain coding for two isoforms of calreticulin. Biochim Biophys Acta 1202: 7076

Liu Z-G, Smith SW, McLaughlin KA, Schwartz LM, Osborne BA (1994) Apoptotic signals delivered through the TCR of a T cell hybrid require immediate-early gene nur 77. Nature 367: 281–284

Lu J, Willis AC, Sim RB (1993) A calreticulin-like protein co-purifies with a "60kD" component of Ro/SSA, but is not recognized by antibodies in Sjögren's syndrome sera. Clin Exp Immunol 94: 429–434
Luis BF, Xu WX, Otwinowski Z, Freedman LP, Yamamoto KR, Sigler PB (1991) Crystallographic analysis of the interaction of the glucocorticoid receptor with DNA. Nature 352: 497–505
Lux FA, McCauliffe DP, Buttner DW, Lucius R, Capra JD, Sontheimer RD, Lieu T-S (1992) Serological cross-reactivity between a human Ro autoantigen (calreticulin) and the lambda Ral-1 antigen of *Onchocerca volvulus*. J Clin Invest 89: 1945–1951
Mason MJ, Mahaut-Smith MP, Grinstein S (1991) The role of intracellular Ca^{2+} in the regulation of the plasma membrane Ca^{2+} permeability of unstimulated rat lymphocytes. J Biol Chem 266: 10872–10879
Mazzarella RA, Gold P, Cunningham M, Green M (1992) Determination of the sequence of an expressible cDNA clone encoding ERp60/calregulin by the use of a novel nest set method. Gene 120: 217–225
McCauliffe DP, Lux FA, Lieu TS, Sanz I, Hanke J, Newkirk MM, Bachinski LL, Itoh Y, Siciliano MJ, Reichlin M, Sontheimer RD, Capra JD (1990a) Molecular cloning, expression and chromosome 19 localization of human Ro autoantigen. J Clin Invest 85: 1379–1391
McCauliffe DP, Zappi E, Lieu TS, Michalak M, Sontheimer RD, Capra JD (1990b) A human Ro/SS-A autoantigen is the homologue of calreticulin and is highly homologous with Onchocercal RAL-1 antigen and an Aplysia "memory molecule". J Clin Invest 86: 332–335
Meilof JF, Van der Lelij A, Rokeach LA, Hoch SO, Smeenk RJT (1993) Autoimmunity and Filariasis. J Immunol 151: 5800–5809
Michalak M, Baksh S, Opas M (1991) Identification and immunolocalization of calreticulin in pancreatic cells: No evidence for calciosomes. Exp Cell Res 197: 91–99
Michalak M, Milner RE, Burns K, Opas M (1992) Calreticulin. Biochem J 285: 681–692
Murthy KK, Banville D, Srikant CB, Carrier F, Holmes C, Bell A, Patel YC (1990) Structural homology between the rat calreticulin gene product and the *Onchocerca volvulus* antigen Ral-1. Nucleic Acids Res 18: 4933
Nakamura M, Moriya M, Baba T, Michikawa Y, Yamanobe T, Arai K, Okinaga S, Kobayashi T (1993) An endoplasmic reticulum protein, calreticulin, is transported into the acrosome of rat sperm. Exp Cell Res 205: 101–110
Nigam SK, Goldberg AL, Ho S, Rohde MF, Bush KT, Sherman MY (1994) A set of endoplasmic reticulum proteins possessing properties of molecular chaperons includes Ca^{2+}-binding proteins and members of the thioredoxin superfamily. J Biol Chem 269: 1744–1749
Opas M, Dziak E, Fliegel L, Michalak M (1991) Regulation of expression and intracellular distribution of calreticulin, a major calcium binding protein of nonmuscle cells. J Cell Physiol 149: 160–171
Ostwald TJ, MacLennan DH (1974) Isolation of a high affinity calcium binding protein from sarcoplasmic reticulum. J Biol Chem 249: 974–979
Ostwald TJ. MacLennan DH, Dorrington KJ (1974) Effects of cation binding on the conformation of calsequestrin and the high affinity calcium-binding protein of sarcoplasmic reticulum, J Biol Chem 249: 5867–5871
Ou WJ, Thomas DY, Bell A, Bergeron JJH (1992) Casein kinase II phosphorylation of signal sequence receptor a and the associated membrane chaperone calnexin. J Biol Chem 267: 23789–23796
Ou WJ, Camron PH, Thomas DY, Bergeron JJH (1993) Association of folding intermediates of glycoproteins with calnexin during protein maturation. Nature 364: 771–776
Pelham HRB (1989) Control of protein exit from the endoplasmic reticulum. Annu Rev Cell Biol 5: 1–23
Putney JW Jr (1990) Capacitative calcium entry revisited. Cell Calcium 11: 611–624
Rojiani MV, Finlay BB, Gray V, Dedhar S (1991) In vitro interaction of a polypeptide homologous to human Ro/SS-A autoantigen (calreticulin) with a highly conserved amino acid sequence in the cytoplasmic domain of integrin α subunit. Biochemistry 30: 9859–9865
Rokeach LA, Haselby JA, Meilof JF, Smeek RJT, Unnasch TR, Greene BM, Hoch SO (1991) Characterization of the autoantigen calreticulin. J Immunol 147: 3031–3039
Schwartzman RA, Cidlowski JA (1993) Apoptosis: the biochemistry and molecular biology of programmed cell death. Endocr Rev 14: 133–151
Smith MJ (1992) A *C. elegans* gene encodes a protein homologous to mammalian calreticulin. DNA Seq 2: 235–240
Smith MJ, Koch GLE (1989) Multiple zones in the sequence of calreticulin (CRP55, calregulin, HACBP), a major calcium binding ER/SR protein. EMBO J 8: 3581–3586
Sueyoshi T, McMullen BA, Marnell LL, Clos TWD, Kisiel W (1991) A new procedure for separation of protein Z, prothrombin fragment 1.2 and calreticulin from human plasma. Thromb Res 63: 569–575

Tan EM (1989) Antinuclear antibodies: diagnostic markers for autoimmune diseases and probes for cell biology. Adv Immunol 44:93–101
Taylor MV, Metcalfe JC, Hesketh RB, Smith GA, Moore JP (1984) Mitogens increase phosphorylation of phosphoinositides in thymocytes. Nature 312:462–465
Tharin S, Dziak E, Michalak M, Opas M (1992) Widespread tissue distribution of rabbit calreticulin, a non-muscle functional analogue of calsequestrin. Cell Tissue Res 269: 29–37
Treves S, Zorzato F, Pozzan T (1992) Identification of calreticulin isoforms in the central nervous system. Biochem J 287: 579–581
Truss M, Beato M (1993) Steroid hormone receptors: interaction with deoxyribonucleic acid and transcriptional factors. Endocr Rev 14: 459–479
Unnasch TR, Gallin MY, Soboslay PT, Erttmann KD, Greene BM (1988) Isolation and characterization of expression cDNA clones encoding antigens of *Onchocerca volvulus* infectious larvae. J Clin Invest 82: 262–269
Wada I, Rindress D, Cameron PH, Ou W-J, Doherty JJ, Louvard D, Bell AW, Dignard D, Thomas DY, Bergeron JJM (1991) SSR α and associated calnexin are major calcium binding proteins of the endoplasmic reticulum membrane. J Biol Chem 266: 19599–19610
Waisman DM, Salimath BP, Anderson MJ (1985) Isolation and characterization of CAB-63, a novel calcium-binding protein. J Biol Chem 260: 1652–1660
Waterston R, Martin C, Craxton M, Huynh C, Coulson A, Hillier L, Durbin R, Green P, Shownkeen R, Halloran N, Metzstein M, Hawkins T, Wilson R, Berks M, Du Z, Thomas K, Thierry-Mieg J, Sulston J (1992) A survey of expressed genes in *Caenorhabditis elegans.* Nature Genet 1: 114–123
Yoshimasa Y, Paul JI, Whittaker J, Steiner DF (1990) Effects of amino acid replacements within the tetrabasic cleavage site on the processing of the human insulin receptor precursor expressed in chinese hamster ovary cells. J Biol Chem 265: 17230–17237

A New Function for an Old Enzyme: The Role of DNase I in Apoptosis

H.G. MANNHERZ[1,3], M.C. PEITSCH[2], S. ZANOTTI[1], R. PADDENBERG[1], and B. POLZAR[1]

1 Introduction

Programmed cell death or apoptosis (Greek for the falling of the leafs in autumn) is an event by which cells are deliberately eliminated (recently reviewed by UCKER 1991, WILLIAMS et al. 1992 and COHEN 1993). It occurs during embryogenesis, during the formation of digits from a limb bud, during the selection of immunocompetent B or autoreactive T cells, and in many tissues to provide a stable balance of cellular mass. The morphologic events during apoptosis differ from those of necrosis, defined as the dying of cells by plasma membrane injury. During apoptosis, the nuclear chromatin rapidly condenses to form crescent-shaped deposits along the nuclear envelope. The nucleus convolutes and fragments, while the cytoplasmic membrane forms protuberances. These are subsequently released forming so-called apoptotic bodies containing highly condensed DNA. The whole cell may disintegrate into a large number of membrane-sealed apoptotic bodies which are immediately phagocytosed by neighboring cells or macrophages. Apoptosis affects only single cells dispersed in a given tissue and there is no release of cytoplasmic contents into the extracellular space and therefore no inflammatory reaction is induced. In contrast, necrotic cell death usually affects a large number of neighboring cells. Due to the membrane injury,

[1] Institute of Cytobiology and Cytopathology, Robert-Koch-Strasse 5, 35033 Marburg, Germany
[2] Glaxo Institute of Molecular Biology, 1021 Plan-Les-Ouates, Switzerland
[3] Institute of Anatomy and Embryology, Ruhr-Universität, 44780 Bochum, Germany

cytoplasmic contents leak into the extracellular space and an inflammatory reaction ensues.

One of the hallmarks of apoptosis is the enzymatic internucleosomal degradation of chromatin which can be demonstrated by the formation of the so-called DNA ladder, which consists of multiples of about 180 bp (base pairs), after electrophoretic separation of the extracted DNA on agarose gels. It most probably constitutes the irreversible step during apoptosis, since the numerous strand breaks will stop transcription.

Until recently, very little was known about the nature of the enzyme catalyzing the DNA cleavage. Recent evidence now suggests that at least one of the endonucleases responsible for the internucleosomal DNA degradation is identical or very similar to a well known and extensively characterized enzyme, deoxyribonuclease I (DNase I) (PEITSCH et al. 1993a). Here we will describe the distribution and the structural and enzymatic properties of DNase I, presenting evidence in favor of its involvement in apoptotic DNA breakdown.

2 Structure and Enzymatic Activity of DNase I

One hundred years ago, in 1894, Albrecht KOSSEL (Professor of Anatomy and Physiology at the Universities of Marburg and Heidelberg) reported the isolation of two types of nucleic acid from bovine thymus: a highly viscous α-form and a β-form of low viscosity. He suspected that the β-form had been generated from the α-form and a long search started for the factor causing this transition, as reviewed by KUNITZ (1950).

DNase I is an endonuclease that hydrolyzes double-stranded DNA down to 5'-phospho- (tri -and/or tetra-) oligonucleotides. Its pH optimum is around pH 7.5 and its full enzymatic activity requires divalent cations: Ca^{2+} and Mg^{2+} or Mn^{2+} in submillimolar concentrations (10^{-4} *M*). Its enzymatic activity is inhibited by chelating agents such as EDTA and by Zn^{2+} ions. In these characteristics it differs from DNase II, which has a pH optimum of about 5.0, is active in a cation-independent manner, and is harbored in lysosomes of many cell types.

The mode of action and the specificity of DNase I depend on the nature of the divalent cations present: in the presence of Ca^{2+} and Mg^{2+} it causes single-stranded cuts of duplex DNA, while double-stranded breaks in purified DNA are formed in the presence of Mn^{2+} (CAMPBELL and JACKSON 1980). It does not cleave DNA randomly, although it does not exhibit base- or sequence specificity. It preferentially cleaves at the 5'-end of pyrimidines and it preferentially cuts bonds with a high degree of local twist (LOMMOSOFF et al. 1981).

The primary structure of bovine pancreatic DNase I was described by LIAO et al. (1973). Subsequent structural analysis to 2 Å resolution led to correction of the sequence (LAHM and SUCK 1991). DNase I is composed of a single polypeptide chain of 260 residues and a carbohydrate moiety of the high mannose type linked to Asn-18. Isoforms of DNase I from the same species can differ in the length (normally eight hexose units) and the composition of the attached glycan. DNase I contains two disulfide bridges between Cys-98–Cys-101 and Cys-170–Cys-206. Reduction of the latter (essential) disulfide bridge leads to the loss of enzymatic activity. Two tightly bound (K_d = 10^{-5} M) structural Ca^{2+} ions protect DNase I against proteolytic degradation at the COOH-terminal end and the essential disulfide bridge from reduction (PRICE et al. 1969).

Bovine pancreatic DNase I was first crystallized by KUNITZ (1940, 1948, 1950). Crystals suitable for three-dimensional structure determination were obtained by SUCK (1983). The three-dimensional structure analysis of bovine pancreatic DNase I showed that its core is made up of two, central, six-stranded β-pleated sheets packed against each other and forming a sandwich-type structure surrounded by eight α-helices and extensive loop regions (SUCK et al. 1988). It is a compact four-layered molecule with extensive hydrophobic interactions both between and along the central β-sheets, explaining its thermal stability and its ability to regain full activity after treatment with chaotropic agents and/or detergents. The active site is formed by a shallow groove rimmed by the two β-sheets on one side.

In a series of elegant investigations SUCK and OEFNER (1986), OEFNER and SUCK (1986), SUCK et al. (1988) and LAHM and SUCK (1991) analyzed the mode of binding of DNase I to DNA and described the mechanism of DNA hydrolysis. The analysis of cocrystals of DNase I with Ca^{2+}-thymidine 3',5'-diphosphate allowed the proposal of a model for the interaction of the enzyme with double-stranded DNA and the hydrolysis event (Fig. 1). This was subsequently confirmed and refined when the determination of the crystal structure of DNase I complexed to a nicked DNA octanucleotide became available at 2 Å resolution.

Structure of a DNase I–DNA Complex

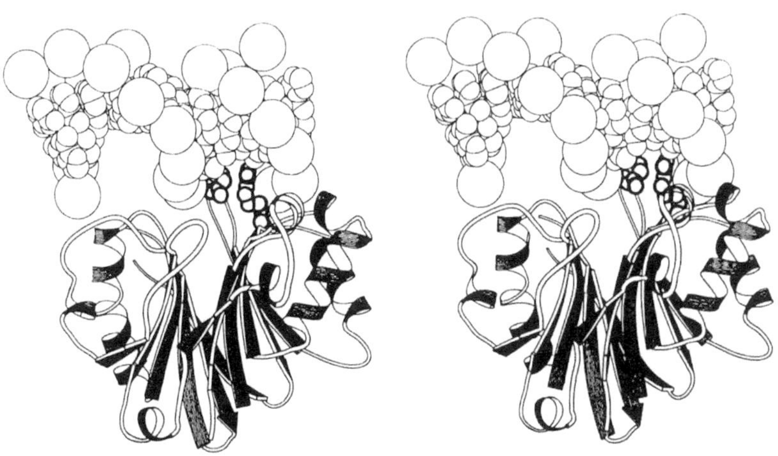

Fig. 1. DNase I-DNA interaction. Stereo picture of a ribbon representation of the 3-D structure of bovine pancreatic DNase I with DNA octamer (space filling). Note the loop implicated in substrate binding (residues 70–74, see text) of DNase I, protruding into the minor groove of the duplex DNA. (From LAHM and SUCK 1991)

According to SUCK and OEFNER (1986), LAHM and SUCK (1991) and SUCK et al. (1988), the loop consisting of residues 70–74 (Arg-Asn-Ser-Tyr-Lys) and the helix-turn region from residue Val-43 to Pro-57 tightly bind to both strands of the minor groove of B-DNA along six base pairs, with two adjacent phosphates forming hydrogen bonds and van der Waal's contacts. The DNA is distorted towards the major groove away from the DNase I, whose conformation remains apparently unchanged by substrate interaction.

The phosphodiester bond to be split is anchored by an additional Ca^{2+} ion. A proton acceptor-donor chain composed of Glu-75–His-131 produces the nucleophilic hydroxyl ion from water which cleaves the 3'-P-O bond (SUCK et al. 1988). After soaking the cocrystal of DNase I and the nicked octanucleotide with Mn^{2+} ions, a second cut appears that was taken as evidence that DNase I is able to cut duplex DNA in a single hit in the presence of Mn^{2+} ions (LAHM and SUCK 1991). The width of the minor groove seems to define the activity of DNase I. It is narrower in A-T and wider in G-C rich regions, which might explain the lower hydrolysis rates of the latter. However, DNA flexibility also seems to be critical in determining DNase I activity.

3 Actin Inhibits DNase I Activity

The existence of a protein in many cells that strongly and specifically inhibits the activity of DNase I was first discovered by DRABOWSKA et al. (1949) and later confirmed by COOPER et al. (1950) and FESTY and PAOLETTI (1963). LINDBERG (1967, 1974) succeeded in the purification of this proteinaceous DNase I inhibitor. LINDBERG (1967) demonstrated that DNase I is inhibited by the stoichiometric high affinity binding of the inhibitory protein. In 1974, LAZARIDES and LINDBERG identified actin as the naturally occurring inhibitor of DNase I, stimulating the interest of scientists involved in research on cytoskeletal proteins, and in particular in the intracellular regulation of the supramolecular organization of actin.

Actin is an ubiquitous protein present in all eukaryotic cells, where it participates in cytoskeletal functions and motile events. Nonmuscle cells contain almost equally large pools of unpolymerized monomeric (G-)actin and polymerized filamentous (F-)actin. DNase I is able to depolymerize F-actin (MANNHERZ et al. 1975; HITCHCOCK et al. 1976), leading to the formation of stable 1:1 complexes of actin and DNase I. In this complex DNase I activity is inhibited (by about 95% at equimolar ratios), and the actin is unable to repolymerize even in the presence of high salt and other actin binding proteins known to induce actin polymerization. The binding constant of skeletal muscle actin to bovine pancreatic DNase I was determined to be $K_a = 10^9\ M^{-1}$, MANNHERZ et al. (1980).

Crystals of the 1:1 complex of skeletal muscle actin and bovine pancreatic DNase I suitable for structural analysis were obtained by MANNHERZ et al. (1977), and the analysis of the atomic structure of this complex was completed by KABSCH

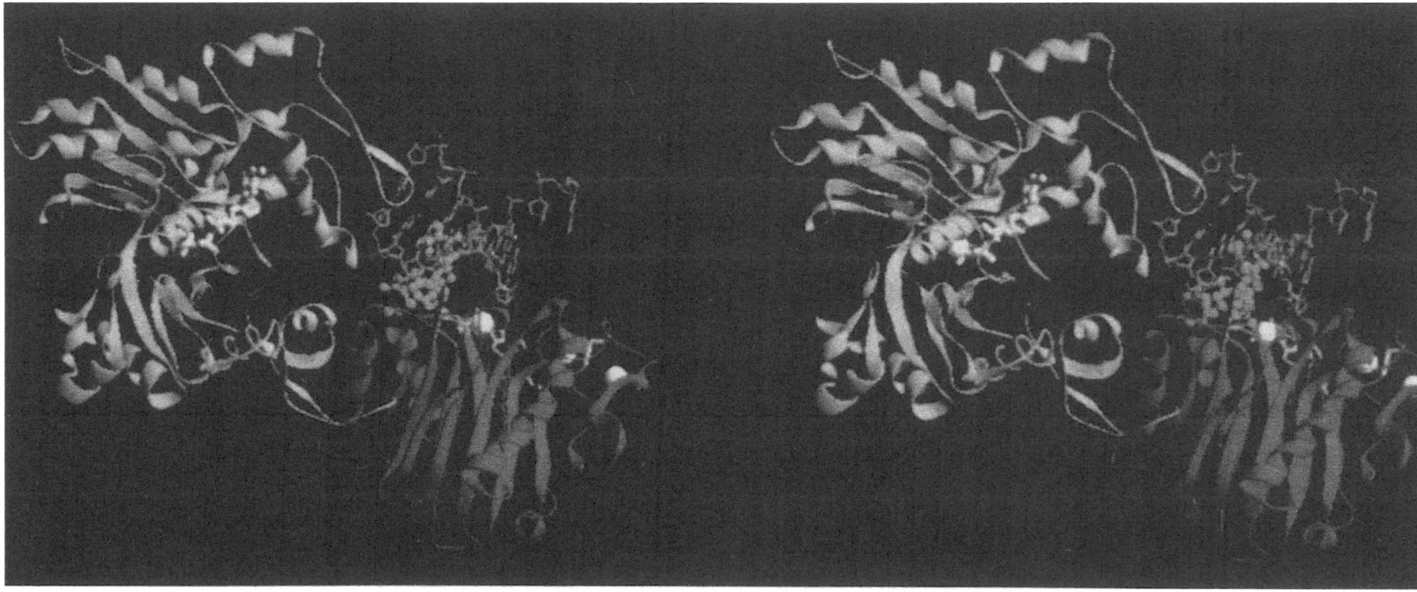

Fig. 2. Stereo representation of the combination of the DNase I-actin coordinate set (Brookhaven Data Bank entry 1ATN) with the DNase I octamer data (kindly provided by Dr. D. Suck, EMBL, Heidelberg). The coordinates of the DNase I model from the DNase I octamer data were superimposed onto the coordinates of the DNase I model in the DNase I-actin complex (1ATN); the same transformation matrix was then applied to the DNA octamer coordinates. The actin molecule is depicted at the left upper corner containing the bound ATP molecule and the calcium atom. DNase I is localized at the right lower corner, the side chains involved in DNA-hydrolysis (Glu 75 and His 131) are highlighted. The carbohydrate chain is attached to Asp 18 and the location of the essential disulfide bridge (Cys 170 to Cys 206) is given. The calcium atoms attached to DNase I are shown in white. Two of them are close to the essential disulfide bridge; these are believed to be essential for the stabilisation of the DNase I conformation. The third calcium atom is directly involved in DNA hydrolysis. Actin interacts with DNase I by its subdomain 2 (*lower inner loop*) and its subdomain 4 (*upper inner loop*). Note the sterical incompatibility between actin and the DNA octamer: the upper external loop of the actin model (residues 242–245) occupies the same portion of space as part of the DNA octamer duplex

et al. in 1990. Figure 2, which was constructed by combining the coordinates of the DNase I-actin complex with those of the DNase I-DNA octamer data, shows that the main contact of DNase I to actin is to its subdomain 2. This contact comprises a hydrophobic loop (residues Gly-42, Val-43 and Met-44) of actin which forms an antiparallel β-sheet with a loop of DNase I comprising residues Tyr-65, Val-66 and Val-67. This contact is very tight and results in a distortion of the α-carbon (C_α) positions of this stretch of the DNase I by 1.8 Å. A second and ionic contact of DNase I is to subdomain 4 : residues Thr-203 and Glu-205 of actin to Glu-13 and His-44 of DNase I. Glu-13 and Ser-14 of DNase I are also involved in binding to DNA. Thus, the binding of actin to DNase I can sterically affect the interaction with DNA (OEFNER and SUCK 1986; KABSCH et al. 1990). The main contact region of DNase I to DNA and the residues involved in DNA hydrolysis mentioned above are not covered by actin.

A number of species-specific variants of DNase I have been purified and characterized (PRICE et al. 1969; KREUDER et al. 1984). The inhibition of its DNA degrading activity by monomeric actin is a reliable criterium for its classification as DNase I. However, not all DNases I interact with monomeric actin with the same high affinity as the bovine enzyme. For instance, the rat parotid DNase I binds to G-actin with an affinity three orders of magnitude lower (KREUDER et al. 1984). This enzyme is therefore not able to shift the equilibrium between F- and G-actin, i.e., unable to depolymerize F-actin. The change in affinity was later attributed to a change of amino acid residues, Glu-13 to Asp and Val-67 to Ile (POLZAR and MANNHERZ 1990). Thus, the high affinity binding of DNase I to actin is not an absolute structural requirement which has been maintained during evolution.

Its highly specific interaction with actin remains enigmatic, although the phenomenon has been used by many scientists to test for the presence of monomeric actin, and to purify actin by affinity chromatography from various sources, BLICKSTAD et al. (1978). The confinement of DNase I to secretory granules appeared to preclude its direct contact with the microfilament system. Thus, reports on the presence of DNase I or DNase I-like enzymes in nonsecretory cells had little impact, since no functional role could be ascribed to the enzyme.

4 DNase I Is an Ubiquitous Protein

The presence of DNase I outside of exocrine tissues of the digestive tract has been reported. For instance, DNase I occurs in the liver of various species, cells of the peripheral blood, and in leukemia L1210 lymphocytes (MALICKA BLASZKIEWICZ and ROTH 1983; MALICKA BLASZKIEWICZ 1986, 1990). Indeed, the internucleosomal DNA degradation (ladder formation) was first observed in isolated liver cell nuclei after their exposure to Ca^{2+} and Mg^{2+} ions (HEWISH and BURGYONE 1973). Recently, the cDNA of rat parotid DNase I was cloned and sequenced (POLZAR and MANNHERZ 1990). This cDNA has been used as a probe to test in a number of tissues for the presence of specific mRNA by northern blot analysis (POLZAR et al. 1994). In agreement with an earlier enzymatic analysis using the so-called zymogram technique (analysis of tissue homogenates by electrophoresis on SDS-gels containing copolymerized DNA; LACKS 1981), the highest-levels of DNase I activity were observed in the parotid gland, kidney, and small intestine (POLZAR et al. 1994). As will be discussed later, these are also the tissues, with the exception of the kidney, which are the site of high levels of cellular turnover. Lower levels were detected in rat liver, thymus and heart, (POLZAR et al. 1994). Using rat DNase I-specific primers, the presence of specific mRNA was furthermore detected by PCR analysis in testis, seminal vesicle, brain, and stomach (POLZAR et al. 1994). Furthermore, DNase I-specific mRNA was shown to be up-regulated in the rat prostate after induction of apoptosis of the secretory epithelium by testosterone withdrawal (BACHER et al. 1993).

5 Intracellular Function of DNase I: Involvement in Apoptosis

So far, DNase I has been regarded as a digestive enzyme secreted by the pancreas or parotid gland and released into the digestive tract, where it fulfils its presumed function, to degrade alimentary DNA. Indeed, its presence in pancreatic secretion was repeatedly demonstrated (ROHR and MANNHERZ 1978). Later it was found that DNase I is also present at high concentration in murine and human serum and urine (CHITRABAMRUNG et al. 1981; KISHI et al. 1990; PEITSCH et al. 1992). Recent evidence now reveals the presence of DNase I within many organs and cells and places DNase I in a central role in an important regulatory event, namely, apoptosis (PEITSCH et al. 1993a)

During apoptosis, the nucleus shrinks and the chromatin condenses, the nuclear material collapses into patches and dissociates into many lobes. Often these nuclear changes are accompanied by the destruction of the nuclear DNA into fragments of oligonucleosomal size, due to DNA hydrolysis in the nucleosomal linker regions. To explain the nuclear DNA breakdown, a Ca^{2+} and Mg^{2+}-dependent apoptosis-specific endonuclease was postulated by WYLLIE (1980). According to this notion, the endonuclease was localized within the nucleus or was newly synthesized and translocated into the nucleus after induction of apoptosis.

Some 20 years ago, HEWISH and BURGOYNE (1973) showed that nuclei isolated from liver cells degrade their nuclear DNA into nucleosomal fragments in a Ca^{2+}, Mg^{2+}-dependent fashion, indistinguishable from the fragments generated during apoptosis. More recently, isolated nuclei of thymocytes and other cells (mostly of hematopoietic origin) were shown to undergo a similar autodigestion when exposed to the divalent cations. Using nuclei isolated from a cell line (RG-17 T cell hybridoma) which harbors only limited amounts of the endogeneous endonuclease, a transfer test was designed to enzymatically characterize the apoptosis-associated endonuclease (PEITSCH et al. 1993a). Endonuclease activity

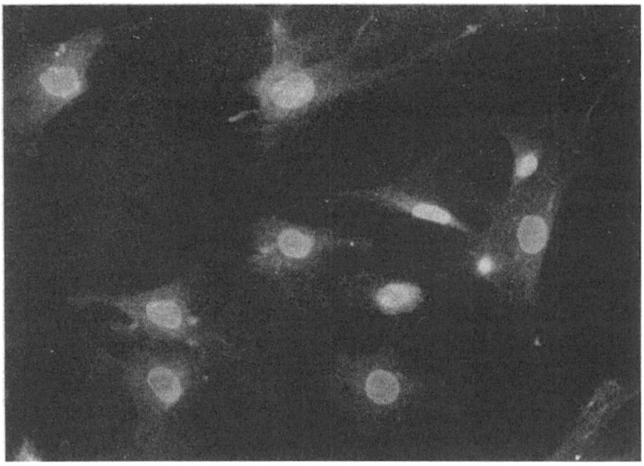

Fig. 3. Intracellular distribution of DNase I in primary rat fibroblasts. Cells were isolated, fixed and stained with an affinity-purified polyclonal anti-DNase I antibody and analyzed by immunofluorescence. Weak staining is observed in the cytoplasm, reflecting the organelles involved in the secretory pathway. A considerable amount of DNase I is found in the perinuclear region

was extracted from thymocytes or lymph node cell nuclei and added to RG-17 nuclei. The chromatin of the RG-17 nuclei was digested, yielding the typical apoptosis ladder, as observed in hematopoietic cells. Analysis of the nucleolytic activity in the nuclear extract revealed that the enzyme was functionally identical to DNase I. Immunoadsorbtion using specific antibodies to rat DNase I effectively reduced the ladder-forming activity in extracts of thymocytes and lymph node cell nuclei. Complete enzymatic inhibition of the apoptotic endonuclease was also obtained in the presence of the actin-gelsolin complex (Peitsch et al. 1992, 1993a). Furthermore, both the culture supernatant and nuclear extracts of COS cells transiently transfected with rat DNase I cDNA induced a clear ladder-forming activity on RG-17 nuclei (Polzar et al. 1993).

In thymocytes, lymphocytes and rat fibroblasts DNase I immunoreactivity was found to be concentrated perinuclearly (Peitsch et al. 1993a). After transfection of COS cells with DNase I cDNA it was also observed in organelles associated with the secretory pathway (Polzar et al. 1993). In rat fibroblasts it is localized perinuclearly (Fig. 3). It thus appears that DNase I is synthesized in the rough endoplasmic reticulum and partly sorted to the secretory pathway, partly stored within the nuclear envelope, in close apposition to the nucleus. This sequence of events was observed in the COS cells transiently transfected with the rat DNase I cDNA (Polzar et al. 1993). Nothing is known about the mechanisms by which DNase I is retained and how it might gain access to the nucleoplasm. Induction of apoptosis is often accompanied by an increase in the intracellular Ca^{2+} concentration (McConkey et al. 1989), which has been reported to induce the dissolution of the endoplasmic reticulum and thereby of the perinuclear envelope (Booth and Koch 1989).

Tissues with a high cellular turnover like the small intestine exhibit increased levels of intracellular DNase I activity and mRNA (Polzar et al. 1994). The

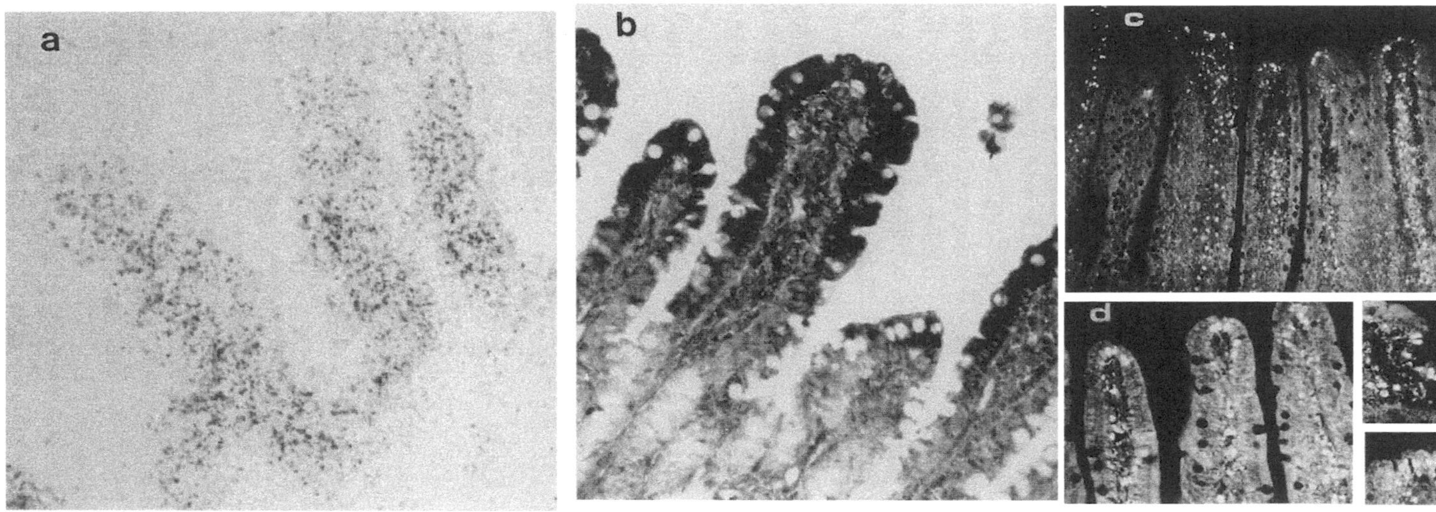

Fig. 4a–d. Expression of DNase I in the small intestine of rats. The enterocytes covering the intestinal villi are regenerated from the crypt and undergo apoptosis at the villar tips **a** In situ hybridization using anti-sense DNase I-specific RNA. Note strong reaction at the base and the middle of the villi, with decreasing intensity towards the villar tips, whereas the crypts are not labeled **b** Immunostaining reveals increasing expression of DNase I-specific immunoreactivity towards the villar tips. **c, d** in situ end-labeling of fragmented DNA using fluorescently labeled dATP. Note nuclear staining of the uppermost cells of the villar tips

enterocytes covering the intestinal villi are regenerated within the crypts of Lieberkühn and shed or retrieved into the lamina propria as dying cells after 2–5 days at the villar tips. Immunostaining again reveals staining and an increasing gradient in DNase I-specific immunoreactivity towards the villar tips (Fig. 4). In situ hybridization with DNase I-specific anti-sense DNA demonstrates a decreasing gradient of specific mRNA from the villar base to the villar tip (Zanotti et al. 1995). Using the procedure of in situ end labeling of fragmented DNA of cells undergoing apoptosis, DNA fragmentation is only observed at the uppermost tips of the villi (Gavrieli et al. 1992; Polzar et al. 1994), indicating that these cells prepare their own elimination by switching on DNase I expression during their migration long before having reached their final location (Fig. 4).

As detailed above, the biological activities of both actin and DNase I are inhibited in the actin-DNase I complex. The concentration of cellular actin in nonmuscle cells is estimated to be about 0.5 mM, 50% of which is in the filamentous form. Most of this actin is in the cytoplasm, although a considerable amount is also found in the nucleoplasm (Fukui and Katsumaru 1979; Nakayasu and Ueda 1983; Valkov et al. 1989). Thus, any DNase I liberated into these compartments, for example during mitosis, will be complexed and inactivated. This might be a mechanism to protect the cell against an untimely liberation or activation of DNase I. Indeed, nuclear extracts of thymocytes contain nucleolytic DNase I activity even before exposure to glucocorticoids. During dexamethasone-induced apoptosis of thymocytes, a translocation of DNase I into the nucleoplasm is observed (Peitsch et al. 1993a). The control mechanisms relieving DNase I inhibition by actin, however, are presently unclear.

In addition, the F-actin depolymerizing activity of DNase I could contribute to the breakdown of the cytoskeleton and cause the shape changes observed during apoptosis (Kolber et al. 1990).

6 Is DNase I the Only Apoptotic Endonuclease?

A number of different candidate endonucleases have been proposed in the past, as reviewed by Peitsch et al. (1994). The involvement of these candidate DNases in apoptotic DNA breakdown has been inferred from the ionic conditions under which DNA ladder formation was observed and these were taken as the enzymatic properties of the particular endonuclease under investigation. In some instances endonucleolytic activity was demonstrated by the zymogram technique. However, a strict correlation or even direct proof of their involvement was not given, since none of these enzymatic entities has been purified nor cloned. Nor have antibodies been raised against any of these enzymes. Furthermore, only some of the proposed endonucleases exhibit ionic requirements or specificities similar to those of DNase I.

We have searched for the presence of apoptotic endonucleases in a number of different cell culture systems using established mouse and human cell lines. In

NIH 3T3 cells apoptosis can be induced by serum withdrawal, in a human pancreatic tumor cell line (PaTu 8902LM) by exposure to drugs inhibiting protein biosynthesis. When these cell lines were screened by the zymogram technique for the presence of endonucleolytic activities after induction of apoptosis, a set of enzymes in the molecular mass region of 45–55 kDa was detected (PADDENBERG et al. 1994; in preparation). Similar enzymes have been observed previously (UCKER et al. 1992; POLZAR et al. 1993). These enzymes are Ca^{2+} and Mg^{2+}-dependent and, after purification, able to degrade the DNA of substrate nuclei internucleosomally (PADDENBERG et al. 1994; in preparation). Using purified plasmid DNA as substrate their enzymatic properties were analyzed. Like DNase I they require Ca^{2+} and Mg^{2+} ions for full enzymatic activity and are inhibited by chelators of divalent cations and Zn^{2+} ions. They are also inhibited by actin fixed in its monomeric state by complexing to gelsolin segment 1 (PADDENBERG et al. 1994). Since there are no sequence data yet available, the nature and relatedness of these enzymes to DNase I or other candidate apoptotic endonucleases are still unclear. It is, however, interesting to note that these enzymes are readily detected in established or permanent cell lines, but absent in organs or primary cell culture systems like freshly isolated thymocytes.

7 Conclusion

There is now an increasing body of evidence that DNase I, in addition to its well-known digestive function, is involved in apoptosis. After induction of apoptosis, DNase I diffuses into the nucleoplasm and attacks the internucleosomal linker regions of chromatin DNA. This can produce a vast number of cuts and nicks—at about every tenth base even of the DNA wound around the histone core (PEITSCH et al. 1993b)—that cannot be repaired, thereby leading to the cessation of transcription. This damage certainly represents an irreversible step in apoptotic cell death, although it has been proposed that DNA breakdown is not a mandatory feature of apoptosis in all cell systems (UCKER et al. 1992; OBERHAMMER et al. 1992).

DNA degradation might also be of great importance in a physiological context. The liberation of high doses of intact DNA from dying cells is potentially harmful. First, high levels of circulating chromatin fragments may cause autoimmune reactions, since anti-chromatin antibodies are associated with such diseases as systemic lupus erythematosus. Second, circulating DNA may transfect other cells leading, for instance, to neoplasmic transformations. The relatively high levels of DNase I found in serum may thus be required to keep the damage of uncontrolled DNA release low. Third, it has been supposed that the phagocytotic activity of normal neighboring cells is facilitated when the chromatin present in the apoptotic bodies is already fragmented.

We are just beginning to appreciate the various roles DNase I may play. Many questions regarding the molecular mechanisms of its regulation and its importance as depicted in Fig. 5 during apoptosis remain to be solved. Does the

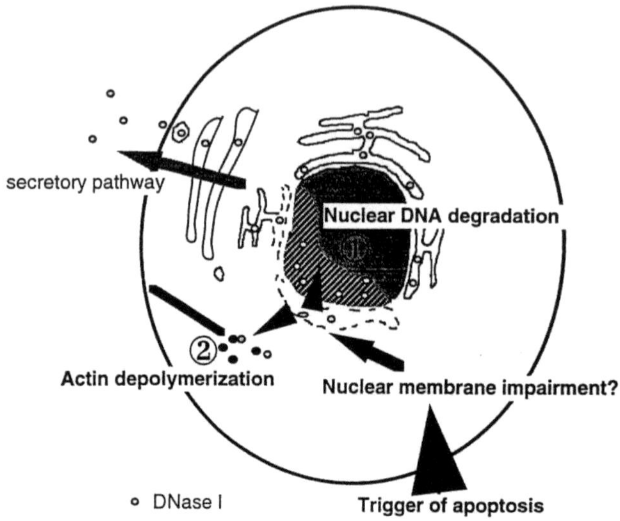

Fig. 5. Hypothetical activation mechanism of DNase I during apoptosis. Activation of an unknown apoptotic signaling pathway leads to the release of DNase I, which is thought to be stored in the perinuclear space in close apposition to the nucleus. After the opening of channels of destruction of the nuclear envelope, the nuclease gains access to the nucleoplasm, leading to the cleavage of DNA at the nucleosomal linker region (route 1). Moreover, the DNase I might also be liberated into the cytoplasm where it will be able to bind to monomeric actin, thereby inducing depolymerization of the actin filaments and thus participate in the induction shape changes observed during apoptosis (route 2)

apototic DNase I also contain a signal peptide, and is it secreted like the pancreatic or parotid DNase I? And if so, how is the access of signal peptide-containing DNase I to the nucleus regulated (route 1 in Fig. 5)? Does DNase I release during apoptosis influence the microfilament integrity leading to the observed membrane blebbing of apoptotic cells (route 2 in Fig. 5)? Is specific up-regulation of DNase I sufficient for the induction of cell death? The challenge ahead will be to find answers to these questions.

Acknowledgement. The continuous financial support of the Deutsche Forschungsgemeinschaft (Bonn) is gratefully noted. Dr. D. Suck (EMBL Heidelberg) kindly supplied the coordinates of the DNase I-DNA octamer complex.

References

Bacher M, Rausch U, Goebel HW, Polzar B, Mannherz HG, Aumüller G (1993) Stromal and epirthelial cells from rat ventral prostate during androgen deprivation and estrogen treatment–regulation of transcription. Exp Clin Endocrinol 101: 78–86

Blikstad I, Markey F, Carlsson L, Persson T, Lindberg U (1978) Selective assay of monomeric and filamentous actin in cell extracts, using inhibition of deoxyribonuclease 1, Cell 15: 935–943

Booth C, Koch GL (1989) Perturbation of cellular calcium induces secretion of luminal ER proteins. Cell 59: 729–737

Campbell WW, Jackson DA (1980) The effect of divalent cations on the mode of action of DNase I. The initial reaction products produced from covalently closed circular DNA. J Biol Chem 255: 3726–3735

Chitrabamrung S, Rubin RL, Tan EM (1981) Serum deoxyribonuclease I and clinical activity in systematic lupus erythematosus. Rheumatol. Int 1: 55–60

Cohen JJ (1993) Apoptosis. Immunol Today 14: 126–130
Cooper EJ, Trautmann ML, Laskowski M (1950) Occurrence and distribution of an inhibitor for deoxyribonuclease in animal tissues. Proc Soc Biol Med 73: 219–222
Drabowska W, Cooper EJ, Laskowski M (1949) A specific inhibitor for deoxyribonuclease J Biol Chem 177: 991–992
Festy B, Paoletti C (1963) Measurement of neutral deoxyribonuclease activity (by liberation of soluble oligonucleotides). Compt Rend H Acad Sci 257: 3682–3685
Fukui Y, Katsumaru H (1979) Nuclear actin bundles in Amoeba Dictyostelium and human HeLa cells induced by dimethyl sulforide. Exp Cell Res 120: 451–455
Gavrieli Y, Sherman Y, Ben-Sasson SA (1992) Identification of programmed cell death in situ via specific labelling of nuclear DNA fragmentation J Cell Biol 119: 493–501
Hewish DR, Burgoyne LA (1973) Chromatin sub-structure. The digestion of chromatin DNA at regularly spaced sites by a nuclear deoxyribonuclease. Biochem Biophys Res Commun 52: 504–510
Hitchcock SE, Carlsson L, Lindberg U (1976) Depolymerization of F-actin by deoxyribonuclease I. Cell 7: 531–542
Kabsch W, Mannherz HG, Suck D, Pai EF, Holmes KC (1990) Atomic structure of actin: DNase I complex. Nature 347: 37–44
Kishi K, Yasuda T, Ikehara Y, Sawazaki K, Sato W, Iida R (1990) Human serum deoxyribonuclease I (DNase I) polymorphism: pattern similarities among isoenzymes from serum, urine, kidney, liver, and pancreas. Am J Hum Genet 47: 121–126
Kolber MA, Broschat KO, Landa-Gonzalez B (1990) Cytochalasin B induces cellular DNA fragmentation. FASEB J 4: 3021–3027
Kossel A (1894) Weitere Beiträge zur Kenntnis der Nucleinsauren. Arch Anat Physiol 195: 1–2
Kreuder V, Dieckhoff J, Sittig M, Mannherz HG (1984) Isolation, characterization and crystallisation of deoxyribonuclease I from bovine and rat parotid gland and its interaction with rabbit skeletal muscle actin. Eur J Biochem 139: 389–400
Kunitz M (1940) Crystalline ribonuclease. J Gen Physiol 24: 15
Kunitz M (1948) Isolation of crystalline deoxyribonuclease from beef pancreas. Science 108: 19–20
Kunitz M (1950) Crystalline deoxyribonuclease I. Isolation and general properties. J Gen Physiol 33: 349–362
Lacks SA (1981) Deoxyribonuclease I in mammalian tissues. Specificity of inhibition by actin. J Biol Chem 256: 2644–2648
Lahm A, Suck D (1991) DNase I induced DNA conformation. 2 Å structure of a DNase I-octamer complex. J Mol Biol 222: 645–667
Lazarides E, Lindberg U (1974) Actin is the naturally occurring inhibor of deoxyribonuclease I. Proc Natl Acad Sci USA 71: 4742–4746
Liao TH, Salnikow J, Moore S, Stein WH (1973) Bovine pancreatic deoxyribonuclease A. Isolation of cynogen bromide peptides; complete covalent structure of the polypeptide chain, J Biol Chem 248: 1489–1495
Lindberg U (1964) Purification of an inhibitor of pancreatic deoxyribonuclease from calf spleen. Biochim Biophys Acta 82: 237–248
Lindberg U (1967) Studies on the complex formation between deoxyribonuclease I and spleen inhibitor II. Biochemistry 6: 343–347
Lommosoff GP, Butler PJG and Klug A (1981) Sequence-dependent variation in the conformation of DNA. J Mol Biol 149: 745–760
Maliska Blaszkiewicz, M, Roth JS (1983) Evidence for the presence of DNase- actin complex in L1210 leukemia cells. FEBS Lett 153: 235–239
Maliska Blaszkiewicz M (1986) DNase I-like activity and actin content in the liver of some vertebrates. Comp Biochem Physiol 84: 207–209
Maliska Blaszkiewicz M (1990) Rat liver DNase I-like activity and its interaction with actin. Z Naturforsch (C) 45: 1165–1170
Mannherz HG, Barrington Leigh J, Leberman R, Pfrang H. (1975) A specific 1:1 G-actin: DNAase I complex formed by the action of DNAase I on F-actin. FEBS Lett 60: 34–38
Mannherz HG, Kabsch W, Leberman R (1977) Crystals of skeletal muscle actin; pancreatic DNAase I complex. FEBS Lett 73: 141–143
Mannherz HG, Goody RS, Konrad M, Nowak E (1980) The interaction of bovine pancreatic deoxyribonuclease I and skeletal muscle actin. Eur J Biochem 104: 367–379
McConkey DJ, Nicotera P, Hartzell P, Bellomo G, Wyllie AH, Orrenius S (1989) Glucocorticoids activate a suicide process in thymocytes through elevation of cytosolic Ca^{2+}- concentration. Arch Biochem Biophys 269: 365–370

Nakayasu H, Ueda K (1983) Association of actin with nuclear matrix from bovine lymphocytes. Exp Cell Res. 143: 55–62

Oberhammer FA, Pavelka M, Sharma S, Tiefenbacher R, Purchino AF, Bursch W, Schulte-Hermann R (1992) Induction of apoptosis in cultured hepatocytes and in regressing liver by transforming growth factor β 1. Proc Natl Acad Sci USA 89: 5408–5412

Oefner C, Suck D (1986) Crystallographic refinement and structure of DNase I at 2 Å resolution. J Mol Biol 192: 605–632

Paddenberg R, Hüttemann B, Wulf S, Mannherz HG (1994) Endonuclease activities of the human pancreatic tumor cell line PaTu 8902LM during cycloheximide induced apoptosis. Eur J Cell Biol 63 [Suppl. 40]: 30

Peitsch MC, Hesterkamp T, Polzar B, Mannherz HG, Tschopp J (1992) Functional characterization of serum DNase I in MRL-lpr/lpr mice. Biochem Biophys Res Commun 186: 739–745

Peitsch MC, Polzar B, Stephan H, Crompton T, MacDonald HR, Mannherz HG, Tschopp J (1993a) Characterization of the endogenous deoxyribonuclease involved in nuclear DNA degradation during apoptosis (programmed cell death). EMBO J 12: 371–377

Peitsch MC, Müller C and Tschopp (1993b) DNA fragmentation during apoptosis is caused by frequent single strand cuts. Nucl Acids Res 18: 4206–4209

Peitsch MC, Mannherz HG, Tschopp J (1994) The apoptosis endonucleases: Cleaning up after cell death. Trends Cell Biol 4: 37–41

Price PA, Stein WH, Moore S (1969) Effect of divalent cations on the reduction and re-formation of the disulfide bonds of deoxyribonuclease, J Biol Chem 244: 929–932

Price PA, Liu TY, Stein WH, Moore S (1969) Properties of chromatographically purified bovine pancreative deoxyribonuclease, J Biol Chem 244: 917–923

Polzar B, Mannherz HG (1990) Nucleotide sequence of a full length cDNA clone encoding the deoxyribonuclease I from the rat parotid gland. Nucleic Acids Res 18: 7151

Polzar B, Peitsch MC, Loos R, Tschopp J, Mannherz HG (1993) Overexpression of deoxyribonuclease I (DNase I) transfected into COS-cells; its distribution during apoptotic death. Eur J Cell Biol 62: 397–405

Polzar B, Zanotti S, Stephan H, Rauch F, Peitsch MC, Irmler M, Tschopp J, Mannherz HG (1994) Distribution of deoxyribonuclease I in rat tissues and its correlation to cellular turnover and apoptosis (programmed cell death). Eur J Cell Biol 64: 200–210

Rohr G, Mannherz HG (1978) Isolation and characterization of secretory actin: DNase I complex from rat pancreatic juice. Eur J Biochem 89: 151–157

Suck D (1982) Crystallisation and preliminary crystallographic data of bovine pancreatic deoxyribonuclease I. J Mol Biol 162: 511–513

Suck D, Oefner C, Kabsch W (1984) Three-dimensional structure of bovine pancreatic DNase I at 2.5 Å resolution, EMBO J 3: 2423–2430

Suck D, Oefner C (1986) Structure of DNase I at 2.0 Å resolution suggests a mechanism for binding to and cutting DNA. Nature 321: 620–625

Suck D, Lahm A, Oefner C (1988) Structure refined to 2 Å of a nicked DNA octanucleotide complex with DNase I. Nature 332: 464–468

Ucker DS (1991) Death by suicide: One way to go in mammalian cellular development? New Biol 3: 103–109

Ucker DS, Obermiller PS, Eckhart W, Apgar JR, Berger NA, Meyers J (1992) Genome digestion is a dispensable consequence of physiological cell death mediated by cytotoxic T lymphocytes. Mol Cell Biol 12: 3060–3069

Valkov NI, Ivanova MI, Uscheva AA, Krachmarov CP (1989) Association of actin with DNA and nuclear matrix from Guerin ascites tumor cells. Mol Cell Biochem 87: 47–56

Williams GT, Smith CA, McCarthy NJ, Grimes EA (1992) Apoptosis: Final control point in cell biology. Trends Cell Biol 2: 263–267

Wyllie AH (1980) Glucocorticoid-induced thymocyte apoptosis is associated with endogenous endonuclease activation. Nature 284: 555–556

Zanotti S, Polzar B, Doll K, Stephan H, Niessing J, Mannherz HG (1995) Localization of deoxyribonuclease I gene transcripts in rat tissues and its correlation to apoptotic cell elimination. Histochem (in press)

The Ligands and Receptors of the Lymphotoxin System

C.F. WARE[1], T.L. VANARSDALE[1], P.D. CROWE[1], and J.L. BROWNING[2]

1 Tumor Necrosis Factor-Related Cytokine and Receptor Families

The tumor necrosis factor (TNF)-related cytokines have emerged over the past 2 years as a large family of pleiotropic mediators of host defense and immune regulation. Members of this family exist in membrane-anchored forms acting locally through cell-to-cell contact, or as secreted proteins capable diffusion to

[1] Division of Biomedical Sciences, University of California, Riverside, CA 92521, USA
[2] Biogen, Inc., Department of Immunology and Inflammation, Fourteen Cambridge Center, Cambridge, MA 02142,USA

more distant targets. A parallel family of receptors signals the presence of these molecules leading to the initiation of cell death or cellular proliferation and differentiation in the target tissue (see SMITH et al. 1994; BANCHEREAU et al. 1994 for reviews). The focus of this review is on two members of this family produced by activated T cells, the original lymphotoxin-α (LT-α, previously referred to as TNF-β), and a new member, lymphotoxin-β (LT-β), and their specific receptors. Initially discovered by cytotoxic activity in vitro, lymphotoxin, as a secreted molecule, was one of the earliest postulated mechanisms used by cytotoxic T lymphocytes (RUDDLE and WAKSMAN 1968; GRANGER and WILLIAMS 1968). The molecular cloning of LT and TNF dramatically revised the view of these cytokines as limited nonspecific cytotoxins and revealed their more intricate role in immunoregulation and host defense. Although once thought to be merely a redundant form of TNF, new findings have indicated that lymphotoxin has a role in immune physiology distinct from TNF and forms a system of secreted and membrane-anchored immunoregulatory molecules.

1.1 Ligands

Defined as a pair of genes in 1984, the TNF family of ligands and receptors has grown to at least nine receptor-ligand pairs (SMITH et al. 1994; BEUTLER and VANHUFFEL 1994). Characteristics of this family of ligands are summarized in Table 1 and a sequence comparison is shown in Fig. 1. It is seen that even the most related cases, i.e., the two LT species and TNF are only about 25%–30% identical although amino acid homology is about 50%. A more careful examination shows that there are several regions of relatedness between the different members and Fig. 1 shows one such alignment that more heavily weights the conservation of cysteine residues. Other alignments have been suggested (SMITH et al. 1993; GOODWIN et al.1993a).

TNF and LT-α are both structured into a sandwich of two anti-parallel β-pleated sheets with the "jelly roll" or Greek key topology (JONES et al. 1989; ECK et al. 1992). The rms deviation between the Cα atoms of the β-strand residues is 0.61 Å suggesting a high degree of similarity in their molecular topography. A structural feature emerging from molecular studies of TNF and LT-α is the propensity to assembly into oligomeric complexes. Intrinsic to the oligomeric structure is the formation of the receptor binding site at the junction between

Fig. 1. The Tumor necrosis factor (*TNF*) ligand family. An alignment of eight members of the family illustrating the variations in the length of the intracellular NH_2-terminal domains and the stalk regions spacing the COOH-terminal receptor binding domain from the transmembrane region (*beginning*, just before the first β strand). All sequences are human except for FasL (rat) and 4-1BB-L (mouse). The alignment weighs cysteine conservation heavily, and due to the poor relatedness between some family members many alternative alignments can be used. The *bars* over the sequences indicate β strand structures in lymphotoxin (*LT*) using the nomenclature of ECK and SPRANG (1989). Canonical N-linked glycosylation sites are *underlined*, as are probable transmembrane sequences and the two disulfide linked cysteines in TNF are marked with *dots*

```
hTNF                                                               MSTESMIRDVELAEEALPKKTGGPQGSRRCLFLSLFSFLIVAGATTLFCLL
hLT-α                                                                        MTPPERLFLPRVCGTTLHLLLLGLLLVLLPGAQGLPGVGLT
hLT-β                                                                     MGALGLEGRGGRLQGRGSLLLAVAGATSLVTLLLAVPITVLAVLAL
rFasL          MQQPVNYPCPQIYWVDSSATSPWAPPGSVFSCPSSGPRGPGQRRPPPPPPPPSPLPPPSQPPPLPPLSPLKKKDNIELWLPVIFFMVLVALVGMGLG
hCD27L                                                                          MPEEGSGCSVRRRPYGCVLRAALVPLVAGLVICLVVC
hCD30L                                                    MDPGLQQALNGMAPPGDTAMHVPAGSVASHLGTTSRSYFYLTTATLALCLVFTVATIMVLV
hCD40L                                                                      MIETYNQTSPRSAATGLPISMKIFMYLLTVFLITQMIGSALFAV
m4-1BBL     MDQHTLDVEDTADARHPAGTSCPSDAALLRDTGLLADAALLSDTVRPTNAALPTDAAYPAVNVRDREAAWPPALNFCSRHPKLYGLVALVLLLLIAACVP

                                                                                                 A                A'
hTNF        HFGVIGPQREEFPRDLSLISPLAQAVRSSSRTP-------------------------------------SDKPVAHVVANPQAEGQ---LQWLNR
hLT-α       PSAAQTARQHPKMHLAHS----------------------------------------------------TLKPAAHLIGDPSK--QNS-LLWRAN
hLT-β       VPQDQGGLVTETADPGAQAQQGLGFQKLPEEEPETDLS--------------------------------PGLPAAHLIGAPLK-GQG--LGWETT
rFasL       MYQLFHLQKELAELREFTNHSLRVSSFEKQIANPSTPSETK-----------------------------KPRSVAHLTGNPRSRSIP--LEWEDT
hCD27L      IQRFAQAQQQLPLESLG-----------------------------------------------------WDVAELQLNHTGPQ--QDPRLYWQGG
hCD30L      VQRTDSIPNSPDNVPLKGGNCSEDLLCILKRAP-------------------------------------FKKSWAYLQVAKHL--NKTKLSWNKD
hCD40L      YLHRRLDKIEDERNLHEDFVFMKTIQRCNTGERSLSLLNCEEIKSQFEGFVKDIMLNKEETKKENSFEMQKGDQNPQIAAHVISEASSKTTSV-LQWAEK
m4-1BBL     IFTRTEPRPALTITTSPNLGTRENNADQVTPVSHIGCPNTTQQ---------------------------GSPVFAKLLAKNQASLCNTTLNWHSQ

              A"      B'      B          C                        D                    E
hTNF        RANALLAN-GVELRD--NQ-LVVPSEGLYLIY-SQVLFKGQGCPSTHVLLTHTISRIAVS------YQTKVN--LLSAIKS------PCQRETPEGAEAK
hLT-α       TDRAFLQD-GFSLS--NNS-LLVPTSGIYFVYNSQVVFSGKAY-SPKAT-SSPLYLAHEVQLFSSQYPFHVP--LLSSQKM---VY-PGLQE--------
hLT-β       KEQAFLT-SGTQFSDAEG--LALPQDGLYYLYTCLVGYRGRAPPGGGDPQGRSVTLRSSLYRAGGAYGPGTPELLLEGAETVTPVLDPARRQGYG-----
rFasL       YGTALI--SGVKYKK-GG--LVINEAGLYFVY-SKVYFRGQSCNSQPLSHKVYMRN------FK--YPGDL---VLMEEKK--LNY--CTTGQ-------
hCD27L      PALGRSFLHGPEL-DKGQ--LRIHRDGIYMVHI-QVTLAI--CSSTTASRHHPTTLAVGICSPASRSIS-----LLRLSFHQG-----CTIV--------
hCD30L      G-----ILHGVRYQD-GN--LVIQFPGLYFII-CQLQFLVQ-CPNNSVDLKLELLINKHIKK--QALVT-----VCESGMQTKHVYQNLSQFL-------
hCD40L      GYYTMSNNLVTL--ENGKQ-LTVKRQGLYYIYA-QVTF----CSNREASSQAPFIASLCLKSPGRFERI-----LLRAAN-THSSAKPCGQQSIH-----
m4-1BBL     DGAGSSYLSQGLRYEEDKKELVVDSPGLYYVFLELKLSPTFTNTGHKVQGWVSLVLQAKP--QVDDFDN-----LAL----TVELF-PCSMENKLVDRS-
                                                     •                                        •

                    F                G              H
hTNF        P-WYEPIY--LGGVFQLEK----GDRLSAEINRPDYLDFAESGQ---VYFGIIAL
hLT-α       P-WLHSMY--HGAAFQLTQ----GDQLSTHTDGIPHLVL--SPS--TVFFGAFAL
hLT-β       PLWYTSVG--FGGLVQLRR----GERVYVNISHPDMVDFAR---GKT-FFGAVMVG
rFasL       -IWAHSSY--LGAVFNLTV----ADHLYVNISQL---SLINFEESKT-FFGLYKL
hCD27L      ----------SQRLTPLAR----GDTLCTNLTGTLLPSR-NTDE--T-FFGVQWVRP
hCD30L      ----------LDYLQVNTTISVNVDTFQYIDTST-FPLE-NVL----SIFLYSNSD
hCD40L      ----------LGGVFELQP----GASVFVNVTDPSQVSHG---TGFTS-FGLLKL
m4-1BBL     --WSQLLL--LKAGHRLSV----GLRAYLHGAQDAYRDWELSYPNTTS-FGLFLVKPDNPWE
```

Table 1. Properties of the ligands in the tumor necrosis factor family

System	Names	Size observed (kDa)	Secreted	Surface	Chromosome		Sources	Reference
					Mouse	Human		
TNF	TNF TNF-α, cachectin	17	yes	yes	17 MHC	6 MHC	Macrophages, T cells, many tissues	PENNICA et al. 1984
LT	LT-α, TNF-β	25	yes	yes	17 MHC	6 MHC	T and B cells	GRAY et al. 1984
	LT-β, p33	33	no	yes	17 MHC	6 MHC	T and B cells	BROWNING et al. 1993
Fas/Apo-1	Fas-L	45	?	yes	1	1q23	T cells, reproductive tissue	SUDA et al. 1993; TAKASHI et al. 1994b
4-1BB	4-1BB-L	50	?	yes	17 Non MHC	19p13	Lymphoid cells and stromal lines	GOODWIN et al. 1993
Ox-40	Ox-40-L	34	?	yes	1	1q25	T cells	BAUM et al. 1994
CD27	CD27-L; CD70	50	?	yes	?	19p13	B cells, thymic stroma, T cells	GOODWIN et al. 1993
CD30	CD30-L	40	?	yes	4	9q33	T cells, monocytes	SMITH et al. 1993
CD40	CD40-L gp39;TBAM;TRAP	39	?	yes	?	X	T cells, mast cells	HOLLENBAUGH et al. 1992; ARMITAGE et al. 1992

TNF, tumor necrosis factor; LT, lymphotoxin.

neighboring subunits creating a multivalent ligand. The quaternary structures of both TNF and LT-α have been shown to exist as trimers by analysis of their crystal structures (Eck and Sprang 1989; Eck et al. 1992; Jones et al. 1992), by chemical cross-linking (Smith and Baglioni 1987; Browning and Ribolini 1989; Goh 1993), by analysis of physical properties (Wingfield et al. 1987; Schoenfeld et al. 1991; Pennica et al. 1993) and by complexation with receptor (Loetscher et al. 1991; Pennica et al. 1993). Many of the amino acids conserved between the different ligands are in stretches of the scaffold β-sheet and these regions have been roughly demarcated in Fig. 1. As portions of these scaffold sequences are conserved across the various family members, it is likely that the basic β sandwich structure is preserved in all of these molecules. Since the subunit conformation is likely to remain similar, it follows that the quaternary structure may also be maintained. The low affinity nerve growth factor receptor (LNGFR) also is composed of cysteine-rich TNF receptor (TNFR-) like domains, yet its ligand, NGF, is not particularly related to TNF. The NGF structure, like TNF, is formed from parallel β-sheets and dimerization of NGF relies on hydrophobic interactions between the faces formed by the β-sheets (McDonald et al. 1991; McDonald and Hendrickson 1993). While TNF also exploits hydrophobic interactions between elements of the β-sheet structure to generate the trimer, the similarity ends at this point. NGF lacks all of the amino acid conservation shown in Fig. 1 and is the exception within this family . On this basis it is reasonable to assume that the TNF structural framework is conserved among the non-NGF family members, yet the surface accessible amino acids will be different. The CD30 (Smith et al. 1993) and 4-1BB (Goodwin et al. 1993b) ligands are the most distant members of this family.

Each ligand gene encodes a short NH_2-terminal domain followed by a hydrophobic domain of varying length. The Fas (Suda et al. 1993) and 4-1BB ligands are distinguished by more extensive intracellular domains of about 80 amino acids opening the possibility of bidirectional signalling or cytoskeletal interactions. Such an event has been described for CD40 where receptor binding induced the internalization of its ligand CD40-L (Yellin et al. 1994). The extracellular receptor binding domain resides in the COOH-terminal portion. This arrangement is typical of type II membrane proteins wherein the COOH-terminal domain is external and the NH_2-terminal region remains in the cytoplasm. All of the genes have a stretch of basic residues preceding the hydrophobic domain which may serve as stop transfer signals.

1.2 Receptors

The defining feature of this family of cytokine receptors is found in the cysteine rich extracellular domain initially revealed by the molecular cloning of two distinct TNF receptors (Smith et al. 1990; Kohno et al. 1990; Loetscher et al. 1990; Schall et al. 1990) (Table 2). This family of genes encodes glycoproteins characteristic of type I transmembrane proteins with an extracellular ligand binding domain, a

Table 2. Properties of the receptors in the tumor necosis factor system

Receptor	Names	Size observed (kDa)	Ligands	Soluble forms	Chromosome		Tissue expression
					Mouse	Human	
CD120a	$TNFR_{60}$; R1;A	55–60	TNF, LT-α LT-$\alpha_2\beta_1$	Shed	6	12p13	Broad
CD120b	$TNFR_{80}$; R2;B	75–84	TNF, LT-α LT-$\alpha_2\beta_1$	Shed	4	1p36	Hematopoietic/broad
LT-βR	TNFrrp	61	LT-$\alpha_1\beta_2$	?	6	12p13	Broad
CD95	Fas; Apo1	43	Fas-L	Alternate splice	19	10	Lymphocytes/broad
4-1BB	ILA(hu)	33	4-1BB-L	Alternate splice	4	1p36	Activated T cells/broad
OX-40	ACT35(hu)	48	OX40-L	?	4	1p36	Activated $CD4^+$ T cells
CD27	tp55	50–55 dimer	CD70; CD27-L	Shed	6[a]	12p13	Resting T cells
CD30	Ki-1	120	CD30-L	?	4[a]	1p36	Hematopoietic/ Hodgkin's lymphoma
CD40	Bp50; p50	43–47	CD40-L; gp39; TBAM; TRAP	Shed	2	20q11–q13	B and T cells/carcinomas
LNGFR	p75	75	NGF; neurotrophins	Shed	11	17q21–22	Nervous system

TNF, Tumor necrosis factor; LT, lymphotoxin; LNGFR, low-affinity nerve growth factor receptor; NGF, nerve growth factor.
[a]Assignment by synteny

single membrane spanning region and a cytoplasmic region involved in activating cellular functions. The cysteine-rich ligand binding region exhibits a tightly knit disulfide-linked core domain, and depending upon the particular family member, that is repeated multiple times with four domains for the most of the receptors, but as few as three for Fas/APO-1 (ITOH et al. 1991; LEITHAUSER et al.1993) and as many as six for CD30 (DURKOP et al. 1992). These receptors are heavily glycosylated with as much 40%–50% of their molecular mass contributed by posttranslational modifications.

The TNF receptor family includes several notable members: (1) LNGFR, which binds NGF and three other related ligands forming the neurotrophin family (BRADSHAW et al.1993; DAVIES 1994). The NGF-related ligands also bind (with high affinity) to the *trk* family of receptor kinases which are unrelated to the TNFR family. (2) CD40, identified initially on B cells and carcinomas, was subsequently shown to be a critical factor for B cell growth and differentiation (STAMENKOVIC et al.1989; BANCHEREAU et al.1994); (3) CD27, a T cell expressed protein which can provide a costimulatory signal for T cell proliferation (CAMERINI et al.1991); (4) CD30, an antigenic marker for Hodgkin's lymphoma (DURKOP et al. 1992); (5) Fas/APO-1 (ITOH et al. 1991; LEITHAUSER et al. 1993), to which antibodies induce apoptosis of lymphocytes and is the gene associated with the murine lymphoproliferative *lpr* disorder (WATANABE-FUKUNAGA et al.1992; ITOH et al.1991). Two additional members are 4-1BB, a cDNA clone isolated from an induced murine T helper clone (KWON and WEISSMAN 1989) and its human homologue (SCHWARZ et al.1993) and OX40, a surface protein expressed on activated rat and human T cells (MALLETT et al.1990). The most recently described member (TNFR-related protein) (BAENS et al.1993) is a receptor for the LT–β subunit of the surface LT complex (CROWE et al.1994b).

The genomic localization of several receptors has revealed two clusters, one located on human 12p13, ordered as LT-βR (P. Marynen, personal communication), CD27 (LOENEN et al.1992b) and $TNFR_{60}$ (DERRE et al. 1991); and the other found on chromosome 1p36 with OX40 (LATZA et al. 1994), 4-1BB (KWON and WEISSMAN 1989), CD30 (SMITH et al.1993) and $TNFR_{80}$ (MILATOVICH et al. 1991; BAKER et al. 1991). The gene encoding human $TNFR_{60}$ spans ~11 kb and contains ten exons; the LT-βR spans ~9 kb with a similar exon-intron arrangement (P. Marynen, personal communication). The reported coding region for $TNFR_{80}$ lies within ten exons spanning ~20 kb (S. Santee and L. Owen-Schaub, personal communication).

In addition to these mammalian genes a subfamily of viral proteins exist with homology to TNFR. The T2 open reading frame (ORF) of the Shope fibroma virus (Poxviridae) encodes soluble $TNFR_{80}$ homologue with TNF/LT-α binding activity (UPTON et al.1987; SMITH et al.1991) and is representative of several similar poxvirus encoded genes (MASSUNG et al.1994). The related gene in the rabbit myxoma virus has been shown by McFadden (UPTON et al.1991) to function as a major virulence factor. Indeed, a T2 homologue has been identified in variola (smallpox) virus that is truncated in attenuated vaccinia suggesting its importance in this former scourge (MASSUNG et al. 1993). A gene encoding a single cysteine repeat in the phytopathogenic fungi *Cladosporium fulurum* has been suggested to represent a primordial receptor (BAZAN 1993). Smith and Farrah have suggested that the receptor family may be categorized into two subgroups based upon tissue expression and function, the immune/inflammatory group encompassing the majority of the members, and the nervous system group that has a single member, LNGFR (FARRAH and SMITH 1992).

The cytoplasmic regions differ considerably within this receptor family and none contain any inherent enzymatic activity. Despite this, a common functional relationship is seen in the ability of these receptors to regulate cell growth and death in different tissues. The clearest case is with $TNFR_{60}$ and Fas which both mediate apoptosis and share limited homology over a short span of ~45 residues (TARTAGLIA et al. 1993a). However, it appears that overlapping but divergent signaling mechanisms are used by these receptors (WONG and GOEDDEL 1994). The ligand for CD30 induces cell death (apoptosis) in a susceptible lymphoma line (SMITH et al. 1993), whereas CD40 signaling blocks apoptosis in B lymphocytes (LEDERMAN et al. 1994); in neuronal cells the unligated LNGFR signals apoptosis (RABIZADEH et al. 1993). $TNFR_{80}$ has been shown to enhance T lymphocyte proliferation (TARTAGLIA et al. 1993b), whereas $TNFR_{60}$ can provide a proliferative signal in fibroblasts (ENGELMANN et al. 1990a). Indeed a theme shared by $TNFR_{80}$, 4-1BB and OX-40 is inducible expression upon activation of T cells and their ability to signal enhancement of T cell proliferation; whether this functional theme is coincident with the clustered arrangement of their genes is intriguing speculation. Interestingly, Fas has been implicated as a signaling receptor for apoptosis mediated by Cytotoxic T lymphocytes (CTLs) (ROUVIER et al. 1993), but this receptor can also be a coactivation signal for T lymphocytes (ALDERSON et al. 1993). Clearly, the ultimate cellular fate initiated by ligand-receptor complexes of this family are highly dependent upon the cell type and state of differentiation in which the signals are generated.

2 The Lymphotoxin Ligands

2.1 Lymphotoxin-α

This ligand, also called TNF-β, was characterized biochemically as a cytotoxic protein secreted from a human B lymphoblastoid line RPMI 1788 (AGGARWAL et al. 1984) and the subsequent molecular cloning of both LT-α and TNF revealed their similarity (GRAY et al. 1984; PENNICA et al. 1984) (reviewed in RUDDLE 1992). In contrast to TNF, the LT saga has remained less well developed. There are several reasons for this lack of progress. First, LT appeared to mirror TNF and since recombinant TNF was more readily available, experimentation centered on this cytokine. Secondly, the mouse LT system has not been well characterized at the protein level and recombinant forms are not available, severely limiting proper work.

Recombinant human LT-α has been produced from many systems including bacteria (SCHOENFELD et al. 1991; SEOW et al. 1989), baculovirus-infected insect cells (CROWE et al. 1994a) and mammalian cells (BROWNING and RIBOLINI 1989; FUKUSHIMA et al. 1993). Human LT-α, in contrast to human TNF, is glycosylated on N and O residues and, depending on the source, considerable sugar-based microheterogeneity in the various preparations has been observed (HAINS and AGGARWAL 1989; LANTZ et al. 1991; BENJAMIN et al. 1992; FUKUSHIMA et al. 1993). The oligosaccharide composition clearly affects in vivo clearance rates (FUKUSHIMA et al. 1993), and this parameter should be carefully considered in future work. Moreover, the nature of the antigenic response of various animals to glycosylated and nonglycosylated forms may differ such that monoclonal antibodies (mAbs) that neutralize bacterially derived LT-α do not necessarily bind to mammalian LT-α. Similar variations in the ability of different sera to neutralize all of the different glycosylated forms derived from human B cell lines were noted (BENJAMIN et al. 1992). As outlined above, there is considerable evidence demonstrating that LT-α exists as a trimer in solution. On the other hand, gel exclusion sizing of recombinant preparations of LT-α often show smaller forms whether derived from bacterial expression (SCHOENFELD et al. 1991) or from insect and mammalian cells (Browning and Meier, unpublished). Interactions with the gel matrices have been postulated to affect attempts to size LT-α accurately; however, it is unlikely that these interactions would lead to distinct heterogeneity (SCHOENFELD et al. 1991). A careful study combining light scattering and sedimentation data found that while TNF behaved primarily as a trimer with 20% dimer, bacterially derived LT-α was a mixture of 50% trimer and 50% dimer and monomer. Thus of these two well studied molecules, LT-α appears to form nontrimeric structures more readily than TNF. This observation is curious since the TNF trimeric structure dissociates readily in detergent as opposed to LT-α which is stable in the nonionic detergent NP-40 (BROWNING and RIBOLINI 1989).

In contrast to TNF, LT-α has never been observed to retain its leader or transmembrane region, even in the surface form (BROWNING et al. 1991). Whether LT-α is actually a type II membrane protein which is very efficiently proteolytically processed or a more classical type I secreted protein is not clear. In fact, this distinction may be simply semantic in nature since the same machinery may be used to import both type I and type II membrane proteins into the endoplasmic reticulum.

2.2 Lymphotoxin-β

A number of groups provided evidence that lymphotoxin may exist in a cell surface form (WARE et al. 1981; ANDERSSON et al. 1989; ABE et al. 1991; YASUKAWA et al. 1993). The importance of these observations began to emerge from studies of a CD4$^+$ T cell hybridoma, II-23.D7, that exhibited high levels of expression (BROWNING et al. 1991; ANDROLEWICZ et al. 1992). Biochemical analyses utilizing

[^{35}S]-cysteine and methionine labeling showed that the surface LT-α form lacked a retained transmembrane region and was indistinguishable from the secreted form (BROWNING et al. 1991). An additional 33 kDa band coimmunoprecipitated with the LT-α and biochemical characterization indicated that it was present as a complex with LT-α and not a separate protein immunologically related to LT-α (ANDROLEWICZ et al. 1992). The association between LT-α and the 33 kDa protein occurred during biosynthesis and not as an association of secreted LT with the 33 kDa surface protein. Amino acid sequencing of the 33 kDa protein led to the cloning of a cDNA from a phorbol myristate acetate (PMA) activated II-23 cDNA library that encoded the associated protein now called LT-β (BROWNING et al. 1993). The LT-β molecule is as related to LT-α as LT-α is to TNF (Fig. 1). LT-β has a single N-linked glycosylation site, unlike human TNF, and has only a single cysteine residue that does not appear to be involved in intersubunit disulfide bond formation. As outlined above, the nature of the conserved amino acids located primarily within the internal β strand regions suggested a LT-β structure similar to LT-α and TNF and thereby presented a rational basis for the heteromeric complex formation.

The gene encoding LT-β has 4 exons in a similar arrangement as LT-α and TNF genes and is located in the MHC ~2 kb from TNF (BROWNING et al. 1993). The mRNA transcript for LT-β in activated T cells is 0.9 kb and codes for a type II transmembrane protein characteristic of the ligand family. At the genetic level, a comparison of the mouse and human LT-β promoter regions shows excellent conservation in the sequences surrounding the putative TATA box up through the coding sequence (Browning, unpublished). An NF-κB enhancer-3 element is conserved in both the human and mouse promoters at about -90 bp and multiple copies of NF-κB enhancers have been described in the TNF and LT-α gene promoters. An AU rich sequence has been characterized in the 3' UTR of TNF and LT-α as well as several other cytokine genes that can decrease mRNA stability and have complex direct effects on translation of TNF (BEUTLER 1990). This element is lacking in the 3' UTR of both mouse and human LT-β genes.

The translational start of human LT-β is unusual (BROWNING et al. 1993). There is an ATG followed by a CTG three codons downstream; however, the first in-frame ATG is very close to the 5' end of the RNA and is probably not utilized very efficiently. Clones possessing only the CTG start were functional and, moreover, amino acid sequencing of LT-β indicated that this CTG encoded the NH_2-terminal amino acid of the mature protein. The GC rich region following this CTG may slow ribosomal scanning sufficiently to allow the CTG to function efficiently (M. KOZAK, personal communication). It is possible that both the ATG and CTG start codons are utilized.

In spite of further confusion that is likely to occur, we have suggested that the original nomenclatures (TNF-α and TNF-β or TNF and the original LT, both of which are still in use) be changed to reflect the biochemical structures involved (BROWNING et al. 1993). Typically, individual proteins within an oligomeric structure are designated with greek letters and accordingly LT-α and LT-β refer to subunits of the complex. The actual name "lymphotoxin" is a remnant of the description of the original cytotoxic activity, yet is certainly no less descriptive than the accepted TNF nomenclature. In light of the phenotype of the LT-α knockout, i.e., loss of lymph nodes (discussed below), one could consider the term lymphotoxin prophetic or perhaps "lymph node trophin" might satisfy cytokine lexicologists.

2.3 The Lymphotoxin Membrane Complex: An Unusual Heteromer

The ratios of LT-α to LT-β molecules within the complex have been of much interest due to the ramifications for receptor signaling. Assuming a trimeric arrangement, there are four conceivable forms of these two molecules; LT-α3, LT-α2/β1, LT-α1/β2 and LT-β3. The bulk of the LT-β within the complex could be cross-linked into a dimeric form and as this complex also contained LT-α, a LTα1/β2 stoichiometry is the minimal assembly capable of accommodating these data (Androlewicz et al. 1992). More recently, panels of anti-LT-α and anti-LT-β mAbs have been exploited to analyze the complex (Browning et al. 1995). Anti-LT-α mAbs were found to be of two major categories, those that could only recognize a minor form of surface LT and a second group that is pantropic. Immunoprecipitation analysis using these reagents were consistent with binding of one group to a small proportion of LTα2/β1 complex, whereas the second pantropic group bound to all forms including the predominant LTα1/β2 type. Thus mAb and interactions of the various receptors with the surface LT forms provide results consistent with this model.

The original work leading to the description of the surface LT complex relied heavily on anti-LT-α antibodies and the question whether LT-β assembles without LT-α or forms complexes with other members of the ligand family could not be answered. Currently, there is no evidence supporting the existence of complexes of LT-β with other cytokines in this family. This conclusion must be qualified because of technical considerations. The TNF trimer is readily dissolved in mild detergents and a complex with this molecule may not survive an immunoprecipitation protocol. In contrast the LT-α oligomer is quite stable to nonionic detergents (Browning and Ribolini 1989) and this stability decreases upon inclusion of LT-β molecules (Browning, unpublished). Therefore it is possible that other LT-β complexes may exist; however, it is felt that such promiscuous assembly would be very chaotic in the activated T cell expressing several members of the family simultaneously.

Whether LT-β trimers are formed has been difficult to address. On the surface of the activated II-23 cell line, essentially all LT-β epitopes can be precleared with antibodies to LT-α or the LT-β R:Fc construct. These experiments taken alone essentially rule out the existence of a major LT-β component. There is one anti-LT-β mAb, however, that appears to preferentially recognize a different LT-β form that does not have LTα1/β2 stoichiometry. The biochemistry underlying these observations is still unclear. The surface LT complex has been completely reconstructed by transfection of CHO cells with both genes. These stable cell lines can display all of the epitopes found on the II-23 cell indicating that these two genes alone are sufficient to encode the complex (Browning, unpublished). Surface complexes cannot be observed on stable lines expressing either gene alone; LT-β can be transiently expressed at high levels on COS cells or insect cells infected with baculovirus encoding full length LT-β. Other studies have shown that a soluble version of LT-β secreted by itself is a highly oligomerized. However, upon coexpression with LT-α, a 70 kDa trimeric-like structure is formed containing both proteins (W. Meier, P. Crowe, J. Browning and C. Ware, unpublished results). These results indicate that high level expression of LT-β alone in vivo is unlikely and may not be compatible with cell survival and growth.

2.4 Assembly of the Lymphotoxin Complex

The assembly of the complex occurs very quickly during biosynthesis. Pulse chase experiments using either LT-α (Androlewicz et al. 1992) or LT-β (Browning unpublished) antibodies show that the addition of oligosaccharrides to LT-β is very fast and nonglycosylated intermediates were not observed. LT-α is glycosylated more slowly and heteromeric complexes with LT-β can be seen prior to addition of the N-linked sugar chain to LT-α. An LT-α form retaining its leader sequence apparently cannot be observed suggesting that the proteolytic processing of the signal sequence is rapid and occurs internally. Thus at least some if not all of the LT complex is formed early during transit through the Golgi apparatus.

Analysis of the supernatants of several cell types has failed to show any cleaved (secreted) LT-β forms either with or without LT-α. Since cleaved LT-β forms may be masked by the 25 kDa secreted LT-α that is present, cysteine metabolic labeling was used to examine exclusively LT-β in the absence of LT-α labeling. This approach also failed to show secreted LT-β. The same method was applied to the cell surface LT forms to ask whether LT complexes with mixed cleaved and full length LT-β molecules were present as is observed with surface TNF forms (Kriegler et al.1988; Perez et al.1990). If LT-β cleavage occurs, it is a rare event in the systems examined to date.

It is interesting to speculate on why nature has produced such a complex ligand. The trimeric structure is obviously an efficient way to aggregate receptors. Overlaying multiple proteins within this trimer can led to very complex signaling scenarios (see below) and may be similar to aspects of PDGF function. In addition to the novel possibilities that emerge from the receptor binding aspects, the two gene paradigm may provide a failsafe system for a critical pathway. The requirement for expression of two proteins simultaneously could prevent abberant activation of the system by mistakes in expression.

3 The Lymphotoxin Receptors

3.1 Ligand Specificity of the Tumor Necrosis Factor Receptors

Two specific high affinity receptors mediate signal transduction responsible for the biological effects of TNF and LT-α. However, TNF may also posses trypanolytic and ion channel activities independent of TNFR binding (LUCAS et al. 1994; KAGAN et al. 1992). Two TNF receptors with predicted molecular mass of 75–80 kDa and 55–60 kDa (referred to here as $TNFR_{60}$ and $TNFR_{80}$, respectively)[3] were initially distinguished by cross-linking of radiolabeled TNF to different cell lines. cDNA clones encoding these two distinct TNF receptors have been isolated from partial sequence of the soluble binding protein ($TNFR_{60}$) (NOPHAR et al. 1990; LOETSCHER et al. 1990; SCHALL et al. 1990) or by direct expression cloning ($TNFR_{80}$) (SMITH et al. 1990). Two distinct soluble TNF binding proteins of 30–40 kDa (OLSSON et al. 1989; ENGELMANN et al. 1990b), now recognized as proteolytically processed forms of the cell surface receptors, retain ligand binding activity and appear important for regulating TNF bioavailability (ADERKA et al. 1992). Homotrimers of TNF and LT-α bind to cell surface $TNFR_{80}$ and $TNFR_{60}$ with high affinity in the low nanomolar range (K_d = 0.1–0.5 n*M*) and both receptors can bind ligand when expressed independently. TNF and LT-α are cross-competitive for binding to both TNFR (HOHMANN et al. 1990b) providing a reason for the similarity in initiating cellular responses. The recombinant receptors exhibit multiple equilibria binding isotherms of high and low affinity (K_d = 20 n*M*) indicative of receptor cooperation in generating high affinity sites (SCHALL et al. 1990; SMITH et al. 1990). Both ligands have been shown to express multiple receptor binding sites, with two to three per trimer at saturation (PENNICA et al. 1992a).

As determined from the crystal structure (BANNER et al. 1993) the ligand binding region of $TNFR_{60}$ is an elongated molecule exhibiting four similar cysteine-rich domains with three disulfide bonds arranged as vertically stacked rungs of a ladder within each domain. Domain 2 in $TNFR_{60}$ is the major ligand binding region and has three disulfide bonds organized to form three loops, one at the lower (NH_2-terminal, loop 1) and upper region (COOH-terminal, loop 2) of each domain and a third loop located in the middle of the domain (Figs. 2, 3). These loops flank the central three cysteines, arranged in CxxCxxC motif. The first loop formed by C54–C70 disulfide has 15 residues. In $TNFR_{60}$, domains 3 and 4 have nearly identical positional arrangements of the cysteine residues creating duplicates of domain 2.

---

[3] The molecular cloning by several groups nearly simultaneously by different methods has led to some confusion concerning the nomenclature for these receptors. The two receptors have been given official designation of CD120a for the 55–60 kDa TNFR and CD120b for the 75–80 kDa TNFR (SCHLOSSMAN et al. 1994) (referred to here as $TNFR_{60}$ and $TNFR_{80}$; previously referred to type A and B, or I and II, respectively)

```
          1                                                    50
LTβR      .....AVPPY ASE.N..QTC RDQEKEYYEP QHRICCSRCP PGTYVSAKC.
TNFR80    LPAQVAFTPY APEPG..STC R..LREYYDQ TAQMCCSKCS PGQHAKVFC.
TNFR60    ......LVPH LGDREKRDSV CPQGKYIHPQ NNSICCTKCH KGTYLYNDCP
                               ---------------D1---------------
          51                                                  100
          SRIRDTVCAT CAENSYNEHW NYLTICQLCR PCDPVMGLEE IAPCTSKRKT
          TKTSDTVCDS CEDSTYTQLW NWVPEC.LSC GSRCSSDQVE TQACTREQNR
          GPGQDTDCRE CESGSFTASE NHLRHCLSCS KCRKEMGQVE ISSCTVDRDT
          ---------------------------D2--------------------------
          101                                                 150
          QCRCQPGMFC AAWALE.CTH CELLSDCPPG TEAELKDEVG KGNNHCVPCK
          ICTCRPGWYC ALSKQEGCRL CAPLRKCRPG FGVA.RPGTE TSDVVCKPCA
          VCGCRKNQYR HYWSENLFQC F....NCSLC LNGTVHLSCQ EKQNTVCTCH
          ---------------------------D3--------------------------
          151                                                 200
          AGHF..QNTS SPSARCQPHT RCENQGLV.. EAAPGTAQSD TTCKN.....
          PGTF..SNTT SSTDICRPHQ ICNVVAIP.. GNASMDAVCT STSPTRSMAP
          AGFFLRENEC VSCSNCKKSL ECTKLCLPQI ENVKGTEDSG TT........
          ---------------D4---------------
          201                                            246
          .......... .......... .......... ..PLEPLPPE MSGT..
          GAVHLPQPVS TRSQHTQPTP EPSTAPSTSF LLPMGPSPPA EGSTGD
a         .......... .......... .......... .......... ......
```

```
          1                                                    50
Fas       VKGFVRKNGV NEAKIDEIKN DNVQDTAEQK VQLLRNWHQL HGKKEA.YDT
TNFR60    WKEFVRRLGL SDHEIDRLEL QNGRCLREAQ YSMLATWRRR TPRREATLEL
LTβR      ........GL STPHQED... GKAWHLAETE HCGATPSNRG PRNQFITHD.

          51                    72
          LIKDLKKANL CTLAEKIQTI IL
          LGRVLRDMDL LGCLEDIEEA LC
b         .......... .......... ..
```

Fig. 2a, b. Sequence alignments of the tumor necrosis factor receptor (*TNFR*) and lymphotoxin-β receptor (*LT-βR*). **a** alignment of the extracellular domains of $TNFR_{80}$, $TNFR_{60}$ and LT-βR using "pileup" program (GCG, 1991). Sequence numbering starts at the NH_2-terminal for predicted mature protein and ends at the start of the transmembrane hydrophobic region. *Lines* indicate the position of the cysteine rich domains (D1–D4). **b** Alignment of the sequences in the death/apoptosis domains for $TNFR_{60}$ and Fas with LT-βR from residue 371 for $TNFR_{60}$; 245 in Fas; 398 in LT-βR. Alignment weighted on GLS (gap wt 1.5), where V → N in Fas results in the *lpr^cg^* mutation. *Shaded regions* indicate identity of residues

The cytoplasmic domains of $TNFR_{60}$ and $TNFR_{80}$ differ significantly. Putative sites of serine\threonine phosphorylation are found in both receptors and $TNFR_{80}$ has been shown to be constitutively phosphorylated (Hohmann et al. 1992; Pennica et al. 1992b; Crowe et al. 1993). $TNFR_{80}$ phosphorylation is found primarily on serine residues with a minor amount of threonine phosphorylation. Functional consequences of $TNFR_{80}$ phosphorylation in signaling TNF responses have not been demonstrated. One common characteristic of both receptor cytoplasmic domains is the presence of a high number of serine, threonine, glutamic acid, and proline residues. This is characteristic of so-called PEST sequences thought to promote high rates of protein turnover and limit surface expression (Rogers et al. 1986).

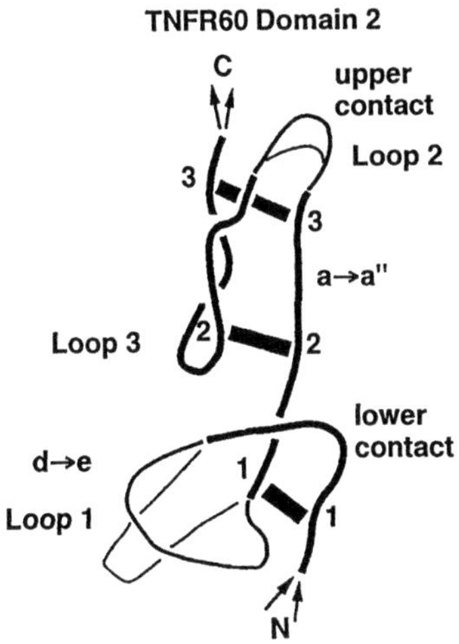

Fig. 3. Disulfide bonds and loop scaffold of the second cysteine-rich domain of the tumor necrosis factor receptor ($TNFR_{60}$) showing the general positions of contacts with lymphotoxin (LT-α) (a → a"and d → e loops). Number *1* indicates the position of C55–C70 disulfide; *2* C73–C88; *3*, C73–C95 (as numbered in Fig. 2a) (From BANNER et al. 1993)

3.2 Structural Aspects of the Tumor Necrosis Factor Receptor

Mutational analyses and X-ray crystallography studies indicate that the $TNFR_{60}$ binding site on the LT-α trimer is located in the cleft formed between two interacting subunits (GOH and PORTER 1991; ZHANG et al. 1992; BANNER et al. 1993). Mutational analysis of TNF and LT-α suggested that two loops, a' → a" and d → e, which are on opposite sides of the ligand monomer creating an asymmetric subunit, are critical for receptor binding (ZHANG et al. 1992; references therein)[4]. Selected mutations within these loops do not alter trimer formation but decrease receptor interactions. A comparison of the sequences within these loops reveals nonconservative changes among the ligands in this family (Fig. 1). TNF mutants at L29S and R32W have been shown to distinguish binding between $TNFR_{60}$ and $TNFR_{80}$ by selectively knocking out $TNFR_{80}$ binding (VAN OSTADE et al. 1993). Extensive mutational analyses by Loetscher (LOETSCHER et al. 1993) have revealed additional subregions of TNF that are important in providing contacts that separate $TNFR_{60}$ and $TNFR_{80}$ binding sites.

Mutation of D50 in LT-α, adjacent to R51 which contacts residues Y108 and P113 between monomer A and C, indicates that residues involved in intersubunit contacts are also important to receptor binding (GOH et al. 1991). Naturally occurring mutants in other family members confirm the importance of the regions involved in trimer contacts. Interestingly, the W140G mutation in human CD40L does not alter surface expression, but the ligand fails to bind to CD40:Fc (KORTHAUER et al. 1993). This mutation, within the highly conserved a' β strand just prior to be receptor contact loop, is important for β sheet interactions and its position is likely to cause an altered conformation of the a' → a" loop (PEITSCH and

[4] The designation of the β strands is by the convention of ECK et al. (1992)

Jongeneel 1993). This mutation causes a profound immunodeficiency in immunoglobulin heavy chain switching. A number of other mutations in CD40L have confirmed the importance of this region (Hollenbaugh et al. 1992; Allen et al. 1993; Korthauer et al. 1993;. DiSanto et al. 1993; Aruffo et al. 1993). Another example is found in murine Fas-L, in which an F273L mutation, located in the conserved COOH-terminal β-strand (equivalent to h β strand), gives rise to the *gld* phenotype (Takahashi et al. 1994), the ligand equivalent of *lpr*.

Domains 2 and 3 of $TNFR_{60}$ bind at the cleft formed between the A and C subunits of the LT-α trimer (Banner et al. 1993). Each receptor domain has two distinct ligand binding regions, a lower and upper region (Fig. 3). The lower region consists of loop 1 through the middle of the three central cysteines (C73) making contacts with residues in both the A and C subunits of LT-α. The main contacts are in the d → e loop of the C subunit and a → a" loop of the A subunit, typically described as the base of the ligand. Residues in loop 1 of domain 2 show primarily hydrophobic interactions with the d → e loop of LT-α where Y108 is critical; D50 in the a → a" loop is important in binding CxxCxxC region where the primary contacts are polar. The upper binding region of the $TNFR_{60}$ is formed by loop 2 in domain 2 (R77-G81) and loop 1 of domain 3 (W107–C114). These receptor residues make contact at the top of the trimer in the β strands d and e in the C subunit of LT-α (abbreviated (d+e)C). Other contacts are (c+f)A; (g)C; and a → a' loop of the A subunit.

The ligand-receptor complex, as it would exist on the surface of juxtaposing cells, shows the narrow end of the elongated trimeric ligand oriented outward from to the cell membrane. This provides the proper orientation for interaction of cell surface ligands and receptors during cell-cell contacts. Receptors within the complex protrude above and below the ligand and are roughly parallel to each other and to the threefold axis of the ligand.

3.3 Receptor Specificity of Surface Ligands

The ligand-receptor interactions determined by crystallography of the LT-α/$TNFR_{60}$ complex serve as the basis for understanding receptor specificities of the surface LT complex. In addition, a major tool in the discovery of most of the related ligands, and especially important in characterization of the membrane-anchored LT-α β complex, is the receptor:Fc fusion protein, created as chimeras between the Fc region of IgG and extracellular domain of the receptors (Ashkenazi et al. 1991; Peppel et al. 1991). These constructs, with specificity for the ligand, are bivalent and can be utilized in classical immunochemical procedures, such as precipitation and flow cytometry (Crowe et al. 1994b).

We found that $TNFR_{60}$:Fc and $TNFR_{80}$:Fc bound to a minor fraction of the surface LT complex expressed by human T cell hybridoma, II-23 (Crowe et al. 1994b), although both TNFR:Fc reacted well with membrane TNF and secreted LT-α and TNF. This indicated that another receptor may bind the major form of surface LT-α β complex and was further supported by the fact that none of the other members of the receptor family recognized the LT-$\alpha\beta$ complex. Identification of a TNFR-related sequence, discovered by Marynen and coworkers (Baens et al. 1993) using a somatic cell hybrid expressing human chromosome 12, prompted the construction of a receptor:Fc chimera that was subsequently shown to bind to the LT-β subunits in major form of surface LT-α β complex (Crowe et al. 1994b). The specificity of this receptor for LTα/β complex and lack of reactivity with TNF and LT-α which are recognized by $TNFR_{60}$ and $TNFR_{80}$, promoted the designation of LT-βR.

3.4 Structural Aspects of the Lymphotoxin-β Receptor

LT-β R is translated from a 2.2 kb mRNA and when expressed in COS cells as a transfected cDNA has an observed molecular mass of 61 kDa (Crowe, Van Arsdale, Force, and Ware unpublished). Analysis of the LT-βR cDNA sequence encodes a predicted 408 amino acid protein sharing 41% and 46% homology with $TNFR_{60}$ and $TNFR_{80}$ respectively (Fig. 2). LT-βR has a ligand binding domain with four cysteine-rich pseudorepeats followed by a short proline-rich membrane proximal region. Two potential N-glycosylation sites are present in the extracellular region. The ligand binding domain of LT-βR has characteristics of both $TNFR_{60}$ and $TNFR_{80}$ in the positioning of the cysteine residues. Similarity with $TNFR_{60}$ is found in the first and second domains. In contrast, the equivalent of loop 1 in domain 3 of LT-βR is dramatically shortened having only five residues, and in this regard is highly similar to domain 3 of $TNFR_{80}$. The fourth domain of LT-βR resembles $TNFR_{80}$ more closely than $TNFR_{60}$. An additional similarity of LT-βR with $TNFR_{80}$ is the proline rich region proximal to the membrane spanning sequence, suggested to form a "stalk" extending from the cell surface.

The cytoplasmic portion of LT-βR shares limited homology with $TNFR_{60}$ and Fas in a span identified as the apoptosis domain, including the GLS sequence, which in *lpr*^{cg} mutation (I $\rightarrow$ N225) knocks out apoptosis or the cytotoxic function of $TNFR_{60}$ (Fig. 2) (Itoh and Nagata 1993; Tartaglia et al. 1993a). However, LT-βR lacks the consensus homology KEFVR found in both $TNFR_{60}$ and Fas as well as a COOH-terminal homology region that functions as a "apoptosis protective" domain in Fas (Itoh and Nagata 1993). The structural homology of LT-βR cytoplasmic region with $TNFR_{60}$ and Fas suggests the possibility that this receptor may have cell viability regulating activity via surface LTα/β ligands, and as discussed later may take on an unexpected role in immune system development.

3.5 Model of the Lymphotoxin-$\alpha\beta$ Receptor Interactions

These results have led us to formulate a working model (Fig. 4) in which LT-α and LT-β can form four distinct ligands consisting of homo- and heteromeric complexes. The model assumes an overall trimeric configuration which is in keeping with the quaternary structure of LT and TNF. It also views the receptor

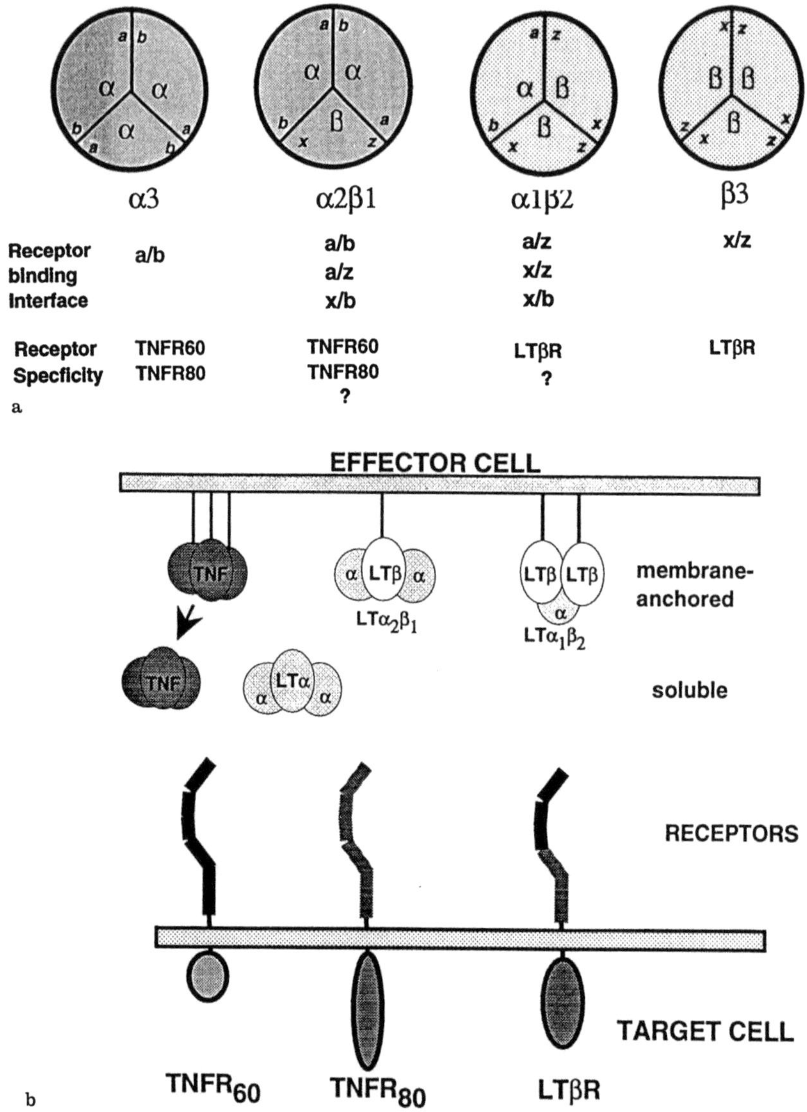

Fig. 4a, b. Model of receptor binding sites in lymphotoxin-αβ (LT-αβ) homo- and heterotrimers. **a** Four distinct ligands are created by interactions of two subunits in a trimeric configuration. Each side of the monomer is designated *a* and *b* for LT-α and *x* and *z* for Lt-β giving rise to different receptor binding faces. **b** The secreted and membrane-anchored ligands and their receptors in the LT system

binding site to be at the interface between two ligand subunits which is consistent with mutational and crystallographic data of LT-α $TNFR_{60}$ complexes (Goh and Porter 1991; Zhang et al. 1992; Banner et al. 1993). Binding to both ligand faces is assumed to be required for productive interaction and moreover, the receptor binding cleft is formed by interactions between the A and C subunits where both subunits contribute to the receptor binding site. For the homotrimers there are three equivalent bind sites per trimer. The asymmetric conformation of each LT subunit forms distinct sides of each receptor binding site, designated *a* and *b* for LT-α; *x* and *z* for LT-β. This model implies that each ligand will have a unique display of receptor binding site(s) potentially inducing distinct signaling pathways and thus a divergent set of cellular responses. The available data suggest that $TNFR_{60}$ and $TNFR_{80}$ bind an α/α cleft and LT-βR binds a β/β cleft in the LT-α1/β2 complex. (However, binding to α/β cleft unique to LT-α1β2 has not been excluded). Whether distinct receptors exist for the two heteromeric clefts (*az* and *xb*) created by α/β interactions is not clear and, as discussed earlier, the existence of LT-β_3 remains to be demonstrated. The mechanism of initiating signal transduction by this family of oligomeric ligands is hypothesized to be through their ability to cluster the respective receptors by multivalent binding. LT-α2β1 and LT-α1β2 complexes are predicted to have a single α/α or β/β receptor binding sites, respectively. These heteromeric ligands would theoretically bind receptor but be unable to directly induce clustering of either $TNFR_{60}$ or LT-βR. One caveat, is that low affinity binding to the heteromeric cleft may be possible if the ligand binding site in the receptor is "sided" and can recognize half the cleft. However, if unique receptors exist for the heteromeric cleft formed by α and β subunits then the LT-α2β1 and LT-α1β2 ligands could induce clustering of multiple receptors including $TNFR_{60}$/$TNFR_{80}$ or LT-βR.

The properties of surface LT on the activated II-23 cell line based on mAb and receptor based immunoprecipitations have led to the following picture of surface forms of LT and TNF. Within 1h following PMA activation significant levels of surface TNF are displayed followed shortly by the appearance of a LTα2/β1 form that reach peak levels by 3–4 h; within 8 h, TNF and the LTα2/β1 form have disappeared from the surface. Starting at 2–4 h, the LTα1/β2 form appears and soon comprises greater than 95% of the total surface forms. This form eventually disappears over 24–48 h. The II-23 cell line has been extensively used in the study of surface LT since the levels of surface LT reach the order of $\geq 10^5$ copies per cell (Ware et al. 1992). These high levels are atypical of both lymphoid tumor lines and primary cells and it is likely that the high expression coupled with the temporarily skewed synthesis of LT-α relative to LT-β forces the appearance of small amounts of alternative forms, i.e., LTα2/β1 and possibly some unassociated LT-β. The LTα2/β1 form has not been found in appreciable levels on most lymphoid tumor lines. In summary the biochemistry points towards the LTα1/β2 form as the dominant form of surface LT on most cells.

4 Biological Roles of the Lymphotoxin-αβ System

4.1 Distribution of Ligands and Receptors

4.1.1 Expression of Surface Lymphotoxin Complex

The expression characteristics of LT-α have been reviewed extensively (PAUL and RUDDLE 1988; RUDDLE and HOMER 1988; FASHENA et al. 1990; RUDDLE 1992). Essentially, LT-α is secreted primarily by activated T and B lymphocytes and NK cells, although a recent report described expression in melanocytes which was the first case of nonlymphoid expression (MELANI et al. 1993). Microglia and T cells in the lesions of multiple sclerosis patients can be stained with anti-LT-α sera (SELMAJ et al. 1991) which is interesting given the previous inability to detect LT-α synthesis in cells of monocytic lineage. It is possible that LT-α synthesis may be more widespread than currently realized.

Surface LT as defined by FACS analysis or immunohistology using anti-LT-α antibodies is primarily expressed by activated T and B cells and NK cells and has been described on human CTL clones (WARE et al. 1992; YASUKAWA et al. 1993), peripheral mononuclear lymphocytes activated with anti-CD3 (ANDERSSON et al. 1989; WARE et al. 1992) or staphylococcal enterotoxin A (ANDERSSON et al. 1989), IL-2 activated peripheral blood lymphocytes (lymphokine activated killer cells) (ABE et al. 1991; WARE et al. 1992;. YASUKAWA et al. 1993), pokeweed activated peripheral B lymphocytes (WARE et al. 1992) and various lymphoid tumors of both T and B lineage (WARE et al. 1992). Engagement of alloantigen bearing target cells specifically induced expression by $CD8^+$ CTL clone (WARE et al. 1992). Additional work is required to fully delineate the expression patterns of surface LT both in normal immune function, in development and in pathogenesis.

The tissue expression of LT-α and LT-β in adult tissues shows essentially exclusive localization to peripheral blood lymphocytes, spleen and thymus in humans. LT-α is much more difficult to observe in these tissues suggesting that LT-β is not simply an activation marker. Likewise, LT-β RNA could be detected in unstimulated human peripheral blood lymphocytes, whereas LT-α expression requires cell activation (BROWNING et al. 1993). LT-β expression in mouse spleen cells has been observed (MILLET and RUDDLE 1994).

Within peripheral blood lymphocyte populations, anti-CD3 activation leads to surface expression and further PMA stimulation resulted in immediate down-regulation (Browning, unpublished). The culturing of lymphocytes in IL-2 to generate LAK cells also induces surface LT (WARE et al.1992; ABE et al. 1991, 1992). The expression of TNF and surface LT have also been analyzed in a number of human T cell clones (WARE et al.1992). As with the tumor cell lines, differing patterns were found, with some clones displaying surface TNF and LT as well as secreting both cytokines, whereas others showed surface expression without LT secretion. A myelin basic protein specific clone was also positive for surface LT, as were three HSV-specific $CD4^+$ CTL lines (YASUKAWA et al.1993). From these experiments and observations of peripheral blood lymphocytes, it is clear that both $CD4^+$ and $CD8^+$ T cells can express surface LT. Weak FACS signals were observed with pokeweed mitogen (PWM) activated peripheral B cells and currently the issue of B cell expression is under further study.

To get an approximation of what cell types are capable of surface LT expression and to provide systems that are easy to study, a panel of tumor lines have been examined (WARE et al. 1992). Several different patterns of expression of the surface LT complex were observed in human lymphoid cell lines; fast induction within 3–6 h by PMA followed by decay (e.g., Il-23 and Jurkat cells), a slow induction requiring 15–25 h to reach maximum (some Burkitt lymphoma lines, e.g., Ramos) and constitutive expression that either increases (Hut 78, Reh) or decreases (Hs445) with PMA treatment. These results suggest that expression patterns in primary cells may depend on the cell type.

With few exceptions, in cell lines in which LT-α had been characterized previously to be secreted, surface LT expression can be observed. Moreover, both LT-α and LT-β mRNA can be detected in every tumor line which was FACS positive for surface LT (Browning, unpublished). An interesting exception is the original B lymphoblastoid line RPMI 1788 from which LT-α was originally purified (AGGARWAL et al. 1984). This line secretes LT-α constitutively and secretion can be further induced with PMA, but the line does not display surface LT under these conditions. Northern analysis was unable to detect LT-β mRNA, again supporting the hypothesis that both genes must be expressed to drive LT to the surface.

4.1.2 Expression of the Tumor Necrosis Factor and Lymphotoxin Receptors

Receptors for TNF/LT-α are broadly distributed in the body, as are many members of the family. Most tissues express both receptors simultaneously although the ratio of $TNFR_{60}/TNFR_{80}$ varies dramatically (Table 3). As a general finding from studies of cell lines $TNFR_{60}$ is usually more abundant than $TNFR_{80}$ in epithelial cells, whereas $TNFR_{80}$ is the predominant form expressed on hematopoietic tissue (WARE et al. 1991; RYFFEL et al. 1991; RYFFEL and MIHATSCH 1993). The number of receptors per cell range from very few to only several thousand, with coexpression of the receptors typical. In the lymphoid tissue, the two receptors are found in discrete regions (RYFFEL et al. 1991). $TNFR_{80}$ is abundant on lymphoblasts in the thymic medullary region and in the interfollicular area of lymph nodes; in contrast $TNFR_{60}$ expression is limited to the germinal centre dendritic reticulum cells. In vitro analysis has shown that $TNFR_{80}$ is inducible on T and B cells following stimulation via their antigen receptor (SCHEURICH et al. 1987; WARE et al. 1991; ERIKSTEIN et al. 1991) and is the major ligand binding receptor on lymphocytes. TNF and IL-2 reciprocally induce the other's receptor suggesting an amplification loop enhancing T cell functions (CRUMP et al. 1990; WARE et al. 1991).

Expression of the receptor for LT-βR has not yet been extensively studied, but mRNA expression is observed in human spleen and thymus and other major

Table 3. Tumor necrosis factor receptor expression on human cell lines

Cell line	Receptor				
	$TNFR_{80}$	$TNFR_{60}$	$TNFR_{80}/TNFR_{60}$	LT-βR	Fas
T cells anti-CD3 activated (PBL)	+	–	>10	–	+
II-23.D7 CD4⁺ T hybridoma	+	+	4	–	+
U937 promonocytic leukemia	+	+	1.6	+	+
K562 erythroleukemia	+	+	0.7	+	–
WI38 normal lung fibroblasts	+	+	0.6	+	+
VA-13 SV40 transformed WI38	–	+	<0.1	+	+
MelJuSo melanoma	–	+	<0.1	+	–
Hep-G2 hepatoma	–	–	<0.1	–	–
ME-180 cervical carcinoma	–	+	<0.1	+	+

TNFR, tumor necrosis factor receptor; LT, lymphotoxin.

organs, but absent in brain (Browning, unpublished). In vitro studies have indicated LT-βR is expressed in a pattern similar to $TNFR_{60}$ (Table 3), and is notably lacking on peripheral blood T cells and T cell lines as detected by flow cytometry (C. Ware, unpublished).

4.1.3 Soluble Receptors

Activation by a variety of different stimuli results in rapid loss of surface $TNFR_{60}$ and $TNFR_{80}$ and concomitant accumulation of soluble TNF-binding proteins in the supernatant (Heller et al. 1990; Kohno et al. 1990; Nophar et al. 1990; Lantz et al. 1990). Pulse-chase analysis of T lymphocytes activated through the T cell receptor (TCR) clearly demonstrated that down-regulation of $TNFR_{80}$ occurs by proteolytic cleavage on the extracellular side and release (shedding) of cell surface receptors (Ware et al. 1991). Similarly, inducible shedding of TNFR has been documented for other cell types (Porteu and Nathan 1990). Low levels of $TNFR_{60}$ and $TNFR_{80}$ are shed spontaneously by many cells but the rate of release is dramatically increased with activation. TNF has been reported to trigger shedding of its own receptor in some cell types (Lantz et al. 1990), but not in others (Brakebusch et al. 1992; Pennica et al. 1992b). More recent evidence suggests that TNF-induced shedding of $TNFR_{80}$ is mediated by $TNFR_{60}$ (Porteu and Hieblot 1994; Higuchi and Aggarwal 1994). Pharmacological agents such as PMA or the serine/threonine phosphatase inhibitor okadaic acid also trigger

TNFR shedding (Higuchi and Aggarwal 1993), whereas protein kinase C-selective inhibitors (H-7, staurosporine) block TNFR shedding suggesting a role for protein phosphorylation in this process (Crowe et al. 1993). Cytoplasmic truncation of $TNFR_{80}$ does not prevent inducible shedding of $TNFR_{60}$ or $TNFR_{80}$ and suggests that phosphorylation of a different protein (e.g. a protease) may be a necessary signal for shedding (Brakebusch et al. 1992; Crowe et al. 1993). The identity of the protease responsible for TNFR shedding is not known. Inhibitors of serine or thiol proteases had no effect on TNFR shedding in neutrophils (Porteu and Nathan 1990). In contrast, serine protease inhibitors were able to block TNFR shedding by THP-1 cells (Hwang et al. 1993). Recent evidence indicates that a specific metalloprotease inhibitor of TNF processing also blocks shedding of TNFR80 in activated T lymphocytes (Crowe et al. 1995). Site-directed mutagenesis of $TNFR_{60}$ revealed that cleavage occurs at Asn-172/Val-173 near the transmembrane domain (Gullberg et al. 1992).

Soluble forms of several other TNFR family members have been described (Loenen et al. 1992a; van Kooten et al. 1994; Josimovic-Alasevic et al. 1989;. DiStefano and Johnson 1988) and appear to be derived by shedding as described above for $TNFR_{60}$ and $TNFR_{80}$ (Table 2). In contrast, soluble forms of Fas and 4-1 BB are derived from alternatively spliced mRNA transcripts which encode proteins lacking a transmembrane region (Goodwin et al. 1993b; Cheng et al. 1994).

Naturally occurring soluble forms of TNFR capable of binding ligand may represent an important control mechanism regulating the bioavailability of TNF in vivo. Soluble TNFR have been detected in urine and serum of healthy individuals and in patients in a variety of pathologic states where TNF levels are elevated. For example, soluble TNFR levels are elevated in sera of cancer patients (Gatanaga et al. 1990; Aderka et al. 1991), in systemic lupus erythematosus (Aderka et al. 1993), individuals infected with HIV-1 (Godfried et al. 1993), in urine of patients with chronic renal failure (Olsson et al. 1989; Brockhaus et al. 1992), and in synovial fluid of persons with active rheumatoid arthritis (Cope et al. 1992; Deleuran et al. 1992). Since soluble TNFR retain ligand binding capacity, they may function to inhibit the biologic activities of TNF by competing for binding of cell surface receptors (Van Zee et al. 1992). Other data, however, have shown that soluble TNFR can prolong TNF bioactivity presumably through stabilization of the TNF trimer structure (Aderka et al. 1992). Thus, it is likely that soluble TNFR play a dual role, as either antagonist or agonist, in regulating local and systemic effects of TNF.

4.2 Receptor Activation and Signal Transduction

Activation of TNFR signaling is thought to occur through close spatial aggregation of multiple receptors as a result of binding trivalent ligand, a process referred to as receptor clustering (Loetscher et al. 1991; Pennica et al. 1992a). Receptor clustering as a means of receptor activation in other systems has been well documented, especially for receptor tyrosine kinases (Ullrich and Schlessinger 1990;

Schlessinger and Ullrich 1992; Kolanus et al. 1993). Direct evidence for TNFR activation through clustering comes primarily through the use of receptor specific antibodies. $TNFR_{60}$-specific rabbit IgG was shown to mimic TNF responses including cytotoxicity (Engelmann et al. 1990a). Monovalent Fab fragments of the anti-$TNFR_{60}$ IgG lacked cytotoxic activity; however, activity was restored upon cross-linking with bivalent anti-rabbit IgG antibodies. This would suggest that at least two receptor molecules need to be in close spatial proximity to initiate TNFR signaling pathways. Structural studies of TNF, LTα, and ligand-receptor complexes further supports clustering as the mechanism of receptor activation (Eck et al. 1988, 1992; Banner et al. 1993). Distances between receptors in the ligand complex are proposed to be within a range allowing interactions between cytoplasmic domains.

Cytoplasmic regions brought together in receptor-ligand complex could therefore create a unique surface for initiation of signaling. Involvement of receptor cytoplasmic domains in the generation of signaling has been shown with cytoplasmic mutations that block TNF induced cellular responses (dominant negative mutations) (Tartaglia and Goeddel 1992; Brakebusch et al. 1992). The use of TNFR-specific antibodies as agonists (or antagonists) have demonstrated that individual TNF receptors signal distinct cellular responses (Naume et al. 1991; Greenblatt and Elias 1992; Thoma et al. 1990; Barna et al. 1993; Mackay et al. 1993; Tartaglia et al. 1991; Higuchi and Aggarwal 1992; Kruppa et al. 1992). The results observed with antibodies have been reproduced with TNF muteins with selective binding to either $TNFR_{60}$ or $TNFR_{80}$ (Tavernier et al. 1990; Loetscher et al. 1993; Van Ostade et al. 1993, 1994). Also, mice have been developed with specific TNF receptors deleted from the genome, allowing evaluation of the function of these receptors in vivo (Pfeffer et al. 1993; Rothe et al. 1993) (M. Moore and D. Goeddel, personal communication). The majority of studies have shown $TNFR_{60}$ to be the receptor responsible for many of the responses to TNF defined by in vitro and in vivo systems. These responses include cytotoxicity (Espevik et al. 1990; Engelmann et al. 1990a; Thoma et al. 1990; Tartaglia et al. 1991, 1993a; Greenblatt and Elias 1992; Pfeffer et al. 1993; Rothe et al. 1993), NF-κB activation (Kruppa et al. 1992; Schutze et al. 1992; Pfeffer et al. 1993), induction of MHC class I and II expression (Thoma et al. 1990), ICAM-1 expression (Mackay et al. 1993), fibroblast growth (Espevik et al. 1990), protection from TNF-mediated cytotoxicity (Engelmann et al. 1990a; Tartaglia et al. 1991) and others (Tartaglia et al. 1991; Engelmann et al. 1990a; Mackay et al. 1993); and in vivo, activation of bacterial pathogen defenses and systemic shock in response to lipopoly saccharide (LPS) (Pfeffer et al.1993; Rothe et al. 1993).

Receptor-specific TNF muteins also suggest that at least two different signaling pathways are initiated by activated $TNFR_{60}$ (Barbara et al.1994). Three TNF muteins, with either single or double amino acid substitutions, show near exclusive binding to $TNFR_{60}$ and retain the cytotoxic activity of wild-type TNF. The muteins, however, display a marked reduction in proinflammatory activities, as measured by the priming of neutrophils, induction of adhesion molecules on endothelial cells, IL-8 secretion and others. This data would suggest that TNF-activated $TNFR_{60}$ initiates at least two distinct signaling pathways individually responsible for generating cytotoxicity and proinflammatory activities.

$TNFR_{80}$ has been shown to signal independently of $TNFR_{60}$ as demonstrated by enhancement of thymocyte and CTL proliferation (NAUME et al. 1991; TARTAGLIA et al. 1991,1993b) and induction of GM-CSF expression (VANDENABEELE et al. 1992). $TNFR_{80}$ has been suggested to signal cytotoxicity (HELLER et al.1992) and NF-κB activation (HOHMANN et al. 1990a) but these reports are controversial (TARTAGLIA et al. 1993d; PFEFFER et al.1993). Understanding of TNF signaling through $TNFR_{80}$, therefore, remains somewhat limited at this time. In addition to direct signaling activity, $TNFR_{80}$ has been shown to cooperate with $TNFR_{60}$-induced responses by enhancing TNF binding to $TNFR_{60}$. This enhancement is thought to occur via the passing of ligand from the higher affinity $TNFR_{80}$ to the lower affinity $TNFR_{60}$ (TARTAGLIA et al. 1993c). The "ligand passing" theory is supported by three observations: (1) antibodies to $TNFR_{80}$ modulate $TNFR_{60}$ signaling; (2) antibody blockade of $TNFR_{80}$ slows the rate of $TNFR_{60}$ ligand occupancy; and the unusually fast rate of ligand dissociation noted for $TNFR_{80}$ ($T_{1/2}$=10 min) supports the possibility that ligand dissociated from $TNFR_{80}$ can serve to increase the local concentration of TNF at the cell surface and augment $TNFR_{60}$ signaling. This receptor activation scenario, initiated by the receptor clustering induced by TNF, is complicated by the simultaneous presence of two receptors with similar ligand specificities but with distinct cytoplasmic domains. The binding of $TNFR_{60}$ and $TNFR_{80}$ simultaneously to a single trimeric ligand (heteromeric receptor clustering) could initiate signals distinct from homomeric receptor-ligand complexes which possess defined activities. This could suggest a role for $TNFR_{80}$ in modulating TNF signaling through formation of heteromeric receptor complexes. As pointed out above heteromeric receptor complexes may be necessary for signaling by the surface forms of LT.

The events following ligand-receptor clustering and downstream processes involved in signal transduction remain incompletely understood due in part to the uniqueness of the cytoplasmic domains that lack homology to other known signaling mechanisms. The signaling events produced by activated TNF receptors involve a number of classical and novel second messenger systems, activation of transcription factors and a plethora of gene regulatory events which produce the many cellular responses characteristic of TNF and related ligands. The second messenger pathways activated by TNF have recently been reviewed (BEYAERT and FIERS 1994) and a detailed discussion goes beyond the focus here. The pathways implicated in TNF signaling cytotoxic responses include GTP binding proteins (IMAMURA et al.1988), phospholipases (KIM et al.1991; SCHUTZE et al.1991; GOPPELT-STRUEBE and REHFELDT 1992; HOECK et al. 1993; DE VALCK et al.1993; BEYAERT et al. 1993a), and various protein kinases (HEPBURN et al. 1988; VIETOR et al. 1993). Recent excitement has been generated by the identification of a TNF-activated sphingomyelinase, generating ceramide and phosphocholine (WIEGMANN et al.1992; KOLESNICK and GOLDE 1994). In turn, ceramide activates a serine/threonine protein kinase (MATHIAS et al. 1991; LIU et al.1994).

4.3 Receptor-Mediated Cytotoxicity and Cytotoxic T Lymphocytes

4.3.1 Tumor Necrosis Factor-Mediated Cytotoxicity

Tumor necrosis factor can induce both necrotic and apoptotic cell death depending upon the target cell. The factor(s) governing which path is followed, however, are not well defined (LASTER et al.1988; REID et al.1991). Characteristics of TNF cytolysis include a relatively slow rate of killing (>4 h) that is, in part, dependent upon occupancy of receptors and cell cycle associated events (DARZYNKIEWICZ et al. 1984; COFFMAN et al.1988). The rate of lysis in murine L929 cells is maximal at ~1%–2% receptor occupancy and appears to occur at M/G2 phase of the cell cycle (COFFMAN et al. 1989b). TNF exhibits a selective cytotoxic effect lysing tumor cells (SUGARMAN et al. 1985) and virus-infected cells which is greatly enhanced by interferon-γ (WILLIAMSON et al.1993), chemotherapeutic agents (ALEXANDER et al. 1987; COFFMAN et al.1989a) or signaling via ligation of Fas with antibodies (WONG and GOEDDEL 1994). An example of tissue selectivity via this receptor family is seen in the disparity in the cytotoxic response between different human target cells to LT and TNF (BROWNING and RIBOLINI 1989). In murine cells TNF and LT are nearly equipotent whereas with many human tumor cells LT is often far less potent (>100-fold) than TNF. The affinity of LT binding is not sufficient to account for the difference in the magnitude of the cytotoxic response. This differential effect between LT and TNF is seen in other systems (ANDREWS et al.1990). The reason for this phenomena is unclear but indicates that LT signaling via $TNFR_{60}$ and $TNFR_{80}$ is complex.

One interesting feature of the signals transduced by TNF and LT is the cytokine's ability to confer resistance to its own cytotoxic effect (WALLACH 1984). Blockade of protein synthesis prior to TNF treatment abolishes induction of the resistant state and many cell lines not directly sensitive to TNF cytotoxicity are killed in the presence of protein synthesis inhibitors. The resistance mechanisms are at least in part associated with TNF inducible genes including Mn^{2+}-superoxide dismutase (WONG et al.1989; BOSS et al.1991), metallothionein (SCIAVOLINO et al.1992) and plasminogen activator II (KUMAR and BAGLIONI 1991). Induction of DNA repair by TNF may also contribute to resistance (LICHTENSTEIN et al.1991). Although controversial, introduction of constitutively expressed bcl-2 may increase resistance of sensitive cells to TNF and Fas cytotoxicity (ITOH et al.1993; HENNET et al. 1993; VANHAESEBROECK et al.1993).

Although no intrinsic enzyme activity is present in $TNFR_{60}$ a recent report suggests a possible mechanisms of coupling between the receptor ligand complex and downstream signaling pathways associated with resistance to cytotoxicity (VANARSDALE and WARE 1994). A serine kinase activity is activated within a minute following TNF binding to U937 cells. The kinase activity coimmunoprecipitates with $TNFR_{60}$ along with three other phosphoproteins. The protein kinase activity is sensitive to staurosporine and this inhibitor enhances the sensitivity of U937 cells to TNF cytotoxicity (BEYAERT et al.1993b) leading to the suggestion that the $TNFR_{60}$ associated kinase may be involved in regulating genes involved in resistance to cytotoxicity.

Cellular resistance to TNF cytotoxicity is abrogated by infection with viruses. Studies of adenovirus by Wold and Gooding illustrate this point well (GOODING et al. 1990; WOLD and GOODING 1991; GOODING 1992). The adenovirus E1A oncogene dramatically induces susceptibility to TNF cytotoxicity (DUERKSEN-HUGHES et al. 1989; AMES et al.1990). However, several additional adenovirus genes in the E3 region block the cytotoxic sensitizing effect of E1A at steps distinct from the ligand-binding process. These mechanisms are quite different from the soluble-TNFR strategy used by the poxvirus (UPTON et al.1991). Thus, viruses have developed distinct mechanisms to specifically alter cellular responses to TNF underscoring the importance of this cytokine family in host defense.

4.3.2 Cytotoxic T Lymphocytes

The last decade of research has clearly shown that CTLs carry a diverse arsenal of molecular mechanisms that utilize both necrotic and apoptotic pathways to eliminate infected cells. In vitro studies have demonstrated two major pathways that induce cell death, the granule-associated exocytosis pathway (GOLSTEIN et al. 1991; PODACK et al. 1991; YAGITA et al. 1992; SCHAERER and TSCHOPP 1993; BERKE 1994) and the receptor-mediated cytotoxic response, best exemplified by Fas and $TNFR_{60}$ (WARE et al. 1990; ROUVIER et al. 1993; RATNER and CLARK 1993). The granule pathway uses the membranolytic molecule perforin and proteases (granzymes) that mediate necrosis and apoptosis of cells. TNF induces a necrotic death via oxidative damage but also triggers apoptotic death depending upon the properties of the target cell. The answer to the question of which pathway is the most important cytotoxic mechanism of CTLs, best viewed from the perspective of the specific pathogen, is the one which confers an effective defense. The use of animals genetically deficient in one or more mechanisms is beginning to illuminate the boundaries of this issue in greater detail and should provide excellent models for defining function in vivo (KAGI et al.1994).

The secreted forms of TNF and LT are apparently not involved in CTL mediated lysis. This has been demonstrated by a variety of evidence including the inability of neutralizing antibodies to block cytolysis (WARE et al. 1981; LIU et al. 1992), the disparity in the spectra of target cells lysed by the cytokines and CTLs, and, most recently, genetic deletion of the LT-α gene (DE TOGNI et al. 1994), or TNFR genes (PFEFFER et al. 1993; ROTHE et al. 1993) (M. Moore and D. Goeddel, personal communication), which did not disrupt classic CTL activity, but were important for host defense.

How might TNF and LT participate in CTL reactions to protect the host? The recognized pleiotropic proinflammatory activities may serve to orchestrate specific steps in the CTL pathway. For instance, TNF acts to enhance antigen recognition by modulation of target cell MHC class I and II molecules (COLLINS et al. 1986); adhesion is promoted through the induction of the target cell ligands, ICAM-1 and VCAM-1, for the integrins, LFA-1 and VLA-4, expressed on CTLs

(Staunton et al. 1988; Osborn et al. 1989). In addition, TNF up-regulates the expression of IL-2 receptors leading to enhanced T cell responses mediated by IL-2 and may stimulate proliferative responses in mixed lymphocyte cultures (Shalaby et al.1988). Additional studies have shown that LAK cells and CD8+ CTL are optimally induced with TNF and, in combination with IL-2, increase expression of CTL granule components (Robinet et al.1990). These studies suggest how a pleiotropic cytokine may function to integrate a specific effector function, such as CTL activity, in cellular immune responses.

Identification of several membrane-anchored ligands related to TNF and their ability to initiate cell death has refocussed attention on these molecules in CTL action. Recent studies of CTL activity has revealed a granule-independent pathway of target cell lysis (Ostergaard and Clark 1989). This alternate pathway was found to be independent of Ca^{2+} ions (which perforin requires), but was specifically and rapidly induced in CTLs following contact with the antigen bearing target cell. Insightful studies by Golstein and coworkers (Rouvier et al.1993) have lead to the hypothesis that Fas and its ligand are involved as mediators of this pathway. Expression and regulation of Fas-L and several of the related ligands are consistent with this hypothesis, including the rapid induction of ligand expression in CTL, and apoptosis of target cells via receptor clustering. It is also interesting that Fas-L is released in a soluble form with lytic activity (Suda and Nagata 1994). Preliminary reports (Doherty 1993) of perforin gene disrupted mice indicate that these animals are significantly impaired in CTL function as are mice deficient in granzyme B (Heusel et al.1994); however, significant CTL activity remains. Further detailed analysis must be performed to reach clear conclusion. The natural knock-out of Fas and its ligand, *lpr* and *gld*, have impaired CTL responses (Davidson et al. 1986; Rouvier et al.1993). Collectively, these data provide further intrigue and complexity to the issue of molecular mechanisms of CTL mediated tissue cytotoxicity and raise further speculation that CTL and other cell types may utilize ligands in the TNF family to induce cell death in tissue-specific processes, or perhaps during developmental events, in addition to their roles in host defense.

4.4 Lymphotoxin in Regulation and Development of the Immune System

The elucidation of the surface LT form and recently the identification of a unique LT-β receptor have now clearly delineated these two cytokines. What then is the biological role of the LT system? In this respect it is instructive to consider the roles of other members of this family. In the well studied case of TNF, roles have been implied in literally hundreds of biological systems; however, the elimination of TNF receptors in mice by gene disruption (Pfeffer et al. 1993; Rothe et al.1993) tends to point towards a primary function as an initiator of acute inflammatory events.

Probably dominant in this process is the activation of endothelium to display altered repertoires of adhesion systems. The apparently normal state of a transgenic mouse constitutively expressing $TNFR_{60}$:Fc, i.e., a TNF inhibitor, would point to a lack of a critical role in development (PEPPEL et al.1993). In principle, this mouse should not be able to mount any surface TNF, secreted TNF or LT-α dependent activities. As the mouse can function reasonably well without a complete immune system, further work will be needed to detect more subtle alterations in these mice. A role for TNF in the immune system in addition to the proinflammatory aspects has often been cited, but tangible in vivo data have remained elusive. The best case exists for involvement of TNF and possibly LT in B cell regulation. TNF is required for the proliferation of purified splenic B cells in response to CD40 signaling (BOUSSIOTIS et al.1994) or, viewed from a different perspective, it can be a costimulatory signal along with CD40 ligand for B cell activation (AVERSA et al.1993). Aspects of this pathway may be responsible for the polyclonal B cell activation accompanying HIV infection of T cells (MACCHIA et al. 1993). LT has been implicated as a growth factor for normal B cells (KEHRL et al. 1987), B cell tumors (SEREGINA et al.1989; ALVEREZ-MON et al.1993) and EBV transformed B cells (Estrov et al.1993), yet not for other B cell tumors (JANSSEN et al.1990). The somewhat conflicting nature of all of these results may derive from several facets. Firstly, while either TNF or LT may drive B cell growth in a pharmacological format (which actually was not the case for EBV transformed B cells (ESTROV et al.1993)), the actual cytokine acting physiologically may depend on the cell populations employed in the culture. It is likely that B cells immortalized by viral infection or other oncogenic events will differ depending upon which growth controlling signals were bypassed upon transformation (JANSSEN et al.1990). Lastly, many of the signaling systems in this family, which appear to be focused on lymphocytes, can act to either inhibit or induce proliferation depending on the state of the cells (SMITH et al.1994) and similar considerations may be affecting B cells. A careful characterization of the roles of TNF and LT in the B cell system will hopefully provide a piece of this puzzle that will complement the information gained by genetic disruption.

Genetic disruption of the LT-α gene provides a direct assessment of the specific functions of LT without the complications associated with the receptor disruptions. In dramatic contrast to the TNF receptor knock-outs, the LT-α knock-out mouse has profound defects in the peripheral immune system (DE TOGNI et al.1994). The immediate question arising is whether these effects are due to secreted LT-α or, as we view more likely, the surface LT-α/β complex. The lack of a similar phenotype in the $TNFR_{60}$ or $TNFR_{80}$ knock-outs or in the $TNFR_{60}$:Fc transgenic mouse suggests that the surface LT form mediates this unique biology. It is still conceivable that additional LT-α-specific receptors exist that could account for the role in the peripheral immune system without invoking LT-β containing complexes. Related defects were seen in mice treated early in development with anti-TNF polyclonal serum, i.e., atrophy of the lymphoid organs. The observation that a mAb directed at mouse TNF could cross-neutralize mouse LT-α (SHEEHAN et al.1989) suggests a combined TNF and LT-α block. In this case, the lymphoid atrophy picture would be consistent with the LT-α knock-out mouse. Another curious observation was made relating to the role of LT in immune regulation. The administration of human LT-α to mice leads to dramatic thymic atrophy, whereas human TNF was much less potent in this assay (Kusumi et al. 1992; Hirahara et al.1994). It is conceivable that either the presence of LT-α in an inappropriate context or complete blockage with an antiserum may both disrupt the LT-dependent system. Studies analyzing the effects of exogenously added murine LT-α and the LT-αβ complex should help in this respect.

The loss of the LT-α gene leads to disrupted development of lymph nodes and mucosal lymphoid tissues and disorganization of the T and B cell zones in the splenic lymphoid follicles. The anatomical phenotype is similar to a recently described mouse mutation called alymphoplasia *aly* (Miyawaki et al. 1994). This mouse is immunodeficient even though most lymphoid cell populations are present. The histological appearance of the spleen and the lack of lymph nodes are essentially identical to the LT-α knock-out mouse although the defect extends into the thymus, which was not observed in the LT-α knock-out. Miyawaki and coworkers utilized bone marrow transfers to address the nature of the defect and concluded that disrupted interactions between lymphocytes and stromal elements in the various lymphoid organs may underlie the phenotype. The lack of both Peyer's patch and peripheral lymph nodes was cited as evidence against a trafficking defect since circulation into these two compartments is mediated by distinct adhesion ligands. Likewise, SCID mice lacking both T and B cells still develop rudimentary lymph nodes and therefore a lack of lymphocyte trafficking alone cannot account for the phenotype. Basically, the defects revealed by the *aly* mutant and LT-α knock-out mice point towards immunological aspects of the immune system that have been only poorly envisioned. Whatever the nature of this signal, it is likely to be central to several aspects of lymphoid follicle and node function. Is LT a trapping signal for lymphocytes as they traffic through the nodes? Does LT serve as a specific signal acting locally to regulate adhesion ligand systems as does TNF? Does LT regulate the communication between follicular dendritic cells and specific lymphocytes? The answers to these questions will certainly lead to new insight into the function and organization of the follicles and lymph nodes.

A number of predictions can be made based on these hypotheses. If the *aly* defect, which maps to a distinct genetic region from the known ligands and receptors (Kuramoto et al. 1994), involves LT-related signaling events then one would expect that the LT-α knock-out mice are also immunodeficient for certain in vivo functional events. Based on our studies, one would predict that the LT-β knock-out should have a similar phenotype to the LT-α disrupted mouse. This experiment as well as the ability to perturb LT related events in vivo in the mouse should advance this field. At the very least these mice will allow one to determine the contribution of the lymph nodes to immune responses. The ability to manipulate specifically lymph node based processes may be useful therapeutically.

Pathology often gives one an excellent view into physiology and the observation of LT staining in the microglia in multiple sclerosis lesions bears consideration (Selmaj et al.1991). The presence of LT and TNF in these lesions is presumed to reflect roles inflammatory processes. Ruddle and coworkers have attempted to analyze this system further using the mouse experimental allergic encephalitis model of multiple sclerosis (RUDDLE et al.1990). T cell clones specific for myelin basic protein secrete both TNF and LT-α (POWELL et al.1990) and likewise a similar human T cell clone displays surface LT (WARE et al.1992). An antibody that blocks both mouse TNF and LT-α was able to prevent transfer of the disease with a T cell clone (RUDDLE et al.1990). It will be of considerable interest to dissect this system with more specific murine reagents.

Within the TNF family, several mutations have been characterized that highlight the general importance of these molecules in immune regulation. Mutations in humans in the CD40 ligand lead to a hyper IgM syndrome and common variable immunodeficiency emphasizing the importance of the CD40 pathway in the control of immunoglobulin class switching and IgG synthesis (CALLARD et al. 1993; MARX 1993; FARRINGTON et al.1994). Likewise two mouse models of autoimmune disease, the *lpr* and *gld* strains, result from disruptive mutations in the Fas receptor and ligand respectively (WATANABE-FUKUNAGA et al.1992; TAKAHASHI et al.1994). It will be interesting to see what pathological events are associated with alterations in the LT system.

4.5 Genetic Aspects of the Lymphotoxin Tumor Necrosis Factor Locus

At the time the LT-β cDNA was identified, it seemed reasonable that the LT-β gene may lie in the class III region of the MHC close to TNF and LT-α. Probing of cosmids spanning about 100 kb of this region quickly showed that this hypothesis was correct (BROWNING et al.1993). The organization of this region of the genome is illustrated in Fig. 5. The mouse LT-β gene has also been completely analyzed and its location within the TNF locus on chromosome 17 is identical to that outlined for the human TNF locus (Browning, unpublished). A parallel organization was expected due to the tight correspondence between these two regions. The region between human LT-α and B144 genes has been completely sequenced and no additional TNF related genes lie within this stretch (Browning, unpublished).

Alterations in the TNF locus have been postulated to be involved with autoimmune disease for some time. As a result the TNF locus has been intensively examined for linkage to disease states both in the mouse and in humans and this topic has been reviewed (JACOB 1992; JACOB and MCDEVITT 1991a,b). In humans, the linkage of TNF and LT-α gene polymorphisms to various haplotypes has been established earlier through RFLP analyses and now short sequence length polymorphisms (SSLP) based on PCR of simple sequence repeats (microsatellites) are able to vastly enhance this analysis (JONGENEEL et al.1991). The production of TNF appears to be linked to certain class II haplotypes such that $DR2^+/DQW1^+$ individuals are low producers wherein that $DR3^+$ and $DR4^+$ individuals produce high levels of TNF (JACOB 1992). SLE individuals with the haplotype associated with low TNF production are predisposed to nephritis.

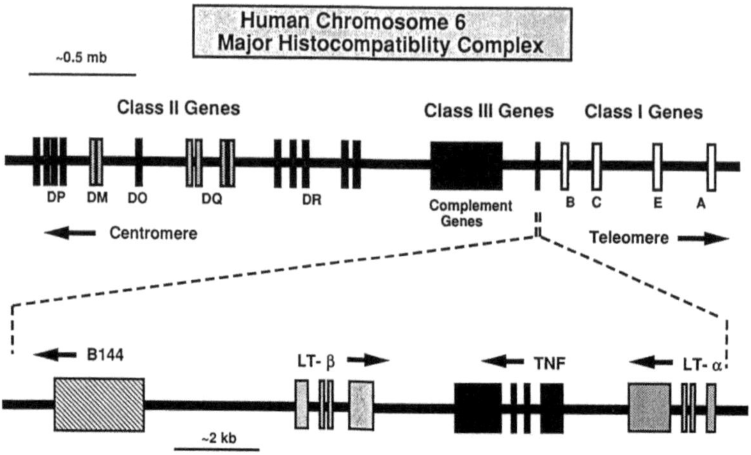

Fig. 5. Genetic map of the tumor necrosis factor/lymphotoxin (*TNF/LT*) locus in the MHC (From BROWNING et al. 1993)

Likewise a polymorphism in the LT-α gene is associated with decreased production with no effects on TNF synthesis (MESSER et al.1991). This particular polymorphism leads to an amino acid substitution, Asn-26 to Thr-26; however, this substitution did not affect complexation with LT-β in a transient COS cell expression assay (Browning, unpublished). Previous work had indicated that both forms were equally active in a standard cytolytic assay (GOH and PORTER 1991).

In the mouse, SSLP techniques have allowed one to ask whether unique stretches of DNA exist in the class III region of mice strains prone to autoimmune disease. These studies indicate that the diabetic mouse, NOD, has a unique stretch of DNA probably extending from I-A through the LT-α gene (JACOB and HWANG 1992; IKEGAMI et al.1993). Other studies have focused on the AU rich 3' untranslated region of TNF that influences mRNA stability and posttranslational events (BEUTLER and BROWN 1993; JACOB and TASHMAN 1993). A distruption of this region in certain mice strains leads to decreased production of TNF and this defect has been postulated to contribute to pathogenicity when present in an autoimmune background, i.e., the (NZBxNZW)Fl mouse (JACOB and TASHMAN 1993).

A fundamental unresolved question concerning the involvement of MHC components in autoimmune disease is whether the class I and II genes are the sole contributing elements due to the nature of the antigens recognized by the binding clefts of the molecules or whether the observed disease linkage also reflects properties of closely linked non-class I or class II genes. In this respect the TNF locus is intriguing because it lies about 200 kb from the HLA B gene in humans and 70 kb from H-2D in mice. It is tantalizing considering the immune regulatory properties of the gene products from this locus to speculate that alterations in their expression or function may influence the impact of other defects such as the aberrant recognition of autoantigen. In this context, the TNF locus genes could be critical susceptibility factors. The challenge for the future is to examine further the function of the TNF/LT locus gene products in both mice and humans. The elucidation of the surface LT complex and its specific receptor calls for a reassessment of this entire area.

Acknowledgements. The authors would like to thank those scientists that provided information prior to publication especially S. VanArsdale, W. Force, C. Benjamin, D. Chaplin, P. Marynen, L. Owen-Schaub, S. Santee, S. Miyawaki, M. Moore, and W. Clark. The dedicated efforts of W. Meier, C. Hession, K. Miakowski, M. Zafari, A. Yampaglia, P. Lawton, and A. Ngam-ek, I. Dougas, P. Bourdon, and B. Ehrenfels were critical to the success of our work. A special thanks to R. Goodwin and C. Smith for their contributions and we apologize to those we missed in citing their work. This work was supported in part by grants (to CFW) from the American Cancer Society (IM663), the Tobacco Surtax Fund of the State of California through the Tobacco-Related Diseases Research Program (RT0261) and the National Institutes of Health (AI33068).

References

Abe Y, Miyake M, Horiuchi A, Kimŭra S, Hitsumoto Y (1991) Expression of membrane-associated lymphotoxin/tumor necrosis factor-β on human lymphokine-activated killer cells. Jpn J Cancer Res 82: 23–26

Abe Y, Horiuchi A, Osuka Y, Kimura S,.Granger GA, Gatanaga T (1992) Studies of membrane-associated and soluble (secreted) lymphotoxin in human lymphokine-activated T-killer cells in vitro. Lymphokine Cytokine Res 11: 115–121

Aderka D, Engelmann H, Hornik V, Skornick Y, Levo Y, Wallach D, Kushtai G (1991) Increased serum levels of soluble receptors for tumor necrosis factor in cancer patients. Cancer Res 51: 5602–5607

Aderka D, Engelmann H, Maor Y, Brakebusch C, Wallach D(1992) Stabilization of the bioactivity of tumor necrosis factor by its soluble receptors. J Exp Med 175: 323–329

Aderka D, Wysenbeek A, Engelmann H, Cope AP, Brennan F, Molad Y, Hornik V, Levo Y, Maini RN, Feldmann M et al. (1993)Correlation between serum levels of soluble tumor necrosis factor receptor and disease activity in systemic lupus erythematosus. Arthritis Rheum 36: 1111–1120

Aggarwal BB, Moffat B, Harkins RN (1984) Lymphotoxin. Production by a lymphoblastoid cell line, purification, and initial characterization. J Biol Chem 259: 686–691

Alderson MR, Armitage RJ, Maraskovsky E, Tough TW, Roux E, Schooley K, Ramsdell F, Lynch DH (1993) Fas transduces activation signals in normal human T lymphocytes. J Exp Med 178: 2231–2235

Alexander RB, Nelson WG, Coffey DS (1987) Synergistic enhancement by tumor necrosis factor of in vitro cytotoxicity from chemotherapeutic drugs targeted at DNA topoisomerase II.Cancer Res 47: 2403–2406

Allen RC, Armitage RJ, Conley ME, Rosenblatt H, Jenkins NA, Copeland NG, Bedell MA, Edelhoff S, Disteche CM, Simoneaux DK et al. (1993) CD40 ligand gene defects responsible for X-linked hyper-IgM syndrome. Science 259: 990–993

Alvarez-Mon M, Garcia-Suarez J, Prieto A, Manzano L, Reyes E, Lorences C, Peraile G, Jorda J, Durantez A (1993) Heterogeneous proliferative effect of tumor necrosis factor-α and lymphotoxin on mitogen-activated B cells from B-chronic lymphocytic leukemia. Am J Hematol 43: 81–85

Ames R, Holskin B, Mitcho M, Shalloway D, Chen M (1990) Induction of sensitivity to the cytotoxic action of TNF α by adenovirus E1A is independent of transformation and transcriptional activation. J Virol 64: 4115–4122

Andersson U, Adolf G, Dohlsten M, Moller G, Sjogren HO (1989) Characterization of individual tumor necrosis factor α-and β-producing cells after polyclonal T cell activation, J Immunol Methods 123: 233–240

Andrews JS, Berger AE, Ware CF (1990) Characterization of the receptor for tumor necrosis factor (TNF) and lymphotoxin (LT) on human T lymphocytes. TNF and LT differ in their receptor binding properties and the induction of MHC class I proteins on a human $CD4^+$ T cell hybridoma (published erratum appears in J Immunol, 1990, 144(12): 4906). J Immunol 144: 2582–2591

Androlewicz MJ, Browning JL, Ware CF (1992) Lymphotoxin is expressed as a heteromeric complex with a distinct 33-kDa glycoprotein on the surface of an activated human T cell hybridoma, J Biol Chem 267: 2542–2547

Armitage RJ, Fanslow WC, Strockbine L, Sato TA, Clifford KN, Macduff BM, Anderson DM, Gimpel SD, Davis-Smith T, Maliszewski CR et al. (1992) Molecular and biological characterisation of a murine ligand for CD40. Nature 357: 80–82

Aruffo A, Farrington M, Hollenbaugh D, Li X, Milatovich A, Nonoyama S, Bajorath J, Grosmaire LS, Stenkamp R, Neubauer M et al. (1993) The CD40 ligand, gp39, is defective in activated T cells from patients with X-linked hyper-IgM syndrome. Cell 72: 291–300

Ashkenazi A, Marsters SA, Capon DJ, Chamow SM, Figari IS, Pennica D, Goeddel DV, Palladino MA, Smith DH (1991) Protection against endotoxic shock by a tumor necrosis factor receptor immunoadhesin. Proc Natl Acad Sci USA 88: 10535–10539

Aversa G, Punnonen J, de Vries JE (1993) The 26-kD transmembrane form of tumor necrosis factor α on activated CD4+ T cell clones provides a costimulatory signal for human B cell activation. J Exp Med 177: 1575–1585

Baens M, Chaffanet M, Cassiman JJ, van den Berghe H, Marynen P (1993) Construction and evaluation of a hnc DNA library of human 12p transcribed sequences derived from a somatic cell hybrid. Genomics 16: 214–218

Baker E, Chen LZ, Smith CA, Callen DF, Goodwin R, Sutherland GR (1991) Chromosomal location of the human tumor necrosis factor receptor genes. Cytogenet Cell Genet 57: 117–118

Banchereau J, Bazan JF, Blanchard D, Briere F, Galizzi JP, Vankooten C, Liu YJ, Rousset F, Saeland S (1994) The CD40 antigen and its ligand. Annu Rev Immunol 12: 881–922

Banner DW, D'Arcy A, Janes W, Gentz R, Schoenfeld HJ, Broger C, Loetscher H, Lesslauer W (1993) Crystal structure of the soluble human 55 kd TNF receptor-human TNF β complex: implications for TNF receptor activation. Cell 73: 431–445

Barbara JA, Smith WB, Gamble JR, Van Ostade X, Vandenabeele P, Tavernier J, Fiers W, Vadas MA, Lopez AF (1994) Dissociation of TNF-α cytotoxic and proinflammatory activities by p55 receptor- and p75 receptor-selective TNF-α mutants. EMBO J 13: 843–850

Barna BP, Barnett GH, Jacobs BS, Estes ML (1993) Divergent responses of human astrocytoma and non-neoplastic astrocytes to tumor necrosis factor α involve the 55 kDa tumor necrosis factor receptor. J Neuroimmunol 43: 185–190

Baum PR, Gayle RB, Ramsdell F, Srinivasan S, Sorensen RA, Watson ML, Seldin MF, Baker E, Sutherland GR, Clifford KN et al. (1994) Molecular characterization of murine and human OX40/OX40 Ligand systems: identification of a human OX40 Ligand as the HTLV-1-regulated protein gp34. EMBO J 13: 3992–4001

Bazan JF (1993) Emerging families of cytokines and receptors. Curr Biol 3: 603–606

Benjamin D, Kofler G, Tschachler E (1992) Human B-cell TNF-β microheterogeneity. Lymphokine Cytokine Res 11: 45–54

Berke G (1994) The binding and lysis of target cells by cytotoxic lymphocytes: molecular and cellular aspects. Annu Rev Immunol 12: 735–773

Beutler B (1990) Regulation of cachectin biosynthesis occurs at multiple levels. Prog Clin Biol Res 349: 229

Beutler B, Brown T (1993) Polymorphism of the mouse TNF-α locus: sequence studies of the 3'-untranslated region and first intron. Gene 129: 279–283

Beutler B, vanHuffel C(1994) Unraveling function in the TNF ligand and receptor families. Science 264: 667–668

Beyaert R, Fiers W (1994) Molecular mechanisms of tumor necrosis factor-induced cytotoxicity. FEBS Lett 340: 9–16

Beyaert R, Heyninck K, De Valck D, Boeykens F, van Roy F, Fiers W (1993a) Enhancement of tumor necrosis factor cytotoxicity by lithium chloride is associated with increased inositol phosphate accumulation. J Immunol 151: 291–300

Bayaert R, Vanhaesebroeck B, Heyninck K, Boone E, De Valck D, Schulze-Osthoff K, Haegeman G, van Roy F, Fiers W (1993b) Sensitization of tumor cells to tumor necrosis factor action by the protein kinase inhibitor staurosporine. Cancer Res 53: 2623–2630

Boss J, Laster S, Gooding L (1991) Sensitivity to tumour necrosis factor-mediated cytolysis is unrelated to manganous superoxide dismutase messenger RNA levels among transformed mouse fibroblasts. Endocrinology 73: 309–315

Boussiotis VA, Nadler LM, Strominger JL, Goldfeld AE (1994) Tumor necrosis factor is an autocrine growth factor for normal human B cells. Proc Natl Acad Sci USA 91: 283–288

Bradshaw RA, Blundell TL, Lapatto R, McDonald NQ, Murray-Rust J (1993) Nerve growth factor revisited. Trends Biochem Sci 18: 48–52

Brakebusch C, Nophar Y, Kemper O, Engelmann H, Wallach D (1992) Cytoplasmic truncation of the p55 tumour necrosis factor (TNF) receptor abolishes signalling, but not induced shedding of the receptor. EMBO J 11: 943–950

Brockhaus M, Bar-Khayim Y, Gurwicz S, Frensdorff A, Haran N (1992) Plasma tumor necrosis factor soluble receptors in chronic renal failure. Kidney Int 42: 663–667

Browning J, Ribolini A (1989) Studies on the differing effects of tumor necrosis factor and lymphotoxin on the growth of several human tumor lines. J Immunol 143: 1859–1867

Browning JL, Androlewicz MJ, Ware CF (1991) Lymphotoxin and an associated 33-kDa glycoprotein are expressed on the surface of an activated human T cell hybridoma. J Immunol 147: 1230–1237

Browning JL, Ngam-ek A, Lawton P, DeMarinis J, Tizard R, Chow EP, Hession C, O'Brine-Greco B, Foley SF, Ware CF (1993) Lymphotoxin β, a novel member of the TNF family that forms a heteromeric complex with lymphotoxin on the cell surface. Cell 72: 847–856

Browning JL, Dougas I, Ngam-ek A, Bourdon P, Ehrenfels B, Miatkowski K, Zafari M, Yampaglia A, Lawton P, Meier W, Benjamin C, Hession C (1995) Characterization of surface lymphotoxin forms: use of specific monoclonal antibodies and soluble receptors. J Immunol (in press)

Callard RE, Armitage RJ, Fanslow WC, Spriggs MK (1993) CD40 ligand and its role in X-linked hyper-IgM syndrome. Immunol Today 14: 559–564

Camerini D, Walz G, Loenen WA, Borst J, Seed B (1991) The T cell activation antigen CD27 is a member of the nerve growth factor/tumor necrosis factor receptor gene family. J Immunol 147: 3165–3169

Cheng J, Zhou T, Liu C, Shapiro JP, Brauer MJ, Kiefer MC, Barr PJ, Mountz JD (1994) Protection from Fas-mediated apoptosis by a soluble form of the Fas molecule. Science 263: 1759–1762

Coffman FD, Green LM, Ware CF (1988) The relationship of receptor occupancy to the kinetics of cell death mediated by tumor necrosis factor. Lymphokine Res 7: 371–383

Coffman FD, Green LM, Godwin A, Ware CF (1989a) Cytotoxicity mediated by tumor necrosis factor in variant subclones of the ME-180 cervical carcinoma line: modulation by specific inhibitors of DNA topoisomerase II. J Cell Biochem 39: 95–105

Coffman FD, Haviland DL, Green LM, Ware CF (1989b) Cytotoxicity by tumor necrosis factor is linked with the cell cycle but does not require DNA synthesis. Growth Factors 1: 357–364

Collins T, Lapierre LA, Fiers W, Strominger JL, Pober JS (1986) Recombinant human tumor necrosis factor increases mRNA levels and surface expression of HLA-A, B antigens in vascular endothelial cells and dermal fibroblasts in vitro. Proc Natl Acad Sci USA 83: 446–450

Cope AP, Aderka D, Doherty M, Engelmann H, Gibbons D, Jones AC, Brennan FM, Maini RN, Wallach D, Feldmann M (1992) Increased levels of soluble tumor necrosis factor receptors in the sera and synovial fluid of patients with rheumatic diseases. Arthritis Rheum 35: 1160–1169

Crowe PD, VanArsdale TL, Goodwin RG, Ware CF (1993) Specific induction of 80-kDa tumor necrosis factor receptor shedding in T lymphocytes involves the cytoplasmic domain and phosphorylation. J Immunol 151: 6882–6890

Crowe PD VanArsdale TL, Walter BN, Dahms KM, Ware CF (1994a) Production of lymphotoxin (LT α) and a soluble dimeric form of its receptor using the baculovirus expression system. J Immunol Methods 168: 79–89

Crowe PD, VanArsdale TL, Walter BN, Ware CF, Hession C, Ehrenfels B, Browning JL, Din WS, Goodwin RG, Smith CA (1994b) A lymphotoxin-β-specific receptor. Science 264: 707–710

Crowe PD, Walter BN, Mohler KM, Otten-Evans C, Black RA, Ware CF (1995) A metalloprotease inhibitor blocks shedding of the 80 kDa TNF receptor and TNF processing in T lymphocytes. J Exp Med (in press)

Crump WL, Owen-Schaub LB, Grimm EA (1990) Molecular identification of the human tumor necrosis factor receptor on interleukin-2-stimulated peripheral blood lymphocytes. Cell Immunol 131: 150–158

Darzynkiewicz Z, Williamson B, Carswell EA, Old LJ (1984) Cell cycle-specific effects of tumor necrosis factor. Cancer Res 44: 83–90

Davidson WF, Dumont FJ, Bedigian HG, Fowlkes BJ, Morse HC (1986) Phenotypic, functional, and molecular genetic comparisons of the abnormal lymphoid cells of C3H-lpr/lpr and C3H-gld/gld mice. J Immunol 136: 4075–4084

Davies AM (1994) Neurobiology. Tracking neurotrophin function (news; comment). Nature 368: 193–194

Deleuran BW, Chu CQ, Field M, Brennan FM, Mitchell T, Feldmann M, Maini RN (1992) Localization of tumor necrosis factor receptors in the synovial tissue and cartilage-pannus junction in patients with rheumatoid arthritis. Implications for local actions of tumor necrosis factor α. Arthritis Rheum 35: 1170–1178

Derre J, Kemper O, Cherif D, Nophar Y, Berger R, Wallach D (1991) The gene for the type 1 tumor necrosis factor receptor (TNF-R1) is localized on band 12p13. Hum Genet 87: 213–233

De Togni P, Goellner J, Ruddle NH, Streeter PR, Fick A, Mariathasan S, Smith SC, Carlson R, Shornick

LP, Strauss-Schoenberger J, Russell JH, Karr R, Chaplin DD (1994) Abnormal development of peripheral lymphoid organs in mice deficient in lymphotoxin. Science 264: 703–706
De Valck D, Beyaert R, van Roy F, Fiers W (1993) Tumor necrosis factor cytotoxicity is associated with phospholipase D activation. Eur J Biochem 212: 491–497
DiSanto JP, Bonnefoy JY, Gauchat JF, Fischer A, de Saint Basile G (1993) CD40 ligand mutations in x-linked immunodeficiency with hyper-IgM. Nature 361: 541–543
DiStefano PS, Johnson EM Jr (1988) Identification of a truncated form of the nerve growth factor receptor. Proc Natl Acad Sci USA 85: 270–274
Doherty PC (1993) Cell-mediated cytotoxicity. Cell 75: 607–612
Duerksen-Hughes P, Wold WS, Gooding LR (1989) Adenovirus E1A renders infected cells sensitive to cytolysis by tumor necrosis factor. J Immunol 143: 4193–4200
Durkop H, Latza U, Hummel M, Eitelbach F, Seed B, Stein H (1992) Molecular cloning and expression of a new member of the nerve growth factor receptor family that is characteristic for Hodgkin's disease. Cell 68: 421–427
Eck MJ, Sprang SR (1989) The structure of tumor necrosis factor-α at 2.6 A resolution. Implications for for receptor binding. J Biol Chem 264: 17595–17605
Eck MJ, Beutler B, Kuo G, Merryweather JP, Sprang SR (1988) Crystallization of trimeric recombinant human tumor necrosis factor (cachectin). J Biol Chem 263: 12816–12819
Eck MJ, Ultsch M, Rinderknecht E, de Vos AM, Sprang SR (1992) The structure of human lymphotoxin (tumor necrosis factor-β) at 1.9-A resolution. J Biol Chem 267: 2119–2122
Engelmann H, Holtmann H, Brakebusch C, Avni YS, Sarov I, Nophar Y, Hadas E, Leitner O, Wallach D (1990a) Antibodies to a soluble form of a tumor necrosis factor (TNF) receptor have TNF-like activity. J Biol Chem 265: 14497–14504
Engelmann H, Novick D, Wallach D (1990b) Two tumor necrosis factor-binding proteins purified from human urine. Evidence for immunological cross-reactivity with cell surface tumor necrosis factor receptors. J Biol Chem 265: 1531–1536
Erikstein BK, Smeland EB, Blomhoff HK, Funderud S, Prydz K, Lesslauer W, Espevik T (1991) Independent regulation of 55-kDa and 75-kDa tumor necrosis factor receptors during activation of human peripheral blood B lymphocytes. Eur J Immunol 21: 1033–1037
Espevik T, Brockhaus M, Loetscher H, Nonstad U, Shalaby R (1990) Characterization of binding and biological effects of monoclonal antibodies against a human tumor necrosis factor receptor. J Exp Med 171: 415–426
Estrov Z, Kurzrock R, Pocsike, Pathak S, Kantarjian HM, Zipf TF, Harris D, Talpaz M, Aggarwal BB (1993) Lymphotoxin is an autocrine growth factor for Epstein-Barr virus-infected B cell lines. J Exp Med 177: 763–774
Farrah T, Smith CA (1992) Emerging cytokine family (letter). Nature 358: 26
Farrington M, Grosmaire LS, Nonoyama S, Fischer SH, Hollenbaugh D, Ledbetter JA, Noelle RJ, Aruffo A, Ochs HD (1994) CD40 ligand expression is defective in a subset of patients with common variable immunodeficiency. Proc Natl Acad Sci USA 91: 1099–1103
Fashena SJ, Tang WL, Sarr T, Ruddle NH (1990) The murine lymphotoxin gene promoter. Characterization and negative regulation. J Immunol 145: 177–183
Fukushima K, Watanabe H, Takeo K, Nomura M, Asahi T, Yamashita K (1993) N-linked sugar chain structure of recombinant human lymphotoxin produced by CHO cells: the functional role of carbohydrate as to its lectin-like character and clearance velocity. Arch Biochem Biophys 304: 144–153
Gatanaga T, Hwang CD, Kohr W, Cappuccini F, Lucci JA, Jeffes EW, Lentz R, Tomich J, Yamamoto RS, Granger GA (1990) Purification and characterization of an inhibitor (soluble tumor necrosis factor receptor) for tumor necrosis factor and lymphotoxin obtained from the serum ultrafiltrates of human cancer patients. Proc Natl Acad Sci USA 87: 8781–8784
Godfried MH, Van der Poll T, Jansen J, Romijin JA, Schattenkerk JK, Endert E, van Deventer SJ, Sauerwein HP (1993) Soluble receptors for tumour necrosis factor: a putative marker of disease progression in HIV infection. AIDS 7: 33–36
Goh C, Porter A (1991) Structural and functional domains in human tumour necrosis factors. Protein Eng 4: 385–389
Goh CR (1993) Protein engineering of tumor necrosis factor-β and its applications in cancer, septicemia and cachexia. Ann Acad Med Singapore 22: 651
Goh CR, Loh CS Porter AG (1991) Aspartic acid 50 and tyrosine 108 are essential for receptor binding and cytotoxic activity of tumour necrosis factor β (lymphotoxin). Protein Eng 4: 785–791
Golstein P, Ojcius DM, Young JD (1991) Cell death mechanisms and the immune system. Immunol Rev 121: 29–65

Gooding LR (1992) Virus proteins that counteract host immune defenses. Cell 71: 5–7
Gooding LR, Sofola IO, Tollefson AE, Duerksen-Hughes P, Wold WS (1990) The adenovirus E3-14.7K protein is a general inhibitor of tumor necrosis factor-mediated cytolysis. J Immunol 145: 3080–3086
Goodwin RG, Alderson MR, Smith CA, Armitage RJ, VandenBos T, Jerzy R, Tough TW, Schoenborn MA, Davis-Smith T, Hennen K et al. (1993a) Molecular and biological characterization of a ligand for CD27 defines a new family of cytokines with homology to tumor necrosis factor. Cell 73: 447–456
Goodwin RG, Din WS, Davis-Smith T, Anderson DM, Gimpel SD, Sato TA, Maliszewski CR, Brannan CI, Copeland NG, Jenkins NA et al. (1993b) Molecular cloning of a ligand for the inducible T cell gene 4-1BB: a member of an emerging family of cytokines with homology to tumor necrosis factor. Eur J Immunol 23: 2631–2641
Goppelt-Struebe M, Rehfeldt W (1992) Glucocorticoids inhibit TNF α-induced cytosolic phospholipase A2 activity. Biochim Biophys Acta 1127: 163–167
Granger GA, Williams TW (1968) Lymphocyte cytotoxicity in vitro: activation and release of a cytotoxic factor. Nature 218: 1253–1254
Gray P, Aggarwal B, Benton C, Bringman T, Henzel W, Jarrett J, Leung D, Moffat B, Ng P, Svedersky L, Palladino M, Nedwin G (1984) Cloning and expression of the cDNA for human lyphotoxin: a lymphokine with tumor necrosis activity. Nature 312: 721–724
Greenblatt MS, Elias L (1992) The type B receptor for tumor necrosis factor-α mediates DNA fragmentation in HL-60 and U937 cells and differentiation in HL-60 cells. Blood 80: 1339–1346
Gullberg U, Lantz M, Lindvall L, Olsson I, Himmler A (1992) Involvement of an Asn/Val cleavage site in the production of a soluble form of a human tumor necrosis factor (TNF) receptor. Site-directed mutagenesis of a putative cleavage site in the p55 TNF receptor chain. Eur J Cell Biol 58: 307–312
Hains JM, Aggarwal BB (1989) Characterization of recombinant human lymphotoxin (tumor necrosis factor-β) produced by a mammalian cell line. Arch Biochem Biophys 274: 417–425
Heller R, Song K, Fan N, Chang D (1992) The p70 tumor necrosis factor receptor mediates cytotoxicity. Cell 70: 47–56
Heller RA, Song K, Onasch MA, Fischer WH, Chang D, Ringold GM (1990) Complementary DNA cloning of a receptor for tumor necrosis factor and demonstration of a shed form of the receptor. Proc Natl Acad Sci USA 87: 6151–6155
Hennet T, Bertoni G, Richter C, Peterhans E (1993) Expression of BCL-2 protein enhances the survival of mouse fibrosarcoid cells in TNF-mediated cytotoxicity. Cancer Res 53: 1456–1460
Hepburn A, Demolle D, Boeynaems JM, Fiers W, Dumont JE (1988) Rapid phosphorylation of a 27 kDa protein induced by tumor necrosis factor. FEBS Lett 227: 175–178
Heusel JW, Wesselschmidt RL, Shresta S, Russell JH, Ley TJ (1994) Cytotoxic lymphocytes require granzyme B for the rapid induction of DNA fragmentation and apoptosis in allogeneic target cells. Cell 76: 977–987
Higuchi M, Aggarwal BB (1992) Modulation of two forms of tumor necrosis factor receptors and their cellular response by soluble receptors and their monoclonal antibodies. J Biol Chem 267: 20892–20899
Higuchi M, Aggarwal BB (1993) Okadaic acid induces down-modulation and shedding of tumor necrosis factor receptors. Comparison with another tumor promoter, phorbol ester. J Biol Chem 268: 5624–5631
Higuchi M, Aggarwal BB (1994) TNF induces internalization of the p60 receptor and shedding of the p80 receptor. J Immunol 152: 3550–3558
Hirahara H, Ogawa M, Kimura M, Iiai T, Tsuchida M, Hanawa H, Watanabe H, Abo T (1994) Glucocorticoid independence of acute thymic involution induced by lymphotoxin and estrogen. Cell Immunol 153: 401–411
Hoeck WG, Ramesha CS, Chang DJ, Fan N, Heller RA (1993) Cytoplasmic phospholipase A2 activity and gene expression are stimulated by tumor necrosis factor: dexamethasone blocks the induced synthesis. Proc Natl Acad Sci USA 90: 4475–4479
Hohmann H, Remy R, Aigners L, Brockhaus M, van Loon APG (1992) Protein kinases negatively affect nuclear factor-kB activation by tumor necrosis factor-α at two different stages in promyelocytic HL60 cells. J Biol Chem 267: 2065–2072
Hohmann HP, Brockhaus M, Baeuerle PA, Remy R, Kolbeck R, van Loon AP (1990a) Expression of the types A and B tumor necrosis factor (TNF) receptors is independently regulated, and both receptors mediate activation of the transcription factor NF-κB. TNF α is not needed for induction of a biological effect via TNF receptors. J Biol Chem 265: 22409–22417
Hohmann HP, Remy R, Poschl B, van Loon AP (1990b) Tumor necrosis factors-α and -β bind to the same

two types of tumor necrosis factor receptors and maximally activate the transcription factor NF-κB at low receptor occupancy and within minutes after receptor binding. J Biol Chem 265: 15183–15188
Hollenbaugh D, Grosmaire L, Kullas C, Chaluphny N, Braesch-Andersen S, Noelle R, Stamenkovic I, Ledbetter J, Aruffo A (1992) The human T cell antigen gp39, a member of the TNF gene family, is a ligand for the CD40 receptor: expression of a soluble form of gp39 with B cell co-stimulatory activity. EMBO J 11: 4313–4321
Hwang C, Gatanaga M, Granger GA, Gatanaga T (1993) Mechanism of release of soluble forms of tumor necrosis factor/lymphotoxin receptor by phorbol myristate acetate-stimulated human THP-1 cells in vitro. J Immunol 151: 5631–5638
Ikegami H, Kawaguchi Y, Ueda H, Fukuda M, Takakawa K, Fujioka Y, Fujisawa T, Uchida K, Ogihara T (1993) MHC-linked diabetogenic gene of the NOD mouse: molecular mapping of the 3' boundary of the diabetogenic region. Biochem Biophys Res Commun 192: 677–682
Imamura K, Sherman ML, Spriggs D, Kufe D (1988) Effect of tumor necrosis factor on GTP binding and GTPase activity in HL-60 and L929 cells. J Biol Chem 263: 10247–10253
Itoh N, Nagata S (1993) A novel protein domain required for apoptosis: mutational analysis of human Fas antigen. J Biol Chem 268: 10932–10937
Itoh N, Yonehara S, Ishii A, Yonehara M, Mizushima S, Sameshima M, Hase A, Seto Y, Nagata S (1991) The polypeptide encoded by the cDNA for human cell surface antigen Fas can mediate apoptosis. Cell 66: 233–243
Itoh N, Tsujimoto Y, Nagata S (1993) Effect of bcl-2 on Fas antigen-mediated cell death. J Immunol 151: 621–627
Jacob CO (1992) Tumor necrosis factor α in autoimmunity: pretty girl or old witch? Immunol Today 13: 122–125
Jacob CO, Hwang F (1992) Definition of microsatellite size variants for Tnfa and Hsp70 in autoimmune and nonautoimmune mouse strains. Immunogenetics 36: 182–188
Jacob CO, McDevitt HO (1991a) Genetic predisposition to autoimmune diseases: the contribution of the major histocompatibility complex. In: Talal N (ed) Molecular autoimmunity. Academic, New York, p 7
Jacob CO, McDevitt HO (1991b) Interferon gamma and tumor necrosis factor in autoimmune disease models: implications for immunoregulation and genetic susceptibility. In: Talal N (ed) Molecular autoimmunity. Academic, New York, p 107
Jacob CO, Tashman NB (1993) Disruption in the AU motif of the mouse TNF-α 3' UTR correlates with reduced TNF production by macrophages in vitro. Nucleic Acids Res 21: 2761–2766
Janssen O, Gillis S, Kabelitz D (1990) In vitro transformation by Epstein-Barr virus induces a switch in growth factor and anti-IgM responsiveness in a human leukemic B cell clone. Eur J Immunol 20: 7–14
Jones EY, Stuart DI, Walker NP (1989) Structure of tumour necrosis factor. Nature 338: 225–228
Jones EY, Stuart DI, Walker NP (1992) Crystal structure of TNF, Immunol Ser 56: 93–127
Jongeneel CV, Briant L, Udalova IA, Sevin A, Nedospasov SA, Cambon-Thomsen A (1991) Extensive genetic polymorphism in the human tumor necrosis factor region and relation to extended HLA haplotypes. Proc Natl Acad Sci USA 88: 9717–9721
Josimovic-Alasevic O, Durkop H, Schwarting R, Backe E, Stein H, Diamantstein T (1989) Ki-1 (CD30) antigen is released by Ki-1-positive tumor cells in vitro and in vivo. I. Partial characterization of soluble Ki-1 antigen and detection of the antigen in cell culture supernatants and in serum by an enzyme-linked immunosorbent assay. Eur J Immunol 19: 157–162
Kagan BL, Baldwin RL, Munoz D, Wisnieski BJ (1992) Formation of ion-permeable channels by tumor necrosis factor-α Science 255: 1427–1430
Kehrl JH, Alvarez-Mon M, Delsing GA, Fauci AS (1987) Lymphotoxin is an important T cell-derived growth factor for human B cells. Science 238: 1144–1146
Kim MY, Linardic C, Obeid L, Hannun Y (1991) Identification of sphingomyelin turnover as an effector mechanism for the action of tumor necrosis factor α and gamma-interferon. Specific role in cell differentiation. J Biol Chem 266: 484–489
Kohno T, Brewer MT, Baker SL, Schwartz PE, King MW, Hale KK, Squires CH, Thompson RC, Kagi D, Vignaux F, Ledermann B, Burki K, Depraetere V, Nagata S, Hengartner H, Golstein P (1994) Fas and perforin pathways as major mechanisms of T cell-mediated cytotoxicity. Science 265: 528–530
Kuramoto T, Mashimo T, Kolke R, Miyawaki S, Yamada J, Miyasaka M, Serikawa T (1994) The alymphoplasia (aly) mutation co-segrates with the intercellular adhesion molecule-2 (ICAM+2) on mouse chromosome II. International Immunol 6: 991–994
Vannice JL (1990) A second tumor necrosis factor receptor gene product can shed a naturally occurring tumor necrosis factor inhibitor. Proc Natl Acad Sci USA 87: 8831–8335
Kolanus W, Romeo C, Seed B (1993) T cell activation by clustered tyrosine kinases. Cell 74: 171–183

Kolesnick R, Golde DW (1994) The sphingomyelin pathway in tumor necrosis factor and interleukin-1 signaling. Cell 77: 325–328

Korthauer U, Graf D, Mages HW, Briere F, Padayachee M, Malcolm S, Ugazio AG, Notarangelo LD, Levinsky RJ, Kroczek RA (1993) Defective expression of T-cell CD40 ligand causes X-linked immunodeficiency with hyper-IgM. Nature 361: 539–541

Kriegler M, Perez C, DeFay K, Albert I, Lu SD (1988) A novel form of TNF/cachectin is a cell surface cytotoxic transmembrane protein: ramifications for the complex physiology of TNF. Cell 53: 45–53

Kruppa G, Thoma B, Machleidt T, Wiegmann K, Kronke M (1992) Inhibition of tumor necrosis factor (TNF)-mediated NF-κB activation by selective blockade of the human 55-kDa TNF receptor. J Immunol 148: 3152–3157

Kumar S, Baglioni C (1991) Protection from tumor necrosis factor-mediated cytolysis by overexpression of plasminogen activator inhibitor type-2. J Biol Chem 266: 20960–20964

Kusumi A, Abo T, Masuda T, Sugiura K, Seki S, Ohteki T, Okuyama R, Kumagai K (1992) Lymphotoxin activates hepatic T cells and simultanously induces profound thymic atrophy. Immunology 77: 177–184

Kwon BS, Weissman SM (1989) cDNA sequences of two inducible T-cell genes. Proc Natl Acad Sci USA 86: 1963–1967

Lantz M, Gullberg U, Nilsson E, Olsson I (1990) Characterization in vitro of a human tumor necrosis factor-binding protein. A soluble form of a tumor necrosis factor receptor. J Clin Invest 86: 1342–1396

Lantz M, Lindvall L, Pantazis P, Olsson I (1991) Lymphotoxin produced by human B- and T-cell lines appears in two distinct forms. Mol Immunol 28: 9–16

Laster SM, Wood JG, Gooding LR (1988) Tumor necrosis factor can induce both apoptic and necrotic forms of cell lysis. J Immunol 141: 2629–2634

Latza U, Durkop H, Schnittger S, Ringeling J, Eitelbach F, Hummel M, Fonatsch C, Stein H (1994) The human OX40 homolog: cDNA structure, expression and chromosomal assignment of the ACT35 antigen. Eur J Immunol 24: 677–683

Lederman S, Yellin MJ, Cleary AM, Pernis A, Inghirami G, Cohn LE, Covey LR, Lee JJ, Rothman P, Chess L (1994) T-BAM/CD40-L on helper T lymphocytes augments lymphokine-induced B cell Ig isotype switch recombination and rescues B cells from programmed cell death. J Immunol 152: 2163–2171

Leithauser F, Dhein J, Mechtersheimer G, Koretz K, Bruderlein S, Henne C, Schmidt A, Debatin KM, Krammer PH, Moller P (1993) Constitutive and induced expression of APO-1, a new member of the nerve growth factor/tumor necrosis factor receptor superfamily, in normal and neoplastic cells. Lab Invest 69: 415–429

Lichtenstein A, Gera JF, Andrews J, Berenson J, Ware CF (1991) Inhibitors of ADP-ribose polymerase decrease the resistance of HER2/neu-expressing cancer cells to the cytotoxic effects of tumor necrosis factor. J Immunol 146: 2052–2058

Liu AY, Miskovsky EP, Stanhope PE, Siliciano RF (1992) Production of transmembrane and secreted forms of tumor necrosis factor (TNF)-α by HIV-1-specific CD4+ cytolytic T lymphocyte clones. Evidence for a TNF-α-independent cytolytic mechanism. J Immunol 148: 3789–3798

Liu J, Mathias S, Yang Z, Kolesnick RN (1994) Renaturation and tumor necrosis factor-α stimulation of a 97-kDa ceramide-activated protein kinase. J Biol Chem 269: 3047–3052

Loenen WA, de Vries E, Gravestein LA, Hintzen RQ, van Lier RA, Borst J (1992a) The CD27 membrane receptor, a lymphocyte-specific member of the nerve growth factor receptor family, gives rise to a soluble form by protein processing that does not involve receptor endocytosis. Eur J Immunol 22: 447–455

Loenen WA, Gravestein LA, Beumer S, Melief CJ, Hagemeijer A, Borst J (1992b) Genomic organization and chromosomal localization of the human CD27 gene. J Immunol 149: 3937–3943

Loetscher H, Pan Y, Lahm H, Gentz R, Brockhaus M, Tabuchi H, Lesslauer W (1990) Molecular cloning and expression of the human 55 kd tumor necrosis factor receptor. Cell 61: 351–359

Loetscher H, Gentz R, Zulauf M, Lustig A, Tabuchi H, Schlaeger EJ, Brockhaus M, Gallati H, Manneberg M, Lesslauer W (1991) Recombinant 55-kDa tumor necrosis factor (TNF) receptor. Stoichiometry of binding to TNF α and TNF β and inhibition of TNF activity. J Biol Chem 266: 18324–18329

Loetscher H, Stueber D, Banner D, Mackay F, Lesslauer W (1993) Human tumor necrosis factor α (TNF-α) mutants with exclusive specificity for the 55-kDa or 75-kDa TNF receptors. J Biol Chem 268: 26350–26357

Lucas R, Magez S, De Leys R, Fransen L, Scheerlinck JP, Rampelberg M, Sablon E, De Baetselier P (1994) Mapping the lectin-like activity of tumor necrosis factor. Science 263: 814–817

Macchia D, Almerigogna F, Parronchi P, Ravina A, Maggi E, Romagnani S (1993) Membrane tumour

necrosis factor-α is involved in the polyclonal B-cell activation induced by HIV-infected human T cells. Nature 363: 464–466
Mackay F, Loetscher H, Stueber D, Gehr G, Lesslauer W (1993) Tumor necrosis factor α (TNF-α)-induced cell adhesion to human endothelial cells is under dominant control of one TNF receptor type, TNF-R55. J Exp Med 177: 1277–1286
Mallett S, Fossum S, Barclay AN (1990) Characterization of the MRC OX40 antigen of activated CD4 positive T lymphocytes—a molecule related to nerve growth factor receptor. EMBO J 9: 1063–1068
Marx J (1993) Cell communication failure leads to immune disorder. Science 259: 896–897
Massung RF, Esposito JJ, Liu LI, Qi J, Utterback TR, Knight JC, Aubin L, Yuran TE, Parsons JM, Loparev VN et al. (1993) Potential virulence determinants in terminal regions of variola smallpox virus genome. Nature 366: 748–751
Massung RF, Liu L, Qi J, Knight JC, Yuran TE, Kerlavage AR, Parsons JM, Venter JC, Esposito JJ Analysis of the complete genome of smallpox variola major virus strain Bangladesh-1975. Virology 201: 215–240
Mathias S, Dressler KA, Kolesnick RN (1991) Characterization of a ceramide-activated protein kinase: stimulation by tumor necrosis factor α. Proc Natl Acad Sci USA 88: 10009–10013
McDonald NQ, Hendrickson WA (1993) A structural superfamily of growth factors containing a cystine knot motif. Cell 73: 421–424
McDonald NQ, Lapatto R, Murrary-Rust J, Gunning J, Wlodawer A, Blundell TL (1991) New protein fold revealed by a 2.3-A resolution crystal structure of nerve growth factor. Nature 354: 411–414
Melani C, Silvani A, Parmiani G, Colombo MP (1993) Lymphotoxin gene expression by melanocytes and melanoma cell lines and persistence of unspliced mRNA. FEBS Lett 335: 114–118
Messer G, Spengler U, Jung MC, Honold G, Blomer K, Pape GR, Riethmuller G, Weiss EH (1991) Polymorphic structure of the tumor necrosis factor (TNF) locus: an NcoI polymorphism in the first intron of the human TNF-β gene correlates with a variant amino acid in position 26 and a reduced level of TNF-β production. J Exp Med 173: 209–219
Milatovich A, Song K, Heller RA, Francke U (1991) Tumor necrosis factor receptor genes, TNFR1 and TNFR2, on human chromosomes 12 and 1. Somat Cell Mol Genet 17: 519–523
Millet I, Ruddle NH (to be published) Differential regulation of lymphotoxin (LT), lymphotoxin-β (LT-β) and tumor necrosis factor (TNF-α) in murine T cell clones activated through the T cell receptor. J Immunol
Miyawaki S, Nakamura Y, Suzuka H, Koba M, Yasumizu R, Ikehara S, Shibata Y (1994) A new mutation, *aly*, that induces a generalized lack of lymph nodes accompanied by immunodeficiency in mice. Eur J Immunol 24: 429–434
Naume B, Shalaby R, Lesslauer W, Espevik T (1991) Involvement of the 55- and 75-kDa tumor necrosis factor receptors in the generation of lymphokine-activated killer cell activity and proliferation of natural killer cells. J Immunol 146: 3045–3048
Nophar Y, Kemper O, Brakebusch C, Englemann H, Zwang R, Aderka D, Holtmann H, Wallach D (1990) Soluble forms of tumor necrosis factor receptors (TNF-Rs). The cDNA for the type I TNF-R, cloned using amino acid sequence data of its soluble form, encodes both the cell surface and a soluble form of the receptor. EMBO J 9: 3269–3278
Olsson L, Lantz M, Nilsson E, Peetre C, Thysell H, Grubb A, Adolf G (1989) Isolation and characterization of a tumor necrosis factor binding protein from urine. Eur J Haematol 42: 270–275
Osborn L, Hession C, Tizard R, Vassallo C, Luhowskyj S, Chi-Rosso G, Lobb R (1989) Direct expression cloning of vascular cell adhesion molecule 1, a cytokine-induced endothelial protein that binds to lymphocytes. Cell 59: 1203–1211
Ostergaard HL, Clark WR (1989) Evidence for multiple lytic pathways used by cytotoxic T lymphocytes. J Immunol 143: 2120–2126
Paul NL, Ruddle NH (1988) Lymphotoxin. Annu Rev Immunol 6: 407–438
Peitsch MC, Jongeneel CV (1993) A 3-D model for the CD40 ligand predicts that it is compact trimer similar to the tumor necrosis factors. Int Immunol 5: 233–238
Pennica D, Nedwin GE, Hayflick JS, Seeburg PH, Derynck R, Palladino MA, Kohr WJ, Aggarwal BB, Goeddel DV (1984) Human tumour necrosis factor: precursor structure, expression and homology to lymphotoxin. Nature 312: 724–729
Pennica D, Kohr W, Fendly B, Shire S, Raab H, Borchardt P, Lewis M, Goeddel D (1992a) Characterization of a recombinant extracellular domain of the type 1 tumor necrosis factor receptor: evidence for tumor necrosis factor-α induced receptor aggregation. Biochemistry 31: 1134–1141
Pennica D, Lam VT, Mize NK, Weber RF, Lewis M, Fendly BM, Lipari MT, Goeddel DV (1992b) Biochemical properties of the 75-kDa tumor necrosis factor receptor. Characterization of ligand binding, internalization, and receptor phosphorylation. J Biol Chem 267: 21172–21178

Pennica D, Lam VT, Weber RF, Kohr WJ, Basa LJ, Spellman MW, Ashkenazi A, Shire SJ, Goeddel DV (1993) Biochemical characterization of the extracellular domain of the 75-kilodalton tumor necrosis factor receptor. Biochemistry 32: 3131–3138
Peppel K, Crawford D, Beutler B (1991) A tumor necrosis factor (TNF) receptor-IgG heavy chain chimeric protein as a bivalent antagonist of TNF activity. J Exp Med 174: 1483–1489
Peppel K, Poltorak A, Melhado I, Jirik F, Beutler B (1993) Expression of a TNF inhibitor in transgenic mice. J Immunol 151: 5699–5703
Perez C, Albert I, DeFay K, Zachariades N, Gooding L, Kriegler M (1990) A nonsecretable cell surface mutant of tumor necrosis fator (TNF) kills by cell-to-cell contact. Cell 63: 251–258
Pfeffer K, Matsuyama T, Kundig TM, Wakeham A, Kishihara K, Shahinian A, Wiegmann K, Ohashi PS, Kronke M, Mak TW (1993) Mice deficient for the 55 kd tumor necrosis factor receptor are resistant to endotoxic shock, yet succumb to L. monocytogenes infection. Cell 73: 457–467
Podack ER, Hengartner H, Lichtenheld MG (1991) A central role of perforin in cytolysis? Annu Rev Immunol 9: 129–157
Porteu F, Hieblot C (1994) Tumor necrosis factor induces a selective shedding of its p75 receptor from human neutrophils. J Biol Chem 269: 2834–2840
Porteu F, Nathan C (1990) Shedding of tumor necrosis factor receptors by activated human neutrophils. J Exp Med 172: 599–607
Powell MB, Mitchell D, Lederman J, Buckmeier J, Zamvil SS, Graham M, Ruddle NH, Steinman L (1990) Lymphotoxin and tumor necrosis factor-α production by myelin basic protein-specific T cell clones correlates with encephalitogenicity. Int Immunol 2: 539–544
Rabizadeh S, Oh J, Zhong LT, Yang J, Bitler CM, Butcher LL, Bredesen DE (1993) Induction of apoptosis by the low-affinity NGF receptor. Science 261: 345–348
Ratner A, Clark WR (1993) Role of TNF-α in $CD8^+$ cytotoxic T lymphocyte-mediated lysis. J Immunol 150: 4303–4314
Reid T, Ramesha C, Ringold G (1991) Resistance to killing by tumor necrosis factor in an adipocyte cell line caused by a defect in arachidonic acid biosynthesis. J Biol Chem 266: 16580–16586
Robinet E, Branellec D, Termijtelen AM, Blay JY, Gay F, Chouaib S (1990) Evidence for tumor necrosis factor-α involvement in the optimal induction of class I allospecific cytotoxic T cells. J Immunol 144: 4555–4561
Rogers S, Wells R, Rechsteiner M (1986) Amino acid sequences common to rapidly degraded proteins: the PEST hypothesis, Science 234: 364–368
Rothe J, Lesslauer W, Lotscher H, Lang Y, Koebel P, Kontgen F, Althage A, Zinkernagel R, Steinmetz M, Bluethmann H (1993) Mice lacking the tumour necrosis factor receptor I are resistant to TNF-mediated toxicity but highly susceptible to infection by Listeria monocytogenes. Nature 364: 798–802
Rouvier E, Luciani MF, Golstein P (1993) Fas involvement in $Ca^{(2+)}$-independent T cell-mediated cytotoxicity. J Exp Med 177: 195–200
Ruddle NH (1992) Tumor necrosis factor (TNF-α) and lymphotoxin (TNF-β). Curr Opin Immunol 4: 327–332
Ruddle NH, Homer R (1988) The role of lymphotoxin in inflammation. Proc Allergy 40: 162
Ruddle NH, Waksman BH (1968) Cytotoxicity mediated by soluble antigen and lymphocytes in delayed hypersensitivity. I. Characterization of the phenomenon. J Exp Med 128: 1237–1254
Ruddle NH, Bergman CM, McGrath KM, Lingenheld EG, Grunnet ML, Padula SJ, Clark RB (1990) An antibody to lymphotoxin and tumor necrosis factor prevents transfer of experimental allergic encephalomyelitis. J Exp Med 172: 1193–1200
Ryffel B, Mihatsch MJ (1993) TNF receptor distribution in human tissues. Int Rev Exp Pathol 34 Pt B: 149–156
Ryffel B, Brockhaus M, Durmuller U, Gudat F (1991) Tumor necrosis factor receptors in lymphoid tissues and lymphomas. Am J Pathol 139: 7–15
Schaerer E, Tschopp J (1993) Cytolytic T cells keep their secrets.Curr Biol 3: 167–169
Schall TJ, Lewis M, Koller KJ, Lee A, Rice GC, Wong GH, Gatanaga T, Granger GA, Lentz R, Raab H (1990) Molecular cloning and expression of a receptor for human tumor necrosis factor. Cell 61: 361–370
Scheurich P, Thoma B, Ucer U, Pfizenmaier K (1987) Immunoregulatory activity of recombinant human tumor necrosis factor (TNF)-α: induction of TNF receptors on human T cells and TNF-α-mediated enhancement of T cell responses. J Immunol 138: 1786–1790
Schlessinger J, Ullrich A (1992) Growth factor signaling by receptor tyrosine kinases. Neuron 9: 383–391
Schlossman SF, Boumsell L, Gilks W, Harlan JM, Kishimoto T, Morimoto C, Ritz J, Shaw S, Silverstein RL, Springer TA, Tedder TF, Todd RF(1994) CD antigen 1993. J Immunol 152: 1–2

Schoenfeld HJ, Poeschl B, Frey JR, Loetscher H, Hunziker W, Lustig A, Zulauf M (1991) Efficient purification of recombinant human tumor necrosis factor β from Escherichia coli yields biologically active protein with a trimeric structure that binds to both tumor necrosis factor receptors. J Biol Chem 266: 3863–3869

Schutze S, Berkovic D, Tomsing O, Unger C, Kronke M (1991) Tumor necrosis factor induces rapid production of 1'2'diaclyglycerol by a phosphatidylcholine-specific phospholipase C. J Exp Med 174: 975–988

Schutze S, Potthoff K, Machleidt T, Berkovic D, Wiegmann K, Kronke M (1992) TNF activates NF-κB by phosphatidylcholine-specific phospholipase C-induced "acidic" sphingomyelin breakdown. Cell 71: 765–776

Schwarz H, Tuckwell J, Lotz M (1993) A receptor induced by lymphocyte activation (ILA): a new member of the human nerve-growth-factor/tumor-necrosis-factor receptor family. Gene 134: 295–298

Sciavolino PJ, Lee TH, Vilcek J (1992) Overexpression of metallothionein confers resistance to the cytotoxic effect of TNF with cadmium in MCF-7 breast carcinoma cells. Lymphokine Cytokine Res 11: 265–270

Selmaj K, Raine CS, Cannella B, Brosnan CF (1991) Identification of lymphotoxin and tumor necrosis factor in multiple sclerosis lesions. J Clin invest 87: 949–954

Seow H-F, Goh CR, Krishnan L, Porter AG (1989) Bacterial expression, facile purification and properties of recombinant human lymphotoxin (TNF-β). Biotechnology 7: 363

Seregina TM, Mekshenkov MI, Turetskaya RL, Nedospasov SA (1989) An autocrine growth factor constitutively produced by a human lymphoblastoid B-cell line is serologically related to lymphotoxin (TNF-β). Mol Immunol 26: 339–342

Shalaby MR, Espevik T, Rice GC, Ammann AJ, Figari IS, Ranges GE, Palladino MA Jr (1988) The involvement of human tumor necrosis factors-α and -β in the mixed lymphocyte reaction. J Immunol 141: 499–503

Sheehan KC, Ruddle NH, Schreiber RD (1989) Generation and characterization of hamster monoclonal antibodies that neutralize murine tumor necrosis factors. J Immunol 142: 3884–3893

Smith CA, Davis T, Anderson D, Solam L, Beckmann MP, Jerzy R, Dower SK, Cosman D, Goodwin RG (1990) A receptor for tumor necrosis factor defines an unusual family of cellular and viral proteins. Science 248: 1019–1023

Smith CA, Davis T, Wignall JM, Din WS, Farrah T, Upton C, McFadden G, Goodwin RG (1991) T2 open reading frame from the Shope fibroma virus encodes a soluble form of the TNF receptor. Biochem Biophys Res Commun 176: 335–342

Smith CA, Gruss HJ, Davis T, Anderson D, Farrah T, Baker E, Sutherland GR, Brannan CI, Copeland NG, Jenkins NA et al. (1993) CD30 antigen, a marker for Hodgkin's lymphoma, is a receptor whose ligand defines an emerging family of cytokines with homology to TNF. Cell 73: 1349–1360

Smith CA, Farrah T, Goodwin RG (1994) The TNF receptor superfamily of cellular and viral proteins: activation, costimulation, and death. Cell 76: 959–962

Smith RA, Baglioni C (1987) The active form of tumor necrosis factor is a trimer. J Biol Chem 262: 6951–6954

Stamenkovic I, Clark EA, Seed B (1989) A B-lymphocyte activation molecule related to the nerve growth factor receptor and induced by cytokines in carcinomas. EMBO J 8: 1403–1410

Staunton DE, Marlin SD, Stratowa C, Dustin ML, Springer TA (1988) Primary structure of ICAM-1 demonstrates interaction between members of the immunoglobulin and integrin supergene families. Cell 52: 925–933

Suda T, Nagata S (1994) Purification and characterization of the Fas-ligand that induces apoptosis. J Exp Med 179: 873–879

Suda T, Takahashi T, Golstein P, Nagata S (1993) Molecular cloning and expression of the Fas ligand, a novel member of the tumor necrosis factor family. Cell 75: 1169–1178

Sugarman BJ, Aggarwal BB, Hass PE, Figari IS, Palladino MA Jr, Shepard HM (1985) Recombinant human tumor necrosis factor-α: effects on proliferation of normal and transformed cells in vitro. Science 230: 943–945

Takahashi T, Tanaka M, Brannan CI, Jankins NA, Copeland NG, Suda T, Nagata S (1994a) Generalized lymphoproliferative disease in mice, caused by a point mutation in the Fas ligand. Cell 76: 969–976

Takahashi T, Tanaka M, Inazawa J, Abe T, Suda T, Nagata S (1994) Human fas ligand: gene structure, chromosomal location and species specificity. Int Immunol 6: 1567–1574

Tartaglia LA, Goeddel DV (1992) Tumor necrosis factor receptor signaling. A dominant negative mutation suppresses the activation of the 55-kDa tumor necrosis factor receptor. J Biol Chem 267: 4304–4307

Tartaglia LA, Weber RF, Figari IS, Reynolds C, Palladino MA Jr, Goeddel DV (1991) The two different receptors for tumor necrosis factor mediate distinct cellular responses.Proc Natl Acad Sci USA 88: 9292–9296
Tartaglia LA, Ayres TM, Wong GH, Goeddel DV (1993a) A novel domain within the 55 kd TNF receptor signals cell death. Cell 74: 845–853
Tartaglia LA, Goeddel DV, Reynolds C, Figari IS, Weber RF, Fendly BM, Palladino MA Jr (1993b) Stimulation of human T-cell proliferation by specific activation of the 75-kDa tumor necrosis factor receptor. J Immunol 151: 4637–4641
Tartaglia LA, Pannica D, Goeddel DV (1993c) Ligand passing: the 75-kDa tumor necrosis factor (TNF) receptor recruits TNF for signaling by the 55-kDa TNF receptor. J Biol Chem 268: 18542–18548
Tartaglia LA, Rothe M, Hu YF, Goeddel DV (1993d) Tumor necrosis factor's cytotoxic activity is signaled by the p55 TNF receptor. Cell 73: 213–216
Tavernier J, Marmenout A, Bauden R, Hauquier G, Van Ostade X, Fiers W (1990) Analysis of the structure-function relationship of tumour necrosis factor. Human/mouse chimeric TNF proteins: general properties and epitope analysis. J Mol Biol 211: 493–501
Thoma B, Grell M, Pfizenmaier K, Scheurich P (1990) Identification of a 60-kD tumor necrosis factor (TNF receptor as the major signal transducing component in TNF responses. J Exp Med 172: 1019–1023
Ullrich A, Schlessinger J (1990) Signal transduction by receptors with tyrosine kinase activity. Cell 61: 203–212
Upton C, DeLange A, McFadden G (1987) Tumorigenic poxviruses: genomic organization and DNA sequence of the telomeric region of the shope fibroma virus genome. Virology 160: 20–30
Upton C, Macen J, Schreiber M, McFadden G (1991) Myxoma virus expresses a secreted protein with homology to the tumor necrosis factor receptor gene family that contributes to viral virulence. Virology 184: 370–382
Van Kooten C, Gaillard C, Galizzi J-P, Hermann P, Fossiez F, Banchereau J, Blanchard D (1994) B cells regulate expression of CD40 ligand on activated T cells by lowering mRNA level and through the release of soluble CD40. Eur J Immunol 24: 787–792
Van Ostade X, Vandenabeele P, Everaerdt B, Loetscher H, Gentz R, Brockhaus M, Lesslauer W, Tavernier J, Brouckaert P, Fiers W (1993) Human TNF mutants with selective activity on the p55 receptor. Nature 361: 266–269
Van Ostade X, Vandenabeele P, Tavernier J, Fiers W (1994) Human tumor necrosis factor mutants with preferential binding to and activity on either the R55 or R75 receptor. Eur J Biochem 220: 771–779
Van Zee KJ, Kohno T, Fischer E, Rock CS, Moldawer LL, Lowry SF (1992) Tumor necrosis factor soluble receptors circulate during experimental and clinical inflammation and can protect against excessive tumor necrosis factor α in vitro and in vivo. Proc Natl Acad Sci USA 89: 4845–4849
VanArsdale TL, Ware CF (1994) TNF receptor signal transduction: ligand-dependent stimulation of a serine protein kinase activity associated with TNFR60. J Immunol 153: 3043–3050
Vandenabeele P, Declercq W, Vercammen D, Van de Craen M, Grooten J, Loetscher H, Brockhaus M, Lesslauer W, Fiers W (1992) Functional Characterization of the human tumor necrosis factor receptor p75 in a transfected rat/mouse T cell hybridoma. J Exp Med 176: 1015–1024
Vanhaesebroeck B, Reed JC, De Valck D, Grooten J, Miyashita T, Tanakla S, Beyaert R, van Roy F, Fiers W (1993) Effect of bcl-2 proto-oncogene expression on cellular sensitivity to tumor necrosis factor-mediated cytotoxicity. Oncogene 8: 1075–1081
Vietor I, Schwenger P, Li W, Schlessinger J, Vilcek J (1993) Tumor necrosis factor-induced activation and increased tyrosine phosphorylation of mitogen-activated protein (MAP) kinase in human fibroblasts. J Biol Chem 268: 18994–18999
Wallach D (1984) Preparations of lymphotoxin induce resistance to their own cytotoxic effect. J Immunol 132: 2464–2469
Ware CF, Harris PC, Granger GA (1981) Mechanisms of lymphocyte-mediated cytotoxicity. II. biochemical and serologic identification of a precursor lymphotoxin form (pre-LT) produced by MLC-sensitized human T lymphocytes in vitro. J Immunol 126: 1927–1933
Ware CF, Andrews JS, Shamansky LM, Grayson MH (1990) Regulation of the CTL lytic pathway by tumor necrosis factor. In: Lotze MT, Finn OJ (eds) Cellular immunity and the immunotherapy of cancer. Wiley-Liss, New York, pp 121–128
Ware CF, Crowe PD, VanArsdale TL, Andrews JL, Grayson MH, Jerzy R, Smith CA, Goodwin RG (1991) Tumor necrosis factor (TNF) receptor expression in T lymphocytes. Differential regulation of the type I TNF receptor during activation of resting and effector T cells. J Immunol 147: 4229–4238
Ware CF, Crowe PD, Grayson MH, Androlewicz MJ, Browning JL (1992) Expression of surface lymphotoxin and tumor necrosis factor on activated T,B, and natural killer cells. J Immunol 149: 3881–3888

Watanabe-Fukunaga R, Brannan C, Copeland N, Jenkins N, Nagata S (1992) Lymphoproliferation disorder in mice explained by defects in Fas antigen that mediates apoptosis. Nature 356: 314–317
Wiegmann K, Schutze S, Kampen E, Himmler A, Machleidt T, Kronke M (1992) Human 55-kDa receptor for tumor necrosis factor coupled to signal transduction cascades. J Biol Chem 267: 17997–18001
Williamson BD, Carswell EA, Rubin BY, Prendergast JS, Old LJ (1983) Human tumor necrosis factor produced by human B-cell lines: synergistic cytotoxic interaction with human interferon. Proc Natl Acad Sci USA 80: 5397–5401
Wingfield P, Pain RH, Craig S (1987) Tumour necrosis factor is compact trimer. FEBS Lett 211: 179–184
Wold WS, Gooding LR (1991) Region E3 of adenovirus: a cassette of genes involved in host immunosurveillance and virus-cell interactions. Virology 184: 1–8
Wong GH, Goeddel DV (1994) Fas antigen and p55 TNF receptor signal apoptosis through distinct pathways. J Immunol 152: 1751–1755
Wong GH, Elwell JH, Oberley LW, Goeddel DV (1989) Manganous superoxide dismutase is essential for cellular resistance to cytotoxicity of tumor necrosis factor. Cell 58: 923–931
Yagita H, Nakata M, Kawasaki A, Shinkai Y, Okumura K (1992) Role of perforin in lymphocyte-mediated cytolysis. Adv Immunol 51: 215–242
Yasukawa M, Yakushijin Y, Hasegawa H, Miyake M, Hitsumoto Y, Kimura S, Takeuchi N, Fujita S (1993) Expression of perforin and membrane-bound lymphotoxin (TNF-β) in virus specific CD4 human cytotoxic T-cell clones. Blood 81: 1527–1534
Yellin MJ, Sippel K, Inghirami G, Covey LR, Lee JJ, Sinning J, Clark EA, Chess L, Lederman S (1994) CD40 molecules induce down-modulation and endocytosis of T cell surface T cell-B cell activating molecule/CD40-L. Potential role in regulating helper effector function. J Immunol 152: 598–608
Zhang XM, Weber I, Chen MJ (1992) Site-directed mutational analysis of human tumor necrosis factor-α receptor binding site and structure-functional relationship. J Biol Chem 267: 24069–24075

Subject Index

Current Topics in Microbiology and Immunology

Volumes published since 1989 (and still available)

Vol. 153: **Lambris, John D. (Ed.):** The Third Component of Complement. Chemistry and Biology. 1989. 38 figs. X, 251 pp. ISBN 3-540-51513-5

Vol. 154: **McDougall, James K. (Ed.):** Cytomegaloviruses. 1990. 58 figs. IX, 286 pp. ISBN 3-540-51514-3

Vol. 155: **Kaufmann, Stefan H. E. (Ed.):** T-Cell Paradigms in Parasitic and Bacterial Infections. 1990. 24 figs. IX, 162 pp. ISBN 3-540-51515-1

Vol. 156: **Dyrberg, Thomas (Ed.):** The Role of Viruses and the Immune System in Diabetes Mellitus. 1990. 15 figs. XI, 142 pp. ISBN 3-540-51918-1

Vol. 157: **Swanstrom, Roland; Vogt, Peter K. (Ed.):** Retroviruses. Strategies of Replication. 1990. 40 figs. XII, 260 pp. ISBN 3-540-51895-9

Vol. 158: **Muzyczka, Nicholas (Ed.):** Viral Expression Vectors. 1992. 20 figs. IX, 176 pp. ISBN 3-540-52431-2

Vol. 159: **Gray, David; Sprent, Jonathan (Ed.):** Immunological Memory. 1990. 38 figs. XII, 156 pp. ISBN 3-540-51921-1

Vol. 160: **Oldstone, Michael B. A.; Koprowski, Hilary (Eds.):** Retrovirus Infections of the Nervous System. 1990. 16 figs. XII, 176 pp. ISBN 3-540-51939-4

Vol. 161: **Racaniello, Vincent R. (Ed.):** Picornaviruses. 1990. 12 figs. X, 194 pp. ISBN 3-540-52429-0

Vol. 162: **Roy, Polly; Gorman, Barry M. (Eds.):** Bluetongue Viruses. 1990. 37 figs. X, 200 pp. ISBN 3-540-51922-X

Vol. 163: **Turner, Peter C.; Moyer, Richard W. (Eds.):** Poxviruses. 1990. 23 figs. X, 210 pp. ISBN 3-540-52430-4

Vol. 164: **Bækkeskov, Steinnun; Hansen, Bruno (Eds.):** Human Diabetes. 1990. 9 figs. X, 198 pp. ISBN 3-540-52652-8

Vol. 165: **Bothwell, Mark (Ed.):** Neuronal Growth Factors. 1991. 14 figs. IX, 173 pp. ISBN 3-540-52654-4

Vol. 166: **Potter, Michael; Melchers, Fritz (Eds.):** Mechanisms in B-Cell Neoplasia. 1990. 143 figs. XIX, 380 pp. ISBN 3-540-52886-5

Vol. 167: **Kaufmann, Stefan H. E. (Ed.):** Heat Shock Proteins and Immune Response. 1991. 18 figs. IX, 214 pp. ISBN 3-540-52857-1

Vol. 168: **Mason, William S.; Seeger, Christoph (Eds.):** Hepadnaviruses. Molecular Biology and Pathogenesis. 1991. 21 figs. X, 206 pp. ISBN 3-540-53060-6

Vol. 169: **Kolakofsky, Daniel (Ed.):** Bunyaviridae. 1991. 34 figs. X, 256 pp. ISBN 3-540-53061-4

Vol. 170: **Compans, Richard W. (Ed.):** Protein Traffic in Eukaryotic Cells. Selected Reviews. 1991. 14 figs. X, 186 pp. ISBN 3-540-53631-0

Vol. 171: **Kung, Hsing-Jien; Vogt, Peter K. (Eds.):** Retroviral Insertion and Oncogene Activation. 1991. 18 figs. X, 179 pp. ISBN 3-540-53857-7

Vol. 172: **Chesebro, Bruce W. (Ed.):** Transmissible Spongiform Encephalopathies. 1991. 48 figs. X, 288 pp. ISBN 3-540-53883-6

Vol. 173: **Pfeffer, Klaus; Heeg, Klaus; Wagner, Hermann; Riethmüller, Gert (Eds.):** Function and Specificity of γ/δ TCells. 1991. 41 figs. XII, 296 pp. ISBN 3-540-53781-3

Vol. 174: **Fleischer, Bernhard; Sjögren, Hans Olov (Eds.):** Superantigens. 1991. 13 figs. IX, 137 pp. ISBN 3-540-54205-1

Vol. 175: **Aktories, Klaus (Ed.):** ADP-Ribosylating Toxins. 1992. 23 figs. IX, 148 pp. ISBN 3-540-54598-0

Vol. 176: **Holland, John J. (Ed.):** Genetic Diversity of RNA Viruses. 1992. 34 figs. IX, 226 pp. ISBN 3-540-54652-9

Vol. 177: **Müller-Sieburg, Christa; Torok-Storb, Beverly; Visser, Jan; Storb, Rainer (Eds.):** Hematopoietic Stem Cells. 1992. 18 figs. XIII, 143 pp. ISBN 3-540-54531-X

Vol. 178: **Parker, Charles J. (Ed.):** Membrane Defenses Against Attack by Complement and Perforins. 1992. 26 figs. VIII, 188 pp. ISBN 3-540-54653-7

Vol. 179: **Rouse, Barry T. (Ed.):** Herpes Simplex Virus. 1992. 9 figs. X, 180 pp. ISBN 3-540-55066-6

Vol. 180: **Sansonetti, P. J. (Ed.):** Pathogenesis of Shigellosis. 1992. 15 figs. X, 143 pp. ISBN 3-540-55058-5

Vol. 181: **Russell, Stephen W.; Gordon, Siamon (Eds.):** Macrophage Biology and Activation. 1992. 42 figs. IX, 299 pp. ISBN 3-540-55293-6

Vol. 182: **Potter, Michael; Melchers, Fritz (Eds.):** Mechanisms in B-Cell Neoplasia. 1992. 188 figs. XX, 499 pp. ISBN 3-540-55658-3

Vol. 183: **Dimmock, Nigel J.:** Neutralization of Animal Viruses. 1993. 10 figs. VII, 149 pp. ISBN 3-540-56030-0

Vol. 184: **Dunon, Dominique; Mackay, Charles R.; Imhof, Beat A. (Eds.):** Adhesion in Leukocyte Homing and Differentiation. 1993. 37 figs. IX, 260 pp. ISBN 3-540-56756-9

Vol. 185: **Ramig, Robert F. (Ed.):** Rotaviruses. 1994. 37 figs. X, 380 pp. ISBN 3-540-56761-5

Vol. 186: **zur Hausen, Harald (Ed.):** Human Pathogenic Papillomaviruses. 1994. 37 figs. XIII, 274 pp. ISBN 3-540-57193-0

Vol. 187: **Rupprecht, Charles E.; Dietzschold, Bernhard; Koprowski, Hilary (Eds.):** Lyssaviruses. 1994. 50 figs. IX, 352 pp. ISBN 3-540-57194-9

Vol. 188: **Letvin, Norman L.; Desrosiers, Ronald C. (Eds.):** Simian Immunodeficiency Virus. 1994. 37 figs. X, 240 pp. ISBN 3-540-57274-0

Vol. 189: **Oldstone, Michael B. A. (Ed.):** Cytotoxic T-Lymphocytes in Human Viral and Malaria Infections. 1994. 37 figs. IX, 210 pp. ISBN 3-540-57259-7

Vol. 190: **Koprowski, Hilary; Lipkin, W. Ian (Eds.):** Borna Disease. 1995. 33 figs. IX, 134 pp. ISBN 3-540-57388-7

Vol. 191: **ter Meulen, Volker; Billeter, Martin A. (Eds.):** Measles Virus. 1995. 23 figs. IX, 196 pp. ISBN 3-540-57389-5

Vol. 192: **Dangl, Jeffrey L. (Ed.):** Bacterial Pathogenesis of Plants and Animals. 1994. 41 figs. IX, 343 pp. ISBN 3-540-57391-7

Vol. 193: **Chen, Irvin S. Y.; Koprowski, Hilary; Srinivasan, Alagarsamy; Vogt, Peter K. (Eds.):** Transacting Functions of Human Retroviruses. 1995. 49 figs. Approx. IX, 240 pp. ISBN 3-540-57901-X

Vol. 194: **Potter, Michael; Melchers, Fritz (Eds.):** Mechanisms in B-cell Neoplasia. 1995. Approx. 152 figs. XXV, 458 pp. ISBN 3-540-58447-1

Vol. 195: **Montecucco, Cesare (Ed.):** Clostridial Neurotoxins. 1995. Approx. 28 figs. Approx. 260 pp. ISBN 3-540-58452-8

Vol. 196: **Koprowski, Hilary; Maeda Hiroshi (Eds.)** : The Role of Nitric Oxide in Physiology and Pathophysiology. 1995. 21 figs. Approx. 100 pp. ISBN 3-540-58214-2

Vol. 197: **Meyer, Peter (Ed.):** Gene Silencing in Higher Plants and Related Phenomena in Other Eukaryotes. 1995. 17 figs. Approx. 240 pp. ISBN 3-540-58236-3

Springer-Verlag and the Environment

We at Springer-Verlag firmly believe that an international science publisher has a special obligation to the environment, and our corporate policies consistently reflect this conviction.

We also expect our business partners – paper mills, printers, packaging manufacturers, etc. – to commit themselves to using environmentally friendly materials and production processes.

The paper in this book is made from low- or no-chlorine pulp and is acid free, in conformance with international standards for paper permanency.

MIX
Papier aus verantwortungsvollen Quellen
Paper from responsible sources
FSC® C105338

FSC
www.fsc.org

If you have any concerns about our products,
you can contact us on
ProductSafety@springernature.com

In case Publisher is established outside the EU,
the EU authorized representative is:
Springer Nature Customer Service Center GmbH
Europaplatz 3, 69115 Heidelberg, Germany

Printed by Libri Plureos GmbH
in Hamburg, Germany